J. Nita Cogburn
Monsanto

Nuclear Imaging in Drug Discovery, Development, and Approval

Nuclear Imaging in Drug Discovery, Development, and Approval

Edited by
H. Donald Burns
Raymond E. Gibson
Robert F. Dannals
Peter K. S. Siegl

With a Foreword by
Henry N. Wagner, Jr.

Birkhäuser
Boston · Basel · Berlin

Library of Congress Cataloging-in-Publication Data

Nuclear imaging in drug discovery, development, and approval / edited
by H. Donald Burns . . . [et al.] : with a foreword by Henry N.
Wagner, Jr.
p. cm.
Includes bibliographical references and index.
ISBN 0-8176-3601-3 (H : alk. paper). -- ISBN 3-7643-3601-3 (H :
alk. paper)
1. Drugs--Resarch--Methodology. 2. Radioisotope scanning.
3. Radioisotope tracers in biochemistry. I. Burns, H. Donald (Hugh
Donald), 1946- .
[DNLM: 1. Autoradiography. 2. Drug Design. 3. Drug Evaluation.
4. Tomography, Emission-Computed. WN 160 N9645]
RM301.25.N8 1993
615.1'072--dc20
DNLM/DLC 92-48930
for Library of Congress CIP

Printed on acid-free paper.

ISBN 0-8176-3601-3
ISBN 3-7643-3601-3

Camera-ready copy prepared by the editors.
Printed and bound by Edwards Brothers, Ann Arbor, MI
Printed in the U.S.A.

9 8 7 6 5 4 3 2 1

Contents

Foreword

It is the purpose and business of the pharmaceutical industry to discover, develop, and make available drugs for the care of the sick. The purpose of universities and national laboratories is to provide people and scientific knowledge that can help in the process. This book presents the combined efforts of scientists from the drug industry, academic laboratories and national laboratories to describe advances in radiotracer technology in studies on experimental animals and living human beings. The authors believe that the technology is now ready for widespread application in the pharmaceutical industry. The goal of this book is to help bring this about.

The field of Nuclear Medicine is based on the concept that, if treatment of disease is chemical, the patient's diagnosis should be chemical. Anatomy and histopathology have been the principle basis for making a diagnosis. Histopathologic data suffer from being descriptive, subjective, not quantifiable, and based on the study of dead tissue. The era of histopathology as the dominant concept in medical practice is coming to an end. Histopathologic findings are often heterogeneous and a single biopsy will at times not reveal the true nature of the disease, such as the grading of malignancy. Far greater accuracy of staging of disease and in the planning of treatment is possible through chemistry, as well as by making possible a more suitable selection of a histological biopsy site. Basic science advances in genetics, molecular biology, oncology, metabolic diseases, and infectious diseases can be extended to medical practice by means of radiotracer technology. The methods go far beyond anatomy into the domains of physiology and *in vivo* biochemistry; these methods are the topic of this book.

Implicit in the development and use of drugs in medical practice is answering the questions: What is wrong? What is going to happen? What can be done about it? Has the treatment been effective? This book proposes that the diagnostic technology for measuring intercellular communication and the planning and monitoring of treatment should become a major focus of drug design and development. A strong relationship exists between the areas of radioisotopic methods of diagnosis and currently developed drugs. I like to refer to the radiotracers as *nanoDx* molecules, and that the basis of treatment will be using *nanoRx* molecules. The prefix *nano-* indicates that nano- or picomolar quantities of radiotracers are involved in in vivo diagnosis by measuring their emissions of ionizing radiation. *NanoRx* molecules correct regional chemical abnormalities that characterize disease.

NanoDx probes, the radiotracers described in this book, measure specific, general or regional metabolic activity or the state of recognition sites, enzymes, or transport processes. For example, if a pituitary tumor contains dopamine receptors (measured with Carbon-11 labeled N-methylspiperone), the receptor can be stimulated by the administration of the dopamine receptor agonist, bromocryptine. This drug activates the dopaminergic system which in turn inhibits the secretion of prolactin by the prolactin-secreting cells of the pituitary tumor. The beneficial effect of the bromocryptine therapy can be demonstrated on the same day that the treatment is begun by measuring its inhibiting effect on glucose or other substrate utilization by means of radiolabeled glucose, amino acids or thymidine. Photon-emitting radiotracers can characterize a lesion, aid in the planning of treatment, and determine whether or not the treatment has been effective, an approach which goes far beyond relying on clinical manifestations or anatomical changes which may take weeks or months before they occur. This same methodology can be used to demonstrate efficacy of new, experimental drugs in much the same way that the beneficial effect of bromocryptine has been shown.

Amines, such as dopamine and carfentanil, and peptides, such as octreotide, are examples of chemical messengers developed by the pharmaceutical industry that were subsequently labeled in the university community with photon-emitting radiotracers for studies of regional chemistry in living humans. Peptides and proteins can be made in large quantities by means of genetic engineering, and their detailed structure characterized by x-ray crystallography. Knowledge of the stereospecific and other determinants of binding can be incorporated into new ratiotracers for use in human studies, employing external imaging techniques. Radiotracers, with appropriate structure, charge and hydrophobic/hydrophilic properties, search out and characterize abnormalities in regional chemistry. They also make possible, studies of pharmacokinetics, structure/activity relationships, and receptor-specific binding of novel molecules.

For example, patients with neuroendocrine and other types of cancer, including cancer of the breast, are characterized by containing somatostatin receptors in measurable quantities. Characterization of the disease by its receptor sites results in specific treatment with a somatostatin receptor-stimulating analogue. Imaging can also be used to show that new drugs do, in fact, interact with somatostatin receptors *in vivo*.

Nuclear medicine does not just provide new tests for old diseases, but new ways of defining and detecting disease. Chemical changes can almost always be detected before clinical signs of disease, and make possible more specific characterization of the disease. For example, if a metastatic breast tumor contains estrogen receptors it can be treated with the estrogen receptor antagonist, tamoxifen. Both the planning of treatment and the response to treatment can be based on regional biochemistry.

The mechanism of action of many drugs involves stimulation or inhibition of recognition sties - enzymes, transport processes or chemoreceptors - all of which have been extensively studied by PET and SPECT. Hundreds of reports have involved the dopaminergic system, first successfully imaged in living human brain in 1983 with Carbon-11 NMSP and Fluorine-18 L-DOPA. It is now possible to image the regional synthesis of dopamine in the human brain, the binding of dopamine to post-synaptic receptors, and the re-uptake of dopamine into vesicles via pre-synaptic transporter sites. Six different dopamine receptor subtypes have been cloned, and are now available for detailed study in pure form.

Defining diseases in terms of brain chemical abnormalities helps in developing new forms of drug treatment, beyond what would be possible on the bases of subjective symptoms or mental performance tests. In AIDS, for example, chemical changes in the brain can be detected before any neurological signs are present. Such findings are important because of the value of early treatment of HIV infections.

This book provides a timely overview of nuclear medicine technology and its application to the discovery and development of new pharmaceuticals. The potential for use of this technology by the pharmaceutical industry has been recognized but just barely tapped. Its use can greatly reduce the cost and time required for chemicals to be transformed into FDA approved drugs. The goal of this book is to provide the pharmaceutical industry with a look at how this technology can contribute to increasing the efficiency of drug development and to shorten the time from initial synthesis to approval of new drugs.

Henry N. Wagner, Jr.

NUCLEAR IMAGING IN DRUG DEVELOPMENT - INTRODUCTION

Leslie Lars Iversen
Neurosciences Research Center
Merck Research Laboratories

INTRODUCTION

The use of radioactive isotopes to generate images has developed in many directions to yield a series of sophisticated technologies with numerous applications in drug discovery and development. Imaging techniques now allow precise quantitative measurements to be made with high spatial resolution and ever improving time resolution. This brief introductory review will illustrate some of the practical applications of this technology to pharmaceutical research, using examples from our own laboratory which focuses on drug discovery and development in the CNS area. Repeated reference will be made to the applications of nuclear imaging to studies of the compound dizocilpine (**1**, MK-801), an antagonist of glutamate receptors of the N-methyl-D-aspartate (NMDA)

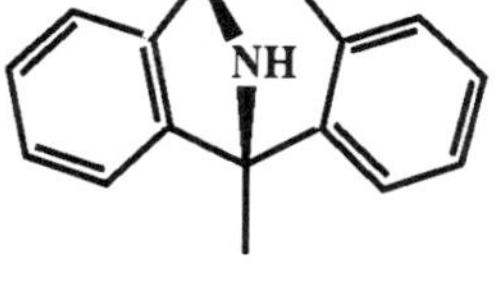

NH

1

Nuclear Imaging in Drug Discovery, Development, and Approval
H.D. Burns, R. Gibson, R. Dannals, Peter Siegl, Editors

subtype. This compound was first discovered as a novel anticonvulsant agent in our US laboratories more than a decade ago (Clineschmidt et al., 1982) and was later found to act as a potent noncompetitive NMDA antagonist (Wong et al., 1986). It has been found to be a powerful neuroprotective agent in a number of animal models of cerebral ischemia, which suggests the potential of NMDA antagonists for use in human cerebral ischemia, e.g., in the acute treatment of stroke (Iversen et al., 1989).

AUTORADIOGRAPHY

The ability to localize radioactively labeled compounds in thin sections of tissue by autoradiography offers the opportunity for studying the microdistribution of drug receptors in tissues, with a spatial resolution considerably better than is possible with *in vivo* imaging techniques. The information thus obtained can be used in the interpretation of Positron Emission Tomography (PET) and Single Photon Emission Computed Tomography (SPECT) studies. Results from *ex vivo* autoradiographic studies provide a valuable tool for validating methods which can subsequently be used in PET and SPECT studies.

Mapping of Drug Receptors

An obvious and powerful application of autoradiographic analysis is to map the distribution of drug receptors in target tissues, using the specific high affinity binding of radiolabeled compounds to these sites. This depends, of course, on the ability to demonstrate such specific receptor binding, which in turn requires that the compounds have high affinity (preferably in the nanomolar range) for these receptors, and that the compounds are not so lipophilic that nonspecific binding to cell membranes obscure the results. Provided the drug acts as a suitable radioligand, autoradiography can yield much valuable information.

In qualitative terms, a high resolution map of the distribution of drug receptors can be created. For example, early in our studies with [^{3}H]MK-801 we were able to demonstrate that the map of drug binding sites in rat brain corresponded precisely with the map of glutamate receptors of the NMDA subtype, created with [^{3}H]L-glutamate (Bowery et al., 1988) (Fig. 1). This provides strong support for the conclusion that the NMDA receptor in brain represented the principal target for MK-801. The use of quantitative autoradiography also permits analysis of the characteristics of drug binding sites in very small regions of the target tissue. By saturation analysis or drug displacement studies it is possible to measure

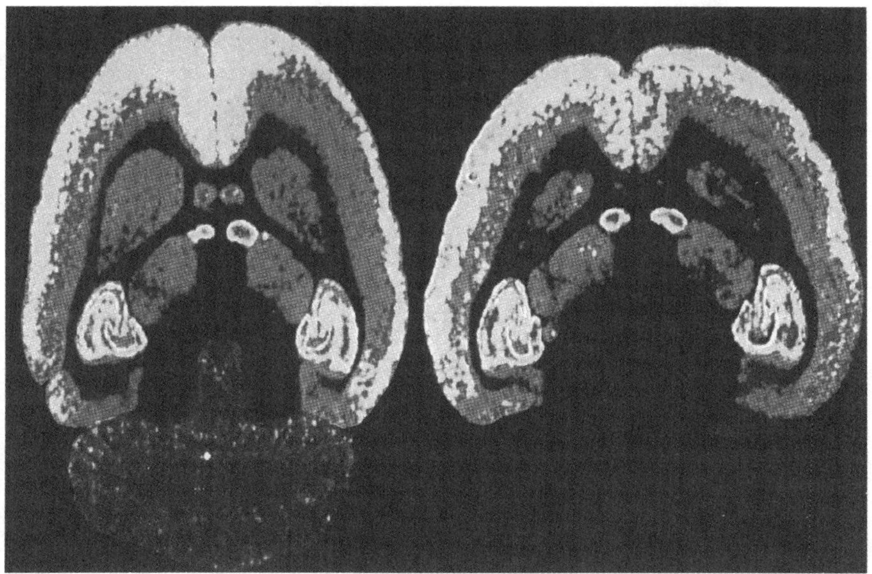

FIGURE 1. Autoradiographic localization of [^{3}H]MK-801 in horizontal sections of rat brain, compared with distribution of NMDA receptors labeled with [^{3}H]glutamate. Unpublished material kindly provided by Dr. A. Young, University of Michigan.

either K_d of the ligand or K_i values for other compounds acting as competitive displacers on a structure which would be far too small for such analysis to be carried out by conventional biochemical studies in tissue homogenates. For example, in our laboratory we found that high affinity binding sites for the radiolabeled peptide [^{125}I]cholecystokinin-8S in brain sections was not uniformly displaced by newly developed nonpeptide drugs which are selective for the cholecystokinin-B (CCK-B) receptor subtype (Hill et al., 1987; Hill et al., 1988; Hill et al., 1990). Although the majority of CCK binding sites in brain do belong to the CCK-B subtype, the binding sites in certain circumscribed regions of rodent and primate brain (i.e., area postrema, interpeduncular nuclei, substantia nigra, substantia gelatinosa, and mammillary body) exhibited the properties of the CCK-A subtype which is found in the alimentary tract. In these brain areas, CCK binding was potently displaced by nonpeptide drugs (e.g., devazepide, MK-329) which are selective for the

CCK-A receptors (Fig. 2).

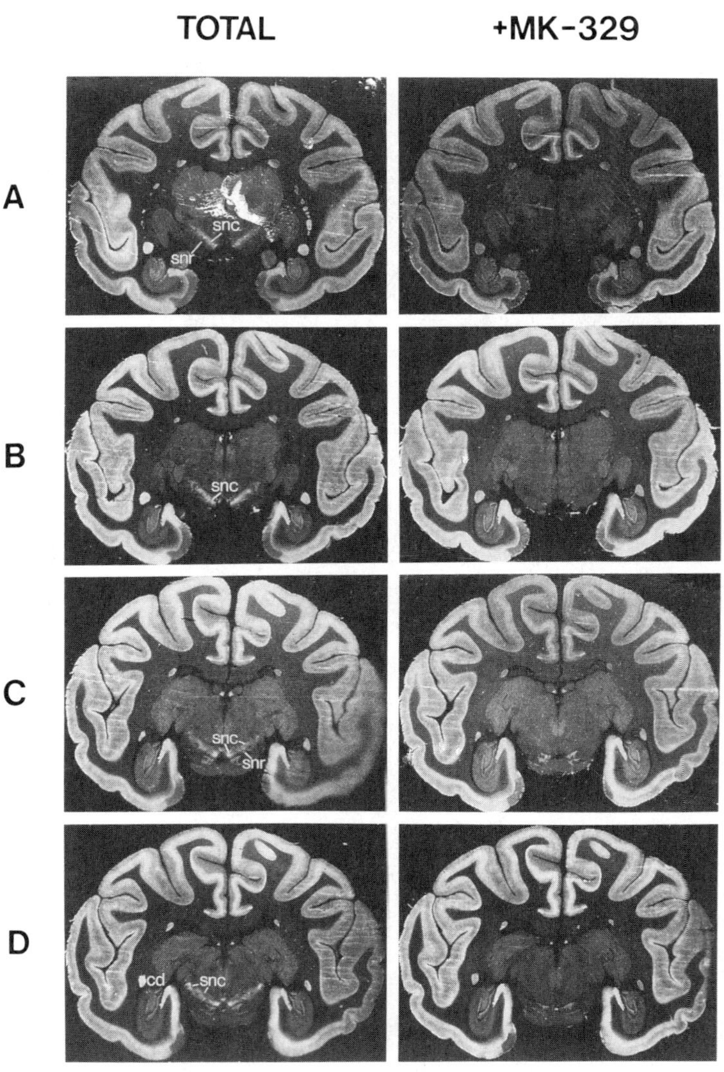

FIGURE 2. Localization of CCK-A receptors in substantia nigra of Cynomolgus monkey brain. CCK-A and CCK-B receptors were labeled with [^{125}I]-Bolton Hunter-CCK, and CCK-A sites discriminated by selective displacement with the CCK-A selective compound MK-329 (10 nM). Sections taken in progressively caudal direction through the substantia nigra compacta (snc) and reticulata (snr) show CCK-A sites, whilst the binding sites in cortex and caudate nucleus (cd) were not sensitive to MK-329. From Hill et al. (1990), with permission.

Distribution of Drug In Vivo

Whole body autoradiography is long established as a powerful technique for studying the disposition of drugs *in vivo*, and forms part of most companies' repertoire of techniques for studies of drug metabolism and distribution. In the case of CNS drugs, it is particularly valuable to have information on the penetration of drugs through the blood-brain-barrier. Such elementary information can be supplemented, however, in more sophisticated studies. For example, Wallace et al. (1992) used quantitative autoradiography to study the uptake and retention of [^{3}H]MK-801 in rat brain in which one cerebral hemisphere was made ischemic by permanent occlusion of the middle cerebral artery 15 min prior to administration of the radioligand. The results showed that at early time intervals (15 min), as expected, the level of [^{3}H]MK-801 was greatly reduced in the poorly perfused ischemic hemisphere but, surprisingly, there was more labelled drug in the ischemic tissue than in the normal hemisphere when measured 60 min after drug administration (Fig. 3). The early uptake of MK-801 into the CNS appears to be limited by the level of cerebral blood flow, whereas at later times the

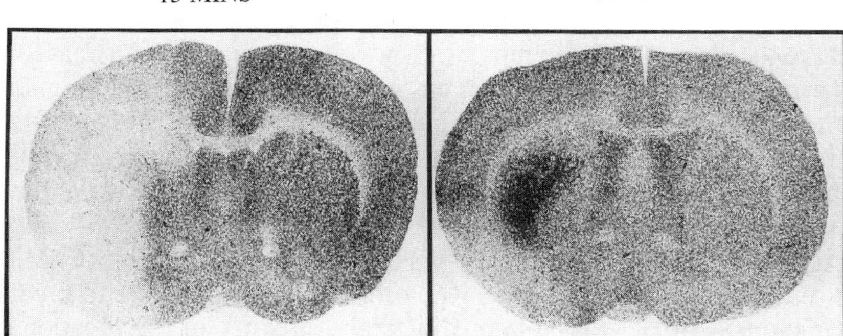

FIGURE 3. Autoradiographic localization of [^{3}H]MK-801 in the anesthetized rat after middle cerebral artery occlusion. [^{3}H]MK-801 was administered intravenously 15 min after artery occlusion. Fifteen min later (left) the uptake of tracer into the ischemic hemisphere is dictated by the low levels of cerebral blood flow. Sixty min after drug administration higher levels of tracer were observed in the ischemic hemisphere, putatively because of glutamate-enhanced binding to NMDA receptors. From McCulloch and Iversen (1991) adapted from Wallace et al. (1992), with permission.

enhancement of MK-801 binding to the NMDA receptor by elevated glutamate concentrations in the ischemic tissue appears to be a more important factor.

The evaluation of radioligands suitable for use in PET or SPECT imaging will frequently involve preliminary studies with compounds labeled with Tritium or Iodine-125 as these may be more easily prepared and are less expensive than the PET and SPECT isotopes. Only if a satisfactory level of "specific" versus "non-specific" binding can be demonstrated *in vivo* should the positron or gamma-emitting ligands be prepared.

Functional Studies

Autoradiography can also provide information on the functional consequences of drug receptor interactions. The powerful technique of [^{14}C]deoxyglucose accumulation as a measure of local rates of glucose metabolism, first introduced by L. Sokoloff and colleagues, has proved particularly valuable in studying the effects of drugs on cerebral metabolism (for review, see McCulloch, 1982). In studies with MK-801 McCulloch and colleagues (Kurumaji et al., 1989) demonstrated how this approach can provide valuable and unexpected information. When conscious rats were treated with anticonvulsant doses of MK-801, the deoxyglucose method revealed that while rates of glucose metabolism were reduced in some brain regions, they were paradoxically increased in other regions - notably in some areas of limbic forebrain; e.g. cingulate cortex, subiculum, and hippocampus. The areas exhibiting greatest increases in cerebral metabolism were also those brain regions in which transient vacuolization has been observed in some of the large neurons, following drug treatment (Olney et al.,1989; McCulloch and Iversen, 1991). Without use of a non-prejudicial approach such as that offered by the autoradiographic method, it would have been impossible to predict such a result from what is known of brain circuitry or the location of NMDA receptors - as the areas involved did not correspond to those possessing the highest density of NMDA sites. Indeed the magnitude and relatively widespread distribution of alterations in glucose utilization in brain after MK-801 are among the most striking ever described with the [^{14}C]deoxyglucose technique. The importance of being able to study conscious animals was emphasized by the observation that MK-801 had little or no effect on local rates of cerebral glucose use in the brains of anesthetized animals (McCulloch and Iversen, 1991).

Another technology pioneered by Sokoloff and his associates is the use of autoradiographic detection of the brain uptake of the freely diffusible tracer [^{14}C]iodoantipyrine to measure local cerebral blood flow. McCulloch and colleagues also applied this method to studies of MK-801 (McCulloch and Iversen, 1991; Park et al., 1989; Nehls et al., 1990). In conscious rats, MK-801 produced marked alterations in CBF in about half of the brain regions examined. Significant increases were observed in the limbic system and several cortical areas. Although in most brain areas examined CBF was tightly coupled to local oxidative metabolism (as measured by the [^{14}C]deoxyglucose method), there was evidence that CBF and glucose use were "uncoupled" in some areas after drug treatment, e.g., neocortex, caudate nucleus, and cingulate cortex.

OTHER NUCLEAR IMAGING TECHNIQUES

Determination of Drug Receptor Occupancy in Man

One of the most difficult problems in drug development is to establish the appropriate human dose for a new drug. If the compound acts on well defined peripheral receptors to cause a readily measurable response (e.g., change in blood pressure), such responses can be used by clinical pharmacologists to determine a suitable dose. In the case of drugs acting on CNS, however, matters are rarely that simple. CNS drugs often target receptors that are found only within the CNS, and are not present in peripheral tissues (e.g., $GABA_A$, glutamate and dopamine receptors). Alternatively, the drug binding sites may be present both in CNS and in peripheral tissues but there is no well defined response that can be measured (e.g. sigma binding sites). If a suitable PET ligand can be developed for use in human subjects, this can be used to establish how much drug is needed to compete for occupancy at the CNS receptors. This is a very important advance in drug development technology, and is beginning to be used more and more widely. For example, a range of clinically used neuroleptics have been studied in man, using the potent dopamine D_2 receptor radioligand [^{11}C]raclopride (Farde et al., 1989). These studies show that, by trial and error, clinicians have arrived at therapeutically effective doses of these neuroleptic drugs in the treatment of schizophrenic illness which, although they ranged from a few mg per day to more than 1 g per day, were remarkably similar in providing 70-85% occupancy of dopamine D_2 sites in brain. This information provided a valuable guide to establishing the correct dosages for the novel Astra/Merck substituted benzamide neuroleptic remoxipride (Farde and von Bahr, 1990).

Functional Studies

A number of authors in this volume will describe the application of a variety of nuclear imaging techniques (PET, SPECT, etc.) to studies of functional effects of drugs in human subjects. Such studies can provide important information on the effects of new compounds on visceral and GI function, secretory activity, as well as basic information on drug delivery and absorption.

CONCLUSION

This book provides a timely review of a fast growing area of medical science, which is of direct relevance to the processes of drug discovery and development. To those of us who work in the CNS area, nuclear imaging technology has proved particularly valuable in providing a window into what goes on in the intact brain, and in offering techniques with the resolution needed to examine drug mechanisms in the extraordinarily complex microarchitecture of the CNS.

REFERENCES

Bowery NG, Wong EHF and Hudson AL (1988): Quantitative autoradiography of [^{3}H]-MK-801 binding sites in mammalian brain. *Br J Pharmac* 93: 944-954.

Clineschmidt BV, Martin GE and Bunting PR (1982): Anticonvulsant activity of (+)-5-methyl-10,11-dihydro-5H-dibenzo[a,d]cyclohepten-5,10-imine (MK-801), a substance with potent anticonvulsant, central sympathomimetic and apparent anxiolytic properties. *Drug Dev Res* 2: 123-133.

Farde L and von Bahr C (1990): Distribution of remoxipride to the human brain and central D2-dopamine receptor binding examined by PET. *Acta Psychiat Scand* 82 (Suppl 358): 67-71.

Farde L, Wiesel FA, Nordstrom AL and Sedvall G (1989): D1 and D2-dopamine receptor occupancy during treatment with conventional and atypical neuroleptics. *Psychopharmacol* 99: S28-S31.

Hill DR, Campbell NJ, Shaw TM and Woodruff GN (1987): Autoradiographic localization and biochemical characterization of peripheral type CCK receptors in rat CNS using highly selective nonpeptide CCK antagonists. *J Neurosci* 7: 2967-2976.

Hill DR, Shaw TM and Woodruff GN (1988): Binding sites for [^{125}I]-cholecystokinin in primate spinal cord are of the CCK-A subclass. *Neurosci Letters* 89: 133-139.

Hill DR, Shaw TM, Graham W and Woodruff GN (1990): Autoradiographic detection of cholecystokinin-A receptors in primate brain using [^{125}I]-Bolton Hunter CCK-8 and [^{3}H]-MK-329. *J Neurosci* 10: 1070-1081.

Iversen LL, Woodruff GN, Kemp JA, Foster AC, McKernan R, Gill R and Wong EHF (1989): Non-competitive NMDA antagonists as drugs. In: *The NMDA Receptor*, Watkins JC and Collingridge GL, eds. Oxford: IRL Press.

Kurumaji A, Nehls DG, Park CK and McCulloch J (1989): Effects of NMDA antagonists, MK-801 and CPP, upon local cerebral glucose use. *Brain Res* 496: 268-284.

McCulloch J (1982): Mapping functional alterations in the CNS with [^{14}C]-deoxyglucose. In: *Handbook of Psychopharmacology, Vol 15*, Iversen LL, Iversen SD and Snyder SH, eds. New York: Plenum Press.

McCulloch J and Iversen LL (1991): Autoradiographic assessment of the effects of N-methyl-D-aspartate (NMDA) receptor antagonists *in vivo*. *Neurochem Res* 9: 951-963.

Nehls DG, Park CK, MacCormack AG and McCulloch J (1990): The effects of N-methyl-D-aspartate receptor blockade with MK-801 upon the relationship between cerebral blood flow and glucose utilization. *Brain Res* 511: 271-279.

Olney JW, Labruyere J and Price MT (1989): Pathological changes induced in cerebrocortical neurons by phencyclidine and related drugs. *Science* 244: 1360-1362.

Park CK, Nehls DG, Teasdale GM and McCulloch J (1989): Effect of the NMDA antagonist MK-801 on local cerebral blood flow in focal ischemia in the rat. *J Cereb Blood Flow Metab* 9: 617-622.

Wallace MC, Teasdale GM and McCulloch J (1992): Autoradiographic analysis of [^{3}H]-MK-801 (dizocilpine) *in vivo* uptake and *in vitro* binding after focal cerebral ischemia in the rat. *J Neurosurg* 76: 127-133.

Wong EHF, Kemp JA, Priestley T, Knight AR, Woodruff GN and Iversen LL (1986): The anticonvulsant MK-801 is a potent N-methyl-D-aspartate antagonist. *Proc Nat Acad Sci* 83: 7104-7108.

NUCLEAR MEDICINE PHYSICS, INSTRUMENTATION, AND DATA PROCESSING IN PHARMACEUTICAL RESEARCH

Jonathan M. Links
Department of Radiology
Johns Hopkins Medical Institutions

INTRODUCTION

Nuclear medicine techniques can play an important role in pharmaceutical research. In order to optimize the quantitative information which can be obtained from such studies, an appreciation of basic nuclear medicine physics, instrumentation, and data processing is required. The techniques to be considered here include both planar and tomographic imaging, the latter including both Single Photon Emission Computed Tomography (SPECT) and Positron Emission Tomography (PET). Since image and data processing are so central to quantification in nuclear medicine, these will be considered as well. In addition to this introduction to the basics of nuclear medicine instrumentation and data processing, several good books are available on this topic (Hendee, 1973; Sorenson and Phelps, 1987; Swenberg and Conklin, 1988).

Nuclear Imaging in Drug Discovery, Development, and Approval
H.D. Burns, R. Gibson, R. Dannals, Peter Siegl, Editors

IMAGING BASICS

In our discussions, *imaging* will be frequently used as a verb, and *image* will be used as a noun. In this context, *imaging* will refer to the acquisition of an *image*, where an image is a pictoral representation of anatomy, physiology, or biochemistry. Typically, the underlying *object*, from which the image is obtained, possesses four-dimensional information (three spatial dimensions plus time). In practice, a single image usually refers to a two-dimensional depiction of a subset of data from the ongoing four-dimensional processes in the object. In this regard, the anatomy or structural information can be regarded as a slow or static process, where the three-dimensional spatial information is most important, while physiological or biochemical processes require all four dimensions of information.

Nuclear Medicine Imaging

Nuclear medicine is that discipline of medicine which makes use of internally administered radioactive substances. These *radiotracers* are drugs or naturally occurring compounds whose biological behavior (i.e., distribution in space and time) in the body we wish to study or *trace*. The molecules are *labeled* with radioactive elements via inorganic or organic chemical syntheses. The radiotracer is then injected into a peripheral arm vein of the patient, or the patient inhales or ingests the tracer. Specialized instrumentation produces images of the internal distribution of radioactivity, which is assumed to mirror the distribution of the compound of interest. These images are compared with known distributions in different disease states. Because there are hundreds of different radiotracers routinely available in nuclear medicine, there are literally hundreds of different nuclear medicine studies, each of which assesses the function of a different physiologic process or organ system within the body.

The technology of nuclear medicine permits direct measurements to be made of body processes in humans that in the past could be examined only in experimental animals. In order to most accurately depict the biodistribution of radiotracers, instrumentation which emphasizes high image quality and quantiative accuracy has been developed. Furthermore, digital image processing plays an important role in nuclear medicine, not only in image enhancement but also in extraction of quantitative information about physiological function.

Planar and Tomographic Imaging

Since each individual image is a two-dimensional depiction, the other two dimensions are either missing or incorporated into the picture. In the case of a static process, such as a study of anatomy, the two-dimensional image can be formed via compression of the third dimension. For example, a conventional chest x-ray radiograph compresses the front-to-back information; underlying and overlapping structures are merged. This type of imaging is known as *planar* imaging.

If, on the other hand, only a slice of information from the three-dimensional object is depicted, this is known as *tomographic* imaging. For example, in x-ray computed tomography, consecutive cross-sectional images of the head or body are produced, much like individual bread slices in a loaf. In practice, most tomographic systems initially produce cross-sectional images, from which slices at arbitrary orientations may be produced.

Static and Dynamic Imaging

In much the same way that planar imaging compresses the third spatial dimension, the finite time over which a single image is acquired compresses the time dimension. In the case of anatomic imaging, where the object is unchanging, this time compression is of no concern, and a single *static* image is obtained. In most nuclear imaging, where the biodistribution of the radiotracer is changing with time, this time compression leads to blurring, and must be considered. Indeed, the entire purpose of the imaging study may be to assess the kinetic behavior of the radiotracer. In such a situation, time compression is unacceptable, and a series of images over time must be acquired. Such *dynamic* imaging is commonplace in nuclear medicine.

Analog and Digital Imaging

A photograph is an *analog* image, with continuous sampling of the object. In nuclear imaging, the equivalent *cameras* (to be described below) were also analog, and produced black-and-white (more correctly gray-scale) images. Today, most nuclear imaging employs digital camera-computer systems. In such a system, an image is formed from discrete sampling, and is represented as a matrix composed of individual cells or *picture elements* (often called *pixels*). Each pixel contains one number, representing the value of the image at that spatial location. Digital imaging is required for accurate quantitative studies.

Qualitative and Quantitative Analyses

Many clinical nuclear medicine studies are interpreted *qualitatively.* A trained physician visually assesses the images in a study, and compares the radiotracer's distribution with known patterns in health and disease. In addition, many studies are analyzed *quantitatively*, with the aid of computer software. Such quantitative approaches are particularly important in drug development, because often the values of physiologic parameters are desired.

Measures of Imaging System Performance

The goal of the nuclear imaging equipment and techniques described below is to accurately and precisely depict the four-dimensional distribution of radioactivity within the object. Many measures of different aspects of imaging system performance are required to assure that the equipment meets this goal. One of the most important parameters of system performance is *resolution.* In our discussion, resolution will be taken to mean the ability to distinguish two separate events in space, time, or energy. The higher or better the resoluition of the system, the better its discriminating abilities. Other important parameters, to be discussed below, include uniformity of response, linearity, sensitivity, and count-rate performance.

RADIATION PHYSICS

Radioactive decay is the spontaneous nuclear transformation of an atom into a more stable (lower inherent energy) combination of protons and neutrons. The term *radioactivity* itself refers to the number of radioactive atoms undergoing decay per unit time. In the process of radioactive decay, *radiation* is emitted from the nucleus, which carries off the excess energy. There are several different modes of decay, including alpha decay, beta decay, and isomeric transitions. These modes of decay result in the emission of either *particulate* or *electromagnetic radiation.* Alphas and betas are particulate radiation. *Alphas* are composed of two protons and two neutrons (i.e., a helium nucleus); *betas* are essentially electrons, differing only in their origin. X rays and gamma rays are electromagnetic radiation. They differ only in their origin: *x rays* come from the electron cloud, while *gamma rays* are emitted during a form of radioactive decay termed isomeric transition. Particulate radiations have ranges in human tissues on the order of millimeters; such radiation, emitted internally, would never leave the body to be detected externally. X rays and gamma rays are much more penetrating (see below); radionuclides which emit electromagnetic radiation are used in nuclear medicine.

Electromagnetic radiation interacts with materials (such as human tissues or radiation detectors) by three main mechanisms: the *photoelectric effect*, *Compton scattering*, and *pair production*. In the photoelectric effect, the radiation interacts with the electromagnetic field of the nucleus of atoms in the tissue; in the process, the radiation's energy is transferred to an electron in the atom (usually a K shell electron). The electron is ejected, ionizing the atom, with kinetic energy equal to the incoming radiation's energy minus the electron binding energy. In Compton scatter, the radiation collides with an unbound electron, and both are scattered in different directions. The amount of energy transferred to the electron depends on scatter angle; the greater the angle by which the incoming radiation is scattered, the greater the fraction transferred to the electron. In pair production, the incoming radiation interacts with the atom, and a positron-electron pair appears in the photon's place. Since the mass equivalent to the positron-electron pair is 1.02 MeV (from $E = mc^2$), electromagnetic radiation with less than this energy cannot interact by pair production. (An *electron volt*, eV, is the amount of energy an electron gains when accelerated across a potential of one volt.) Energy in excess of 1.02 MeV is given as kinetic energy to the positron-electron pair.

The *intensity* of x rays or gamma rays passing through a particular material is given by:

$$I = I_o e^{-\mu x}$$

where I is the intensity of electromagnetic radiation passing through a thickness 'x' of some material given an initial intensity I_o. 'μ' is the *linear attenuation coefficient*, which describes the attenuating ability per unit distance of the material; it is a function of both the material and the energy of the impinging electromagnetic radiation.

Most radionuclides used in nuclear medicine emit electromagnetic radiation in the 70 - 400 KeV range, and thus interact (in either body tissues or radiation detectors) by a combination of photoelectric interactions (which result in complete absorption of the x or gamma rays) or Compton scattering (which results in partial transfer of the radiation's energy). An intrinsic problem in nuclear imaging is that the measurement is a function of two variables, the radioactivity distribution (which is what we wish to know) and the (usually unknown amount of) attenuation by intervening tissues.

PLANAR NUCLEAR IMAGING

The most basic image acquisition is that of a two-dimensional picture. In nuclear medicine, several different approaches have been developed over the past 25 years. These include multi-wire proportional counters, multi-crystal cameras, and the Anger scintillation camera. Today, the scintillation camera dominates planar nuclear imaging.

Anger Scintillation Camera

The Anger scintillation camera was invented by Hal Anger at Donner Laboratory, University of California, Berkeley in the late 1950s. It is the most commonly used imaging instrument in nuclear medicine today. The complete camera system consists of a multi-hole lead collimator, a 10 to 25 inch circular, square, or rectangular sodium iodide scintillation crystal, an array of photomultiplier tubes on the crystal, a positioning logic network, a pulse height analyzer, a scaler-timer, and a cathode ray tube display (Richardson, 1977; Sorenson and Phelps, 1987; Links, 1989).

Collimators

The collimator is a 1 to 2 inch thick slab of lead the same dimensions as the scintillation crystal, with a geometric array of holes in it. The lead in-between each hole is called a *septum*; collectively the lead represent *septa*. The collimator provides an interface between the patient and scintillation crystal, by only allowing those photons traveling in an appropriate direction (i.e., those that can pass through the holes without being absorbed in the lead) to interact with the crystal. Thus, the collimator discriminates against photons based on their direction of travel, and restricts the field-of-view of the crystal. There are several types of collimators used with Anger cameras: parallel-hole, converging, diverging, and pinhole.

The most commonly used collimator is the parallel hole collimator, which consists of an array of parallel holes, essentially perpendicular to the crystal face, and thus presents a real-size image to the crystal face. The resolution of a parallel-hole collimator is best at the collimator surface. The sensitivity is independent of the distance between the source and the collimator.

Converging collimators have an array of tapered holes that aim at a point at some distance in front of the collimator. The image that is presented to the crystal is a magnified version of the real object. Converging collimators have their best resolution at the surface of the collimator. The sensitivity of a converging collimator slowly increases as the source is moved from the collimator face back to the focal plane (the plane parallel to the collimator face that passes through the point in space at which all the holes aim), and then decreases.

Diverging collimators are essentially upside-down converging collimators. They have an array of tapered holes that diverge from a hypothetical focal point behind the crystal. The image presented to the crystal face is a minified image of the real object. Since converging and diverging collimators are simply flipped versions of each other, some collimators may have an insert that can be flipped either way, in effect producing two collimators in one.

Pinhole collimators are thick conical collimators with a single 2 - 5 mm hole in the bottom center. As a source is moved away from the surface of a pinhole collimator, the camera image gets smaller. However, the camera image is magnified (that is, larger than real size) from the collimator face to a distance equal to the length of the collimator, and is then progressively minified at larger distances.

Crystal

The basic detector used in an Anger camera is a scintillation crystal. This detector emits a flash of light when radiation interacts with crystal molecules and transfers energy. These flashes of light are recorded by photomultiplier tubes, and converted into electrical pulses. Crystals used in Anger cameras are extremely sensitive to moisture, and are sealed in an aluminum housing. In addition, they are sensitive to temperature, especially rapid changes in temperature, which can produce cracking. The crystals used in Anger cameras vary from 7 to 20 inches in diameter, and one-quarter to one-half inches thick. The thicker the crystal, the higher the probability that an incoming photon will interact, deposit its energy, and be detected. This yields a camera with higher sensitivity. However, the thicker the crystal, the poorer the spatial resolution, due to the complex interaction between the crystal, the photomultiplier tubes, and the light pipe that is generally used to optically couple the two. One-quarter inch thick crystals have about 1 mm better intrinsic resolution than one-half inch thick crystals. When counting low energy radionuclides such as Thallium-201, there is no difference in sensitivity. However, when counting Technetium-99m, one-quarter inch thick crystals have 15% less sensitivity than one-half inch thick crystals. At higher energies, the difference in sensitivity is even more significant. Crystals with a thickness of three-eighths to one-half inch are required to efficiently detect gamma rays above 200 keV.

Event Positioning Logic

Anger cameras have an array of 19 to 91 photomultiplier tubes optically coupled to the back of the scintillation crystal. These numbers are determined by the size and shape of both the crystal and each individual photomultiplier tube; there is no fundamental reason why 19, 37, and 61 tubes are so common. The more photomultiplier tubes the better the spatial resolution and linearity. Early photomultiplier tubes had a round cross-section. Current tubes often have a hexagonal cross section to cover more of the crystal area for more efficient detection of scintillation photons.

When a scintillation event occurs, each photomultiplier tube produces an output pulse. The amplitude of the pulse from a given photomultiplier tube is directly proportional to the amount of light (number of scintillation photons) its photocathode has received. Those photomultiplier tubes closest to the scintillation event have the largest output pulses. If only that tube with the largest pulse were used for positioning, the spatial resolution of the camera would be equivalent to the cross-sectional size of each tube. By combining the pulses from each photomultiplier tube through a resistor or capacitor network, a higher resolution X,Y-coordinate of the gamma ray location can be generated, based on a centroid approach.

Energy Discrimination

The desired goal of the Anger camera is to create an image that portrays the distribution (i.e., sites and numbers of radioactive atoms) of radioactivity within the patient. Because the collimator only allows those photons traveling in predetermined directions to interact in the crystal, a line drawn from the scintillation event in the crystal through the nearest collimator hole is presumed to intersect the site of origin of the photon (i.e., the radioactive atom it came from) in the patient. If the photon has been scattered in the patient, a line drawn through its direction of flight will not intersect its site of origin, only the site of the Compton interaction. Thus, photons scattered into the field-of-view could be falsely attributed to activity at the sites of Compton interactions in the patient. It is clearly not desirable to have these scattered photons contribute to the final image, as they significantly degrade resolution and contrast. It is important to note that a large percentage of photons striking the crystal have been scattered in the patient.

The Z pulse is used by the pulse height analyzer to discriminate against these scattered photons. The pulse height analyzer is used to set a window around the photopeak. Because the window has a finite width, some scattered photons may still be accepted (those which scattered through a small angle, and thus retained most of their energy). For example, 140 keV photons can scatter by as much as 50° and still be accepted by the often-used 20% window. In practice, proper window setting is vital, as a window which is not centered around the photopeak (an offset window) can degrade field uniformity for many cameras; this is typically a result of the slightly better light collection efficiency directly under each photomultiplier tube. Newer cameras, with microprocessor based correction circuitry (which will be described later), generally maintain good uniformity even with offset pulse height windows. Such cameras may be purposely peaked to the high side of the photopeak to further reduce scatter, by eliminating any Compton scattered photons which show up in the lower half of the photopeak (due to every camera's less than perfect energy resolution). Some cameras have two or three separate pulse height windows, to simultaneously image the multiple emissions of some radionuclides (e.g., those from Gallium-67). In this way, counts are acquired in a shorter amount of time, as the multiple energy emissions are utilized.

Anger Camera Performance

There are many factors that affect the performance of Anger cameras, including spatial resolution, sensitivity, and field uniformity. Modern Anger cameras contain correction circuitry to improve performance in these areas. Field *uniformity* is the ability of the camera to depict a uniform distribution of activity as uniform. At one time it was thought that non-uniform response arose from changes in sensitivity across the crystal. To correct the non-uniformity, a uniform *flood* or sheet source of radioactivity was imaged and recorded in an electronic memory in the camera. Clinical images were corrected during acquisition by either adding counts to the image in areas where the flood had too few counts relative to the other areas, or by subtracting (i.e., purposely not recording) counts in areas with too many counts. It is currently thought that non-uniformities arise mainly from mispositioning of counts in the image (i.e., from errors and non-linear distortions in the positioning logic), rather than from changes in crystal sensitivity. Reference images are acquired, and digital correction maps generated and stored. Each map contains values which represent X,Y correction shifts. Sophisticated microprocessor circuitry is used to reposition each count in real time during acquisition using these shifts. Variation in the position of a pulse from different areas of the camera within the pulse height window can also produce non-uniformities. This spatially-dependent energy distortion may also be corrected by microprocessor circuitry. Correction of spatial and energy distortion improves spatial resolution and linearity as well as uniformity. (Used in this context, *spatial resolution* refers to the ability of the imaging system to distinguish separate points of radioactivity, and *linearity* refers to the ability of the system to depict a straight line source as straight.)

Spatial resolution and linearity can be assessed with phantoms. These typically consist of a lead sheet with an array of holes, through which a point or sheet source is shined. The image is compared with the known pattern of holes in the phantom.

Count-rate performance is also important. Anger cameras, like other nuclear instrumentation, require a finite amount of time to register one event or pulse. During this period of time, the system may be unresponsive to other radiation. In such a situation, as the radioactivity in the object increases, the count-rate may not proportionately increase; this produces an underestimation of activity at count rates.

Overall image quality is a function of both total effective resolution (which governs accuracy) and statistical precision. In this context, the term *accuracy* refers to the agreement between the depicted distribution of radioactivity and the true distribution within the patient; both spatial resolution and contrast are linked to this concept. The term *precision* refers to the point-to-point variation in observed counts due to image noise within an area of uniform radioactivity (see below). Assuming that each factor which can affect resolution is independent of the other factors, and that the factors act in parallel, the total effective resolution is determined by the intrinsic camera resolution, the collimator resolution, the effects of patient or organ motion, and scatter of radiation within the patient. The statistical precision is determined by the total number of counts acquired in the image. Statistical precision can be expressed by the relative standard deviation (or coefficient of variation). For Poisson statistics, the standard deviation equals the square root of the mean, and the coefficient of variation decreases (i.e., the precision improves) as the observed total counts increase. The statistical precision of an image determines what is often referred to as the "noise" in an image. Noise gives rise to a mottled or "salt-and-pepper" appearance in the image. The larger the total counts in an image, the lower the noise, and the less mottled in appearance.

TOMOGRAPHIC IMAGING

Tomography is the process of producing a picture of a section or slice through an object. In medical imaging, tomography is performed either by transmitting x rays through an object (as in transmission computed tomography - CT scanning), by measuring proton density (as in magnetic resonance imaging - MRI), or by tomographically determining the distribution of radioactivity in a patient as in emission computed tomography.

Emission computed tomography, in its most general use, refers to the process of producing a picture of the distribution of radioactivity in a slice through the patient. The slice can be oriented horizontally (a transaxial slice), vertically (coronal or sagittal slices), or at any arbitrary oblique angle to the body. In the past, emission computed tomography utilized either coded aperature systems (such as "seven-pinhole tomography") or true transaxial tomographic acquisition and reconstruction (as used in CT). Today, only transaxial approaches are in widespread use; these include SPECT and PET.

SPECT

SPECT is generally used today to refer to true transaxial tomography with standard nuclear medicine radiopharmaceuticals (i.e., those that emit a single photon upon decay, as opposed to positron emitters, whose emissions result in two coincident annihilation photons). SPECT is performed with either specialized ring detector systems or rotating Anger cameras. The ring systems consist of an array of individual detectors (usually sodium iodide crystals) that surround the patient. These systems, which produce excellent tomograms, tend to be very expensive. By far the most popular method of doing SPECT is with a rotating Anger camera (or multiple cameras) mounted on a special gantry that allows 360° rotation around the patient.

The essence of emission transaxial tomography is similar to that of CT: an object is viewed at a number of angles between 0° and 360° around it. Images are acquired at many angles, each representing one "projection" of the object. In general, a parallel hole collimator is used; the projections thus have "parallel beam geometry." In some cases, a "fan beam" collimator is used; this collimator has holes which converge in the plane of the slice, but are parallel from slice-to-slice. Projections from these collimators have "fan beam geometry." In order to reconstruct a slice through an object, each projection need only be a one-dimensional linear scan of the object. The use of an Anger camera, which produces two-dimensional images, therefore allows simultaneous acquisition of data for a number of continguous transaxial slices. Note, however, that the data used to reconstruct a given slice come only from that slice. As in CT, a filtered back-projection algorithm is used for slice reconstruction.

It is common in CT to reconstruct images with a "ramp filter." This filter, which is mathematically the theoretically "correct" filter to use in filtered back-projection, produces images with the best resolution. In emission computed tomography, the use of a ramp filter typically results in significant image noise, because the high spatial frequencies, which receive the most boost by the filter, are dominated by noise. In practice, different filters are used to provide different trade-offs between noise reduction and spatial resolution. Typical filters include Butterworth, Hanning, and Shepp-Logan. All of these attenuate high spatial frequencies, thus producing reconstructed images with poorer resolution but less noise than the same images reconstructed with a ramp filter.

Studies have shown that Anger cameras must have significantly better performance for adequate SPECT than for adequate planar imaging. For example, nonuniformity must be reduced to less than 1% in order to avoid reconstruction artifacts. This requires acquisition of a 30 - 120 million count reference image of a uniform field for subsequent computer correction of non-uniformities. The camera image must also be mechanically properly aligned within the computer matrix, or an axis-of-rotation correction made. In addition, there are several other factors that must be accounted for to make SPECT truly quantitative. There is disagreement about the need for attenuation, scatter, and spatial resolution corrections for subjective visual interpretation of SPECT images.

PET

One of the most exciting tomographic techniques is PET scanning. Positron emitting radionuclides are used with this technique. Recall that a positron is an anti-matter electron, and consider a positron-emitting radiopharmaceutical distributed in a patient. As a positron is emitted, it travels several mm in tissue, depositing its kinetic energy. It then meets a free electron in the tissue, and mutual annihilation occurs. From conservation of energy, two 511 keV annihilation photons appear (511 keV is the energy equivalent to the rest mass of an electron or positron); from conservation of momentum, they are emitted 180° back-to-back. An Anger camera could be used to individually detect these 511 keV gamma rays. However, it makes more sense to surround the patient with a ring of detectors, and electronically couple opposing detectors to simultaneously identify the pair of gamma rays.

When two 511 keV gamma rays are detected by opposing detectors in coincidence, an annihilation event must have occured along the line joining the two detectors. The direction of travel of the photons is thus determined without the need for a collimator. Conceptually, the raw PET scan data consist of a number of these coincidence lines. Reconstruction could simply be the drawing of these lines: they would cross and superimpose wherever there was activity in the patient. In practice, the data set is reorganized into projections, and filtered back-projection is used. In some PET scanners, the difference in arrival times of the two photons is used to position the event along the coincidence line. In practice, this "time-of-flight" spatial resolution is limited to several cm because of temporal resolution limitations (hundreds of psec); these limitations are due both to electronics and the decay time of the scintillation detectors used. (Decay time refers to the amount of time required for light to be emitted following deposition of energy in the crystal.) Thus, time-of-flight is used in combination with back-projection reconstruction. Even without time-of-flight, PET differs from SPECT in that the "electronic collimation" of coincidence counting reduces the need for actual lead collimation, thus increasing sensitivity.

The excitement about PET is due to both the chemistry and physics inherent in positron tomography. The most commonly used radionuclides, Carbon-ll, Nitrogen-13, Oxygen-15, and Fluorine-18, are isotopes of elements that occur naturally in organic molecules. (Fluorine usually does not, but is a bioisoteric substitute for hydrogen.) Thus, radiopharmaceutical synthesis is simplified, and the tracer principle (which mandates as small a change in the molecule to be traced as possible) is better satisfied. Indeed, useful PET radiopharmaceuticals are now available to measure *in vivo* such important physiologic and biochemical parameters as (i) blood flow, (ii) oxygen, glucose, and free fatty acid metabolism, (iii) amino acid transport, (iv) pH, and (v) neuroreceptor densities. The short half-lives of the radionuclides (Carbon-ll, 20 min; Nitrogen-13, 10 min; Oxygen-15, 2 min; Fluorine-18, 110 min) permit the acquisition of serial studies on the same day without background activity from prior injections interfering with the measurements. The physics of PET permits greater quantitative accuracy and precision. The use of small, high density crystals improves spatial resolution (about 4 mm in the best commercial PET scanners). The lack of collimation to determine photon direction dramatically increases sensitivity. Finally, coincidence detection allows mathematically accurate attenuation correction.

DATA PROCESSING

Image and data processing play important roles in both planar and tomographic imaging. Specific protocols are necessary to ensure appropriate acquisition and analysis of data, whether the final "product" is an image for visual interpretation or quantitative data. Major factors which influence data quality are filtering and other image processing procedures, and definition of the location of anatomic regions-of-interest for analysis.

Image Processing

Nuclear images are characterized by both poor spatial resolution and high image noise. Digital image processing can play an important role in improving image characteristics. For example, it is common to filter nuclear images, either through a convolution operation or the Fourier transform. Such filters typically are used to reduce the image noise produced by the statistical nature of radioactive decay. These "smoothing" filters usually greatly improve the image signal-to-noise ratio, although at the expense of spatial resolution.

Tomographic reconstruction itself can be viewed as a form of image processing. The "raw" projection data from a PET scan consists of a number of "coincidence lines," representing those pairs of opposing detectors which simultaneously detected annihilation photons, while the projection data from SPECT are usually organized into parallel or fan beam projections. In practice, the PET data are reoriented into similar parallel or fan beam projections; the SPECT or PET image is then reconstructed with a standard filtered back-projection algorithm.

As discussed above, an important consideration is the choice of reconstruction filter. A "sharp" filter, like the ramp, will preserve spatial resolution in the reconstructed image, but will amplify the random noise in the image. Conversely, a "softer" filter, like the Butterworth, will reduce both spatial resolution and statistical noise. The choice of an appropriate filter is dependent on the required spatial resolution and statistical precision of the study. New approaches to filtering include the use of sophisticated Fourier-space analyses, such as Wiener or Metz filtering, which seek to simultaneously improve spatial resolution and reduce noise in the filtered image (King et al., 1984; King et al., 1986).

Definition of Anatomic Regions

In applications relevant to radiopharmaceutical research, image quantification is commonly performed. Typically, knowledge of the radioactivity in specific organs or anatomic regions is desired. The placement of a region-of-interest can be made on the basis of several pieces of data. To begin with, many structures are "visible" in the nuclear images themselves. Indeed, most investigators use the actual images as the primary or only guide to the placement of regions-of-interest. Even so, other investigators argue that this approach to region placement is not adequate, because: (i) only "functionally active" parts of a structure will be visible in the image, (ii) images have insufficient spatial resolution to permit accurate placement, and (iii) manual placement compromises objectivity and reproducibility. These investigators have proposed the use of either matched, registered anatomic images in the same subject (Bergstrom et al., 1981; Mazziotta et al., 1982; Kearfott et al., 1984), or the use of a stereotactic coordinate system (for brain studies) (Fox et al., 1985).

The use of corresponding anatomic and functional images is most advanced in brain studies. Most approaches to registration of anatomic and functional images have relied on head-holders (i.e., pre-acquisition registration). Several investigators have described post-acquisition registration approaches based on two- or three-dimensional matching of data sets, typically by reducing the tomographic data to a series of outlines of brain structures (Levin et al., 1988; Pelizzari et al., 1989; Dann et al., 1989). The use of matched, registered anatomic images from the same subject eliminates the need to assume certain elastic morphometric properties. However, many structures, particularly in the cortex, have little contrast with their surroundings in either CT or MRI images. The use of a standardized stereotactic coordinate system with a predefined template of regions-of-interest eliminates the need for an operator to manually position regions on either SPECT/PET or anatomic images. However, such systems may not be accurate in the setting of gross brain distortion (e.g., atrophy or tumor).

The production of an appropriately sized region-of-interest requires consideration of the size and shape of the anatomic structure, the nature of the surrounding tissues, and the resolution characteristics of the nuclear imaging device. In the absence of noise, the best estimate of the actual radioactivity concentration in a small structure is that single pixel in the middle of the structure with the highest counts (Hoffman et al., 1979; Mazziotta et al., 1981; Kessler et al., 1984). In the presence of noise, a larger region-of-interest must be used to improve precision. In practice, the region is made large enough to reduce statistical noise, yet small enough to limit resolution effects, which dictate a progressive underestimation of radioactivity concentration with increasing region size.

Typically, regions-of-interest are used in one of two ways. In static studies (i.e., those with only one time measurement), regions are generated and individual region values obtained. These can be compared with each other (e.g., to compare radiopharmaceutical uptake in different organs) or between subjects. In dynamic studies (i.e., those with multiple time images of the same subject), the same region can be applied to the series of images, and the resulting values graphed as a time-activity curve. This curve is particularly useful in characterizing tracer pharmacokinetics, and is used in mathematical modeling.

QUANTIFICATION

One of the most important features of nuclear imaging is the ability to quantify the four-dimensional distribution of radioactivity, and to derive useful parameters of tracer behavior or actual physiology. Approaches to quantification require knowledge of, and correction for, various physical factors present in nuclear imaging. In nuclear imaging, quantitative accuracy is at its highest in tomographic imaging. Planar imaging is typically used to obtain overall organ estimates of dose, or to generate time-activity curves. While simple correction schemes for the effects of such physical factors as attenuation exist, they are rarely utilized. For advanced quantification, SPECT and PET are used. Indeed, for many investigators, the major value of SPECT or PET is not the production of an image per se, but the measurement of physiologic and biochemical processes in absolute terms. The conversion of an image into these absolute values is a function of many factors. The first step is to convert observed "count rate" into true radioactivity. Next, an equilibrium or kinetic mathematical model is used to describe the biodistribution of radioactivity in space and time. From this model, the physiologic variables of interest, which are incorporated into the model, can be derived.

From a physics point of view, there are five major factors which affect the quantification of absolute radioactivity. These include: (i) attenuation of photons by tissue, (ii) transaxial spatial resolution and effective slice thickness of the SPECT or PET scanner (Hoffman et al., 1979; Huang et al., 1980; Mazziotta et al., 1981; Hoffman et al., 1982), (iii) detection of scattered photons (Bergstrom et al., 1983; Logan and Bernstein, 1983), (iv) accidental counting of "random" (non-paired) photons in coincidence (Hoffman et al., 1981), applicable only to PET, and (v) "noise" resulting from the statistical nature of radioactive decay (Budinger et al., 1978; Hoffman et al., 1984).

Quantification in Planar Imaging

As stated above, planar imaging, while frequently used, is not typically relied on to provide absolute quantification. Even so, there are applications where it is useful to use planar imaging to determine the fraction of injected dose present in each organ or anatomic region. Such an approach must take attenuation of the radiation by intervening body tissues into account. The photons detected in nuclear imaging are electromagnetic radiation. As such, they undergo two major types of interactions in tissue: the photoelectric effect, first described by Einstein, and Compton scattering. The photoelectric effect results in complete absorption of the photon, reducing count rate, while detection of scattered photons reduces the contrast in the image. Failure to take attenuation into account results in an erroneously low estimate of organ radioactivity content.

The most common approach utilizes data from opposing views. In this geometric mean method, the activity (A) in an organ is given by:

$$A = \text{sqrt}\,[(C_{ant} \times C_{post})/T] \times S$$

where C_{ant} and C_{post} are the region values from opposing views (typically anterior and posterior), T is the fractional transmission of counts through the subject, and S is the imaging system's sensitivity. T is typically obtained by measuring the transmission of photons through the subject, prior to injection of activity, utilizing a sheet source of activity. S is typically obtained by determining the observed count rate from a known activity point source.

Quantification in Tomographic Imaging

As in planar imaging, attenuation is a major factor influencing quantification. The average attenuation is greatest at the center of an object, such as the head or body of a patient. Attenuation produces a gradual, progressive underestimation of radioactivity from the edge to the center of the head or body, by about a factor of three to five. Thus, it is extremely important to correct for its effects.

There are two main approaches to attenuation correction. In the first, attenuation is measured before the SPECT or PET scan begins by placing a ring of single-photon or positron-emitting activity around the patient's head or body, and measuring the transmission of the photons from the ring through the patient. By relating this measurement to a similar one without the patient, which defines the unattenuated number of photons, the attenuation experienced by radioactivity at each point within the head or body can be determined. This approach is truly only analytically correct for PET, and is only infrequently used in SPECT (Budinger et al., 1977).

The second approach to attenuation correction does not require any additional measurements. This approach is typically only used in the head, not for body scans. After the uncorrected image is reconstructed, the computer operator outlines the scalp, usually with an ellipse. An average value for attenuation is then assumed for each point within this outline of the head. (This average value depends on both the tissues involved and the characteristics of the particular SPECT or PET scanner.) The use of a "measured" attenuation correction may be more accurate, but results in propagation of statistical noise from the transmission study into the emission image. It also requires the additional transmission scan. Because of these considerations, most investigators use the "calculated" correction approach, with an ellipse and an assumed attenuation value.

All tomographic scanners have a limited ability to resolve small objects. The spatial resolution of a scanner can be thought of as that distance by which two small point sources of radioactivity must be separated to be distinguished as separate in the reconstructed image. Finite spatial resolution results in two important effects. First, the image is blurred, with the degree of blurring dependent on the spatial resolution. This blurring has a greater impact in head studies than in body studies. This blurring prevents the delineation of edges of larger structures in the brain, and may not allow the visualization of smaller ones as distinct objects. Furthermore, in the cortex, the grey and white matter are smeared and averaged together, reducing the measured value in the areas with greater radioactivity (usually grey matter), and increasing it in the areas with lesser radioactivity (usually white matter).

The second effect is more subtle. Finite spatial resolution produces an underestimation of radioactivity in small structures with progressive underestimation as the structures get smaller. The effect is not eliminated until the object is approximately three times the resolution of the scanner. This underestimation effect is conceptually separate from the blurring effect, though they both arise from finite spatial resolution. Indeed, blurring from surrounding larger structures can partially compensate for the underestimation in a small structure, by smearing counts into the small structure. Thus, the underestimation is greatest for isolated structures, such as the heads of the caudate nuclei in the brain.

Partial volume effects are related to finite spatial resolution effects, but apply to the so-called "z axis," along the long axis of the patient. If the object is smaller than about three times the slice thickness, the averaging and underestimation effects described above will occur. The underestimation can be corrected with a recovery coefficient as well. Orthogonal one-dimensional x, y, and z recovery coefficients could be sequentially applied, or an "effective" three-dimensional coefficient can be used. It is important to note that the term "slice thickness" is really a misnomer, as the profile of the slice is not a square wave with a uniform width, but rather more like a bell-shaped curve. Slice thickness is usually given as the "full-width-at-half-maximum" of this gaussian function. Thus, radioactivity lying outside the "slice thickness" still contributes to the slice (though much less so), and not all the activity "within" the slice is detected with equal sensitivity.

Finite spatial resolution and partial volume effects can be corrected if: (i) the "recovery" characteristics of the SPECT or PET scanner are known in three dimensions, and (ii) the shape, dimensions, and position of a structure relative to the tomographic slice can be determined. The recovery characteristics of the scanner can be empirically determined by using a wedge (one dimensional), cylinder (two dimensional), or sphere (three-dimensional) phantom. In this approach, progressively larger phantoms are scanned, with each phantom containing the same concentration of radioactivity. A plot of average counts per pixel as a function of phantom dimension depicts the quantitative recovery as a function of size. Alternatively, recovery can be estimated from measurements of in-plane and z-axis line spread functions, as the fraction of the total area under the line spread function subtended by the geometric edges of the phantom (or actual structure).

Once the recovery characteristics of the scanner have been defined, subject-specific measurements can be used to correct the observed counts for finite spatial resolution and partial volume effects. This approach requires the use of matched, registered anatomic images. The geometric borders of the structures of interest are defined in the anatomic images with conventional edge detection techniques (e.g., thresholds or derivatives). A three-dimensional wire frame of the object is computed. If the z-axis sampling in the anatomic data is coarse, cubic spline interpolation can be used to compute the border outlines of the structure in the "missing" slices. Because the anatomic images are registered to the SPECT/PET images (see above), the location of the SPECT/PET slice of interest with respect to the three-dimensional wire frame is known. Thus, orthogonal x, y (in-plane) and z recovery coefficients can be estimated, based on the phantom measurements described above.

REFERENCES

Bergstrom M, Boethius J, Eriksson L, Greitz T, Ribbe T and Widen L (1981): Head fixation device for reproducible position alignment in transmission CT and positron emission tomography. *J Comput Assist Tomogr* 5: 136-141.

Bergstrom M, Eriksson L, Bohm C, Blomqvist G and Litton J (1983): Correction for scattered radiation in a ring detector positron camera by integral transformation of the projections. *J Comput Assist Tomogr* 7: 42-50.

Budinger TF, Derenzo SE, Gullberg GT, Greenberg WL and Huesman RH (1977): Emission computer assisted tomography single-photon and positron annihilation photon emitters. *J Comput Assist Tomogr*

1: 131-145.

Budinger TF, Derenzo SE, Greenberg WL, Gullberg GT and Huesman RH (1978): Quantitative potentials of dynamic emission computed tomography. *J Nucl Med* 19: 309-315.

Dann R, Hoford J, Kovacic S, Reivich M and Bajcsy R (1989): Evaluation of elastic matching system for anatomic (CT, MR) and functional (PET) cerebral images. *J Comput Assist Tomogr* 13: 603-611.

Fox PT, Perlmutter JS and Raichle ME (1985): A stereotactic method of anatomical localization for positron emission tomography. *J Comput Assist Tomogr* 9: 141-153.

Hendee WR (1973): Radioactive Isotopes in Biological Research. New York: John Wiley & Sons.

Hoffman EJ, Huang SC and Phelps ME (1979): Quantification in positron emission computed tomography: 1. Effect of object size. *J Comput Assist Tomogr* 3: 299-308.

Hoffman EJ, Huang SC, Phelps ME and Kuhl DE (1981): Quantification in positron emission computed tomography: 4. Effect of accidental coincidences. *J Comput Assist Tomogr* 5: 391-400.

Hoffman EJ, Huang SC, Plummer D and Phelps ME (1982): Quantification in positron emission computed tomography: 6. Effect of nonuniform resolution. *J Comput Assist Tomogr* 6: 987-999.

Hoffman EJ, van der Stee M, Ricci AR and Phelps ME (1984): Prospects for both precision and accuracy in positron emission tomography. *Ann Neurol* 15: S25-34.

Huang SC, Hoffman EJ, Phelps ME and Kuhl DE (1980): Quantification in positron emission computed tomography: 3. Effect of sampling. *J Comput Assist Tomogr* 4: 819-826.

Huang SC, Carson RE, Phelps ME, Hoffman EJ, Schelbert HR and Kuhl DE (1981): A boundary method for attenuation correction in positron computed tomography. *J Nucl Med* 22: 627-637.

Kearfott KJ, Rottenberg DA and Knowles RJR (1984): A new headholder for PET, CT, and NMR imaging. *J Comput Assist Tomogr* 8: 1217-1220.

Kessler RM, Ellis JR and Eden M (1984): Analysis of emission tomographic scan data: Limitations imposed by resolution and background. *J Comput Assist Tomogr* 8: 514-522.

King MA, Schwinger RB, Doherty PW and Penney BC (1984): Two-dimensional filtering of SPECT images using the Metz and Wiener filters. *J Nucl Med* 25: 1234-1240.

King MA, Schwinger RB, Penney BC, Doherty PW and Bianco JA (1986): Digital restoration of indium-111 and iodine-123 SPECT images with optimized Metz filters. *J Nucl Med* 27: 1327-1336.

Levin DN, Pelizzari CA, Chen GTY, Chen CT and Cooper MD (1988): Retrospective geometric correlation of MR, CT, and PET images. *Radiology* 169: 817-823.

Links JM (1989): Instrumentation. In: *Nuclear medicine technology and techniques (2nd ed)*, Bernier DR, Christian PE, Langan JK and Wells LD, eds. St. Louis: The C. V. Mosby Company,

Logan J and Bernstein HJ (1983): A Monte Carlo simulation of Compton scattering in positron emission tomography. *J Comput Assist Tomogr* 7: 316-320.

Mazziotta JC, Phelps ME, Plummer Dand Kuhl DE (1981): Quantification in positron emission computed tomography: 5. Physical- anatomical effects. *J Comput Assist Tomogr* 5: 734-743.

Mazziotta JC, Phelps ME, Meadors AK, Ricci A, Winter J and Bentson JR (1982): Anatomical localization schemes for use in positron computed tomography using a specially designed headholder. *J Comput Assist Tomogr* 6: 848-853.

Pelizzari CA, Chen GTY, Spelbring DR, Weichselbaum RR and Chen CT (1989): Accurate three-dimensional registration of CT, PET, and/or MR images of the brain. *J Comput Assist Tomogr* 13: 20-26.

Richardson RL (1977): Anger scintillation camera. In: *Nuclear medicine physics, instrumentation, and agents*, Rollo FD, ed. St. Louis: The C. V. Mosby Company.

Sorenson JA and Phelps ME (1987): Physics in Nuclear Medicine (2nd ed.). Orlando: Grune & Stratton.

Swenberg CE and Conklin JJ (1988): Imaging Techniques in Biology and Medicine. San Diego: Academic Press.

ACCELERATORS FOR POSITRON EMISSION TOMOGRAPHY

Alfred P. Wolf and David J. Schlyer
Department of Chemistry
Brookhaven National Laboratory

INTRODUCTION

The positron emitters Carbon-11, Nitrogen-13, Oxygen-15 and Fluorine-18 can be conveniently produced using low energy particle accelerators. The requirement is for these accelerators to deliver particle beams of sufficient energy to bring about the required nuclear reactions and with sufficient beam current to bring about practical yields of the required radionuclides. Table 1 lists the positron emitters, their characteristics and the most common nuclear reactions for their production which are in use today.

Cyclotron produced radionuclides for biomedical research were used in the late 1930's for some clinical research and for basic research in biochemistry. During World War II a cyclotron in Boston provided a steady supply of radionuclides for medical purposes. One of the earliest pioneers in the use of positron emitters for biomedical research was Dr. William Myers at Ohio State University. Unfortunately, no detailed written record in the literature of his efforts is available. In the mid 1950's a group at Hammersmith in the United Kingdom put into operation a cyclotron wholly dedicated to radionuclide production. However, their early work on positron emitters was almost exclusively

Nuclear Imaging in Drug Discovery, Development, and Approval
H.D. Burns, R. Gibson, R. Dannals, Peter Siegl, Editors

Nuclide	Half-Life	End Point Positron Energy (MeV)	Reaction	Energy Interval
Oxygen-15	2.01 min	1.72	$^{14}N(d,n)^{15}O$	0.5-12
			$^{14}N(p,n)^{15}O$	3.8-14
			$^{16}O(He\text{-}3,\alpha)^{15}O$	10-45
Nitrogen-13	9.98 min	1.19	$^{12}C(d,n)^{13}N$	0.3-20
			$^{13}C(p,n)^{13}N$	2.8-14
			$^{16}O(p,\alpha)^{13}N$	5.5-20
Carbon-11	20.4 min	0.96	$^{14}N(p,\alpha)^{11}C$	3.1-20
			$^{11}B(p,n)^{11}C$	3.0-30
			$^{10}B(d,n)^{11}C$	3.0-9
			$^{12}C(He\text{-}3,\alpha)^{11}C$	10-45
Fluorine-18	109.8 min	0.64	$^{18}O(p,n)^{18}F$	2.1-14
			$^{16}O(He\text{-}3,n)^{18}F$	10-45
			$^{16}O(He\text{-}4,2n)^{18}Ne\rightarrow^{18}F$	10-46
			$^{20}Ne(d,\alpha)^{18}F$	1.3-16

TABLE 1. Positron Emitting Isotopes.

on Carbon-11 labeled carbon monoxide and carbon dioxide. The major change which came about occurred in the early and mid sixties when work in hot atom chemistry (e.g., the *in situ* chemistry of nucleogenic atoms occurring in a target being bombarded) laid the foundation for the synthesis of organic compounds labeled with positron emitters. A 1966 article by Ter-Pogossian and Wagner focussed on the use of Carbon-11 (Ter-Pogossian and Wagner, 1966). During that decade a few articles began to appear on synthesis with Carbon-11 and Fluorine-18. A major increase in interest in labeled compounds was probably stimulated by the announcement of the first synthesis of ^{18}F-fluorodeoxyglucose (Ido et al., 1978). The book by Clark and Buckingham (1975) was the first to bring together the then understood methods of preparing Carbon-11, Nitrogen-13 and Oxygen-15, primarily using deuteron reactions.

As the field of Positron Emission Tomography (PET) has progressed and expanded rapidly in the last decade, the number of available types of machines with varying characteristics has also expanded. The major class of machines are the positive and negative ion cyclotrons. More recent, innovative machines include superconducting magnet cyclotrons, linacs, and tandem cascade accelerators and helium-particle-only linacs and cyclotrons. These machines are less widespread and, in some cases, not field tested. While traditional Van de Graaff accelerators have been used for radionuclide production, they probably should not be considered for new installations.

The remainder of the chapter (i) describes the basic operation of each of these classes of machines in more detail; (ii) focuses on the considerations pertinent to the choice of an accelerator and the placement of the accelerator in a typical operating environment; and (iii) provides a short summary.

THE ACCELERATORS

Cyclotrons

Cyclotrons are the most commonly used devices for the acceleration of particles to energies sufficient for bringing about the required nuclear reactions. It was the remarkable idea of Ernest O. Lawrence to bend the path of the particles in a linear accelerator into a circle and therefore use the same electrode system over and over again to accelerate the particles. This idea is the basis of all modern cyclotrons and has made the cyclotron the most widely used type of particle accelerator. The first model was built in 1930 with proof of particle acceleration being provided by M. S. Livingston in 1931.

Unfortunately, there is not a rich literature on cyclotrons for medical purposes. The book by Livingood (1961) remains as probably the only general text on cyclotrons and other particle accelerators. Detailed information on advances in cyclotrons and other accelerators is available as a series of symposia papers (Duggan and Morgan, 1991) and in the Proceedings of the XIII International Conference on Cyclotrons and their Applications held in Vancouver, Canada, in 1992 (as well as in previous years). Cyclotrons for biomedical radionuclide production have been reviewed (Wolf and Jones, 1983). It is the purpose of this chapter to focus on particle accelerators purely for the production of the four positron emitters used for PET and in common use today.

Principles of Operation

According to the theory of electrodynamics, the rotational frequency of a charged particle travelling in a magnetic field is independent of the radius of its orbit. The energy of the particle increases as the velocity of the particle increases. The cyclotron utilizes this fact to produce particles of reasonably high energy in a relatively confined space. The acceleration chamber of the cyclotron is placed between the poles of a homogeneous magnetic field as shown in Fig. 1. The magnetic field causes the particles to travel in circular orbits. Ions are produced in an ion source at the center of the machine and accelerated out from the center. The ions are accelerated with a high frequency electric field through two or more hollow electrodes each of which is called a "dee". The ions are accelerated as they pass from one dee to the next through a gap between the dees. Since the rotational frequency of the particles remains constant, as the energy of the particles increases, the diameter of the orbit increases until the particle can be extracted from the outer edge of the machine. The limit on the energy of a particle is determined on a practical basis by the diameter of the magnet pole face. Some very large, high energy cyclotrons have been built, but for the most part, the proton energy is less than 70 MeV.

Radio-Frequency Accelerators

Both the fundamental frequency or higher order harmonics of radio-frequency fields may be used for acceleration of charged particles in a cyclotron. The energy of the particle and the harmonic used will determine the exact structure of the hollow electrodes (dees). When the radio-frequency (rf) field is coupled to the dees, electric field gradients are produced. In each pair, one of the dees is at a ground potential and the other is at a higher voltage. Since the particle will feel an electric field when entering and leaving a dee, the number of gap crossings per revolution equals twice the number of dees. An accelerating structure may have one, two, three or even more dees. The energy gain of a

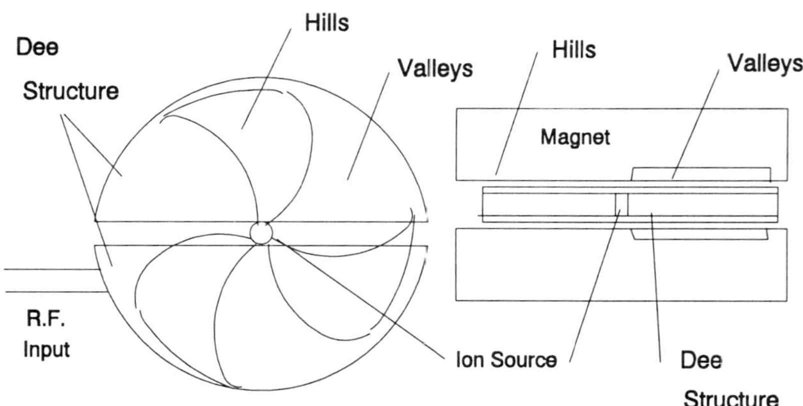

FIGURE 1. Schematic diagram of a cyclotron showing the dee structure, magnet hills and valleys, ion source, and rf input.

particle crossing an accelerating gap depends on the charge, amplitude of the potential difference, the phase relationship of the particle to the electric field and the path length between the electrodes.

The particle must cross the gap between two dees at a given time or it will not be accelerated. For a two dee system operating on the fundamental frequency, the time interval is given by the relation:

$$1/\omega = r/v$$

Where: ω = the orbital frequency.
r = the radius in the median plane.
v = the orbital velocity.

In addition to this acceleration when crossing the gap, the force on the particle has a vertical component. This force is focussing when the electric field is decreasing during the traversal.

The radius of the orbit of a particle in a magnetic field is given by the relation :

$$r = p/qB_z(r)$$

Where: r = the radius of the particle in the median plane.
p = the momentum of the particle.
q = the charge of the particle.
$B_z(r)$ = the magnetic induction in the median plane at the radius r.

This is the stable equilibrium orbit and particles usually oscillate around this orbit. This equation, taken with the equation for the radius of the particle in the magnetic field, are the fundamental equations of cyclotrons. Although in theory the frequency of the orbit is constant, in practice the frequency of the orbit decreases with the radius as a result of relativistic mass increases and the decreasing magnetic field strength with the radius. A particle in phase with the accelerating voltage at the center of the cyclotron will gradually get out of phase and limit the highest energy achievable with this type of cyclotron to about 20 MeV.

The solution to this problem is the use of Azimuthally Varying Field (AVF) isochronous cyclotrons which use the principle of alternating gradient focusing to allow an increase in the magnetic field as the radius increases. The term isochronous refers to the fact that the ions revolve at constant frequency. The alternating gradient is produced by making the gap between the poles of the magnet smaller or larger. The areas where there is a smaller gap (stronger field) are called "hills" and those where there is a larger gap (weaker field) are called "valleys". The use of spiral instead of radial hills and valleys enhances the effect so that cyclotrons of up to 700 MeV may be built (see Fig. 1). Strong magnetic fields will have an effect on the focusing of the beam and this fringing field must be considered in the design of the cyclotron.

Ion Sources

There are several types of ion sources with different operating characteristics. The extraction of positive and negative ions from the ion source and the path in the central region of the cyclotron are determined in part by the ion source characteristics. Cyclotrons may have either an internal or an external source of ions for acceleration. On most commercial cyclotrons, the source is internal. Charged ions are usually produced by an arc discharge in a gas yielding a plasma. The two main types used in cyclotrons are the hot cathode and cold cathode types. In the hot cathode, a heated filament is used to maintain the arc. In the cold cathode type, once the discharge is initiated, no hot filament is used to maintain the plasma during normal operation of the ion source. The spatial distribution and output of an internal source is not well defined. In addition, the ions usually have a broad distribution of energies. These problems are usually overcome by using slits on the ion source which decease the beam intensity but produce a well defined beam profile. Once the ions are out of the ion source, the path of the beam is further defined with a "puller". This is a small slit in a metal plate which accepts only those ions with the proper energy and position to continue past the first orbit. The whole assembly is referred to as the central region. Misalignment of the central region results in a beam which is off center. This can usually be detected and corrected with the use of internal probes (to measure beam current) and ion source positioning and intensity adjustments.

Vacuum Systems

The vacuum inside the cyclotron must be maintained at a fairly high level in order that the particles have a mean free path long enough to be accelerated effectively. In the case of cyclotrons, this translates to a pressure inside the tank of between 1×10^{-5} to 1×10^{-7} mm of Hg depending on the charge of the particle being accelerated and the cyclotron. A higher vacuum (about 10^{-7} mm of Hg) is required for negative ion machines than is required for positive ion machines since gas collisions in the negative ion machine may cause charge stripping. There are two major types of vacuum systems in use for cyclotrons. The first is the oil diffusion pump and the second is the turbomolecular pump. The oil diffusion pump is very reliable and easy to maintain but has the disadvantage of oil vapor which can get back into the cyclotron and cause carbon deposits to build up on some surfaces in spite of in-line trapping systems. The usual failure of diffusion pumps is heater burn out. These are easily replaced and the cyclotron can be back on line in a short time. The turbomolecular pumps are much cleaner and have much higher pumping speeds but suffer from the disadvantage of being susceptible to damage to the vanes by a sudden burst of air in the case of a foil failure on a target; however, there are now turbomolecular pumps on the market which are nominally capable of withstanding a sudden burst of gas with no damage. Turbomolecular pumps are also very sensitive to alignment. They must be protected from small pieces of metal which may be in the cyclotron from ruptured foils or flakes. When these pumps fail it is usually necessary to replace the vane assembly which is an expensive and time consuming proposition.

Extraction Systems

Extraction of the beam depends on whether the charge on the particle is positive or negative. If the particle is positively charged, the extraction is done electrostatically as shown in Fig. 2. If the particle is negatively charged, the extraction is done by stripping the electrons off the negatively charged ion and allowing the magnetic field to reverse the curvature of the resulting positively charged particles' path and push the particles out of the cyclotron as shown in Fig. 2.

Superconducting Cyclotrons

The basic principles of operation of these machines are the same as those for a conventional cyclotron. Using a superconducting magnet, these cyclotrons can be made very compact but must be maintained at liquid helium temperatures. The weight of the magnet is about 1/5 of the weight of a magnet on a conventional cyclotron. Since the magnet is superconducting, the power requirements are also much less. The higher magnetic field causes the orbits to be spaced closer together. The accelerated particle is H^- so the extraction system utilizes a stripper foil.

LINACS

The concept of the linac arose because there is a practical limit to the energy which can be supplied to a particle between two single electrodes. No matter how well the electrodes are insulated, there will be

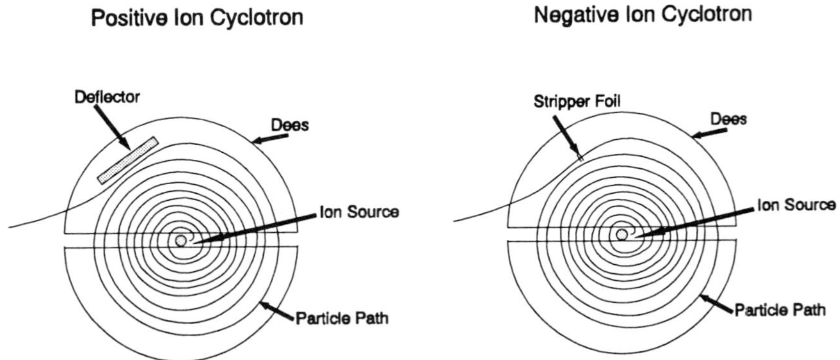

FIGURE 2. Diagram showing the differences between extraction of positive ions and negative ions in a cyclotron.

a discharge of the potential to ground. To overcome this limitation, R. Wideroe devised a system in Germany in the 1930's which would allow the particles to be accelerated in many small steps and would add up to a much greater potential than one giant push.

Principles of Operation

The principle of acceleration used in all accelerators is the fact that a charged particle has its energy changed when it is acted on by an electric field. In the linear accelerator (linac), this change in energy is applied by an alternating potential which must be applied in exactly the proper sequence to keep accelerating the particle. In practice this is achieved with the use of hollow electrodes called drift tubes which allow the particle to drift at constant velocity within the tube and then be accelerated between the tubes. The time the particle spends within the tube allows time for the change in phase of the electric field.

One fact which helps to maintain the timing of the acceleration is what is referred to as phase stability. The potential at each stage of the accelerator can be set so that the maximum potential is applied just after the particles have passed a point. This allows for some margin of error in the timing. If the particle arrives too early, the potential applied will be slightly less than optimum and the particle will traverse the next section more slowly and will be in phase for the next accelerating potential. Likewise, if the particle is moving too slowly, the phase of the accelerator can give it a little extra push and again the particle will move to be in phase at the next section of the accelerator. This will cause the particles to emerge from the end of the linac in bunches. The magnitude of this effect can sometimes be a problem in the design of targetry for the linac since the instantaneous power delivered to a target can be quite high.

Radio-Frequency Acceleration

The radio-frequency power for the current designs of medical linacs is supplied at high frequency (200-500 MHz). The power for the rf system is usually supplied by a bank of power tubes contained within a coaxial cavity. The use of very high frequencies allows the use of shorter sections for the drift tubes and allows the linac to be shorter. This is an extremely important factor since the particles must traverse the length of one drift tube in one cycle of the rf. This implies that the length of the drift tube must be :

$$L = \beta\lambda$$

Where: L = the length of the tube.
β = the fraction of the speed of light for the particle.
λ = the free space wavelength of the radio-frequency.

In order to build a linac with reasonable energies and of reasonable size, a high frequency is essential. Power supplies which can provide high power at high frequency have only recently become available.

A linac design which is typical of the new developments incorporates four sections as shown in Figure 3. The first is an ion source which produces a pulsed proton beam of 30 keV with a single electrode system. In the second section, this beam is bunched and then injected into a radio-frequency quadrupole which is the third section. The energy at the end of this section is on the order of 2 MeV. The beam is then transported to the fourth section where the phase-space of the beam is matched to the phase-space of the drift tube linac and where the protons are accelerated to the final energy of 11 MeV.

Vacuum Systems

In any linear accelerator, a fairly good vacuum must be maintained in order that the charged particles have a mean free path long enough to get from one drift tube to the next without being scattered. In most cases, this is done either with a turbomolecular pump or with a diffusion pump, both of which must be backed with a mechanical pump as is the

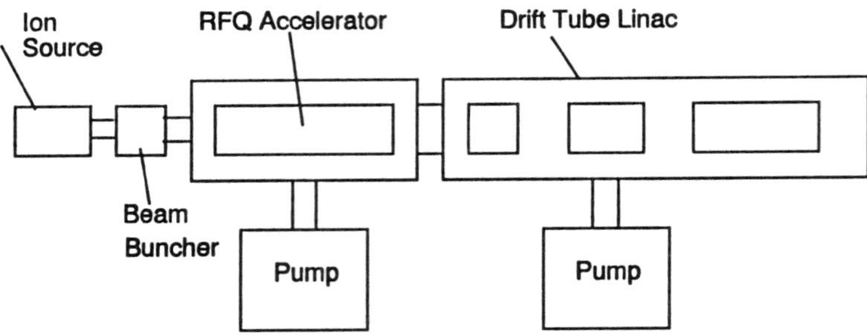

FIGURE 3. Schematic diagram of a PET linac showing the ion source, beam buncher, RFQ accelerator and drift tube linac.

case with cyclotrons. In linacs, this vacuum must be maintained at the level of about 1 x 10^{-6} mm of Hg.

Radio-Frequency Quadrupole

This concept was developed in Russia but was first applied in the United States. The concept is that instead of hollow electrode drift tubes, the sections of the acceleration structure are made of vanes which are set at 90° to each other and to the path of the beam. These vanes are charged by the radio-frequency field and serve the same function as the drift tubes of the traditional linac design.

TANDEM CASCADE ACCELERATORS

The tandem cascade accelerator uses a single large potential to accelerate a charged particle. This type of accelerator was first developed by R. J. Van de Graaff in 1931. The physical principle underlying the machine is that an electric charge placed within a hollow metal conductor must move to the outside surface of the conductor. The accelerator based on the Van de Graaff principle and in use today for PET isotope production is the tandem cascade accelerator. This type of accelerator is composed of four sections. The first is an ion source, the second an acceleration chamber, the third a high voltage terminal with a stripper foil and the fourth a second acceleration chamber. A positive charge is accumulated on the high voltage terminal at the center of the acceleration tube assembly. Negatively charged ions are inserted at ground potential from the ion source at one end of the tube. The ions are accelerated from ground potential to a high voltage at the central high voltage terminal. At this point the charge is stripped off the ion by a thin stripper foil and the repulsive forces accelerate the ion to twice the energy of the high voltage terminal. The acceleration assembly is usually insulated from the outside with sulfur hexafluoride gas which is capable of maintaining a high electric field gradient without ionizing. At the end of the second acceleration tube is the target for the production of the radioisotopes.

OXYGEN-15 ONLY MACHINES

Under this category we are including the Oxygen-15 generators which are now in the field. There are several types. A small deuteron linac has been produced. A compact deuteron cyclotron, and a tandem cascade accelerator have also been devised for Oxygen-15 production. All of these accelerators are capable of producing Oxygen-15 only from the $^{14}N(d,n)^{15}O$ nuclear reaction. All use low energies and are designed for use primarily in hospital environments. The basic operating characteristics are the same as the larger machines of the same type.

PRACTICAL CONSIDERATIONS IN THE CHOICE OF AN ACCELERATOR

The first consideration with any accelerator is whether or not that accelerator is capable of producing sufficient quantities of radionuclides for your particular needs. Regardless of what type of accelerator is installed, it must be kept in mind that the accelerator delivers protons, deuterons or less commonly helium-3 and helium-4 ions. These are the reagents to bring about the needed nuclear reactions. The cross-sections of the nuclear reactions for the production of the common positron emitters are well characterized and the yields as a function of particle energy are known (Austin et al., 1975; Bida et al., 1980; Casella et al., 1980; Ruth and Wolf, 1978; Sajjad et al., 1984; Sajjad et al., 1986). This does not always give an idea about the practical yields which is the determining factor in routine radioisotope production. The practical yield can be quite different from the saturation yield of a radionuclide. Saturation occurs when the production rate of the radionuclide equals the decay rate of the radionuclide. For any radionuclide, $R = N\lambda/(1-e^{-\lambda t})$

where R is the saturation value, N is the number of atoms and λ is the decay constant. In a practical sense, this limit is reached at about 5 half lives. For example, the approximate saturation time for Fluorine-18 production is about 9 hours, which is a totally impractical situation. Oxygen-15 production, on the other hand, reaches saturation at about 10-12 minutes which is practical. Theoretical versus practical yields have been discussed (Wolf, 1984). Today each manufacturer will present the buyer with a "guaranteed" yield of radionuclide under defined conditions. This is a primary consideration especially if clinical application is contemplated.

The chemical form of the radionuclide is also of major interest in considering the attributes of any machine. The target material must withstand the intensely ionizing particle beam and also must be able to withstand the intense radiation field accompanying the particle bombardment. The question of chemical form coming from the target and indeed utilizing the target material and the bombardment conditions has been a matter of research since the early fifties. The chemistry of the positron emitting nucleogenic atoms and the relation to the preparation of labeled compounds has been reviewed (Ferrieri and Wolf, 1983).

Another aspect relating to chemistry is the choice between proton only machines or two particle proton, deuteron machines. If a proton only machine is chosen (Table 1) enriched isotopes are needed to produce three of the four positron emitters: Nitrogen-13, Oxygen-15 and Fluorine-18. To place this in proper perspective, it should be noted that the use of Oxygen-18 water is universal, although recent perturbations in the market and lack of availability of Oxygen-18 water has forced the use of less efficient means of Fluorine-18 production, e.g., the neon $^{20}Ne(d,\alpha)^{18}F$ reaction or the Helium-3 or Helium-4 reactions on natural water (assuming deuterons or Helium-3 or Helium-4 beams are available). Hopefully, the Oxygen-18 dilemma will soon be resolved.

THE ACCELERATORS

Each of the accelerators which have been mentioned have certain advantages and disadvantages. A summary of the basic types of accelerators and some of the characteristics are presented in Table 2. We have tried not to use a particular model to determine the advantages and disadvantages but to give some characteristics which are universal with a particular type of accelerator. This is of course not always possible since in some instances there is only one manufacturer of a particular type of accelerator.

Accelerator	Advantages	Disadvantages
Positive Ion Cyclotron	Proven record Versatility Ease of Maintenance	High Cost High Activation
Negative Ion Cyclotron	Extraction Efficiency Low Activation Beam Uniformity	High Cost High Vacuum Requirements Maintenance of Stripper Foil
Superconducting Cylotron	Compact Size Low Power	Liquid Helium Maintenance
Linac	Stable operation Low Power	Targetry Size
Tandem Cascade	Low Power Low Cost	Targetry
Helium-3 Linac	Low Power Stable Operation	Low Specific Activity Targetry

TABLE 2. Accelerator Advantages and Disadvantages.

Cyclotrons

The cyclotron is by far the most widely used accelerator for the production of positron emitting radioisotopes. The cyclotron has had a long history of reliable isotope production. Once a cyclotron has been chosen as the accelerator, the major choices are one versus two particle and positive versus negative ion. The advantages of the two particle cyclotron is that it does not require the use of isotopically enriched target materials while the one particle cyclotron does. The one particle cyclotron is a little simpler in design and therefore in principle cheaper. It also does not require changing particles which can affect the time between runs and the dose to the operators. The principle differences between the positive ion cyclotron and the negative ion cyclotron are the extraction of the beam and the vacuum characteristics required. The advantages and disadvantages of positive versus negative ion cyclotrons are an important consideration. The difference in the extraction techniques has implications in terms of the vacuum requirements and the uniformity of the internal and extracted beams. In general, the positive ion cyclotrons have a lower vacuum requirement than the negative ion cyclotrons. The use of a septum and deflector means that there is a greater amount of activation when maintenance of the cyclotron becomes necessary. The extraction efficiency of the positive ion cyclotrons are usually lower and can vary from 40 to 95% on various machines. The negative ion machines have a more uniform beam shape, higher extraction efficiency and less activation of the interior of the machine. Negative ion cyclotrons also require a high vacuum to be maintained during operation. The foils which strip the charge from the negatively charged hydrogen ion to yield a positively charged hydrogen ion must be changed on a routine basis. In both types of machines, the ion source filament must be changed routinely. Newer positive ion cyclotrons using a split septum require much less maintenance than the older cyclotrons using a notched septum. Negative ion cyclotrons cannot support Helium-3 or Helium-4 beams.

Superconducting Cyclotrons

The major advantages of the superconducting cyclotrons are the lower weight and the lower power consumption. The major disadvantage is the cost of maintaining the magnet at liquid helium temperatures and the necessity of warming the magnet if repairs are required and then cooling back to liquid helium temperatures. Provisions must also be made for a magnet quenching event in the design of the room where the cyclotron is kept since large amounts of helium could be vaporized in a short time.

Linacs

There are only a few linacs currently in routine use for the production of radionuclides and none are in routine use for the production of PET isotopes other than Oxygen-15. The high cost of the traditional linacs has prevented them from being used extensively. The new technology currently under development with respect to linacs will bring the cost and space requirements down to the levels where they may be used for the production of PET radioisotopes in hospitals. The advantages are the lower shielding requirements, the stability of the beam and low maintenance requirements. The disadvantages are the somewhat larger space requirements, and the pulsed nature of the beam which puts additional constraints on the targetry available.

Tandem Cascade Accelerators

The advantages of this type of accelerator are the low initial cost, the lower power consumption and the lower size and weight of the accelerator over a cyclotron. The disadvantages are the lower yield of the radioisotope which means a longer irradiation time, the high beam current which limits the available targetry, and the restrictions on the number of isotopes which may be produced.

SHIELDING FOR ACCELERATOR AND TARGETS

The shielding for an accelerator of any type is one of the most important considerations in the total initial cost. If an accelerator requires an extensive amount of shielding, the construction cost will be significantly higher than for a room with standard walls. The amount of shielding an accelerator requires depends both on the type of accelerator and on the particular model.

It is beyond the scope of this article to give great detail concerning the current regulations for radiation safety. However, some general definitions are in order. A controlled area is one where the dose rate is less than 5 mR/hr but the annual dose of a person working in the area would be greater than 100 mR. A radiation area is an area where the dose rate is between 5 and 100 mR/hr. It is desirable, if not imperative, to keep the occupied areas in the vicinity of the accelerator at the level of a controlled area.

The major types of radiation found in an accelerator environment are neutrons, betas and gammas. These types of radiation are produced during the operation of the accelerator and from the activated materials surrounding the accelerator. Betas are not normally a shielding problem since they are stopped by light shielding and will not be considered. The attenuation of both neutrons and gammas is an exponential function. The dose rate can be given by the equation:

$$P = P_o e^{(-\mu h)}$$

Where: P = the attenuated dose rate.
P_o = the unattenuated dose rate.
μ = a constant related to energy and material.
h = the thickness.

The simplest way to approach the estimation of the shielding is by using the concept of half-thickness. A half-thickness, as the name implies, is the thickness of material required to reduce the dose rate by a factor of two.

Neutrons are usually produced only during the operation of the accelerator. The most common material for the attenuation of neutrons is concrete, although other more compact materials have been used depending on the neutron energy spectrum. A table of half-thicknesses of concrete versus neutron energy spectrum and water content of the concrete is given in Table 3 (Aglinktlsev, 1965). As an example, if it is necessary to decrease the neutron dose rate by a factor of 1000, then this will require 10 half-thicknesses (2^{10} = 1024) or 99 cm if the water content of the concrete is 7.6%. The gamma dose rate will arise from both the activated materials around the accelerator and from the operation of the accelerator. A table of values for some common materials is given in Table 4 (Brodsky and Beard, 1965). Energies higher or lower may be estimated from these values. These tables are meant to give an order of magnitude estimate for the thickness of the shielding so that some crude cost estimates can be made. There are several references which can supply additional information concerning the shielding of accelerators in general (Patterson and Thomas, 1973).

Manufacturers should supply a map of the radiation environment of the accelerator derived from actual measurement at the factory. Using this map, it is possible to get some rough ideas concerning the shielding required.

Neutron Energy Distribution	% By Weight Water Content	Half-thickness (cm)[a]
Thermal and 18.8 to 4.5 MeV	7.6	9.9
Thermal	7.6	9.1
Thermal	10.0	10.4

a) Half-thickness is the thickness of a particular shielding material which will reduce the radiation dose to one-half of the value when there is no shielding.

TABLE 3. Neutron Attenuation of Half-Thicknesses.

Photon Energy (MeV)	Half-Thickness (cm)[a]			
	Water ($\rho = 1.0$)[β]	Concrete ($\rho = 2.3$)	Iron ($\rho = 7.9$)	Lead ($\rho = 11.3$)
0.2	6.8	3.3	1.0	0.2
0.5	10.3	4.5	1.5	0.5
1.0	14.0	7.0	2.2	1.0
1.5	16.1	8.3	2.4	1.3
2.0	19.5	9.5	2.8	1.5
3.0	26.4	12.4	3.2	1.7

a) Half-thickness is the thickness of a particular shielding material which will reduce the radiation dose to one-half of the value when there is no shielding.

b) ρ is the density in g/cm^3.

TABLE 4. Gamma Attenuation Half-Thicknesses.

HEAT REMOVAL FROM THE ACCELERATOR AND TARGETS

One aspect of accelerator operations which is sometimes overlooked in the planning of the facility is the cooling of the accelerator, power supplies and targetry. There must be a separate deionized closed system water supply for the accelerator (primary cooling water) which passes through a heat exchanger. The water from the secondary coil of the heat exchanger must be either monitored or checked on a routine basis for radioactivity before it is discharged into the open environment. The power supplies required for the accelerator may also need a separate water supply for cooling of power tubes or transistors. The manufacturer should supply all the heat load specifications for the accelerator, power supplies and targets. The manufacturer will usually supply the primary cooling system and the heat exchanger but the purchaser in most cases is responsible for the secondary cooling system and piping. Accommodations must be made for this equipment in the vicinity of the accelerator.

RADIOISOTOPE EMISSIONS

The emission of radioactive gases from the facility is a serious problem and provisions must be made to ensure that short-lived radioactive gases can be held in the facility until they have decayed.

There is some activation of the air in the immediate vicinity of most accelerators which must be contained during the irradiation. The more serious problem is the case of a catastrophic failure of a target foil with the possible release of curie amounts of gaseous activity. Some typical concentration limits for unrestricted areas outside the facility are given in Table 5. Of course the chemical form of the radioisotope is important and cognizance is taken of this fact by making the limits for $[^{11}C]CO_2$ higher than for $[^{11}C]CO$. Estimates can be made based on the production capabilities of the accelerator and target gas processing systems.

MAINTENANCE OF ACCELERATOR AND TARGETS

The maintenance of accelerators is a very important consideration in the choice of an accelerator. One of the problems with the new accelerator designs is that there is no history of how much maintenance will be required. One of the factors to be determined by talking to the manufacturer and more importantly to people who have the type of accelerator you are contemplating buying, is the maintenance history. It is also essential that the design of the accelerator allows easy access to parts which need to be changed on a routine basis such as the ion source filament, stripping foils and target foils. The radiation dose received by

Radionuclide	Limit (μCi/cc)	Reference
^{13}N (N_2)	4.0×10^{-6}	ICRP 53[a]
^{15}O (CO)	2.0×10^{-6}	ICRP 53
^{15}O (CO_2)	4.0×10^{-6}	ICRP 53
^{15}O (O_2)	4.0×10^{-6}	ICRP 53
^{11}C (CO)	4.0×10^{-7}	ICRP 53
^{11}C (CO_2)	2.0×10^{-6}	ICRP 53
^{11}C (other)	6.0×10^{-7}	10 CFR 20[b]
^{18}F (HF)	1.0×10^{-7}	10 CFR 20

a) ICRP 53 is the International Commission on Radiological Protection Report 53 Radiation Dose to Patients from Radiopharmaceuticals, 1988.

b) 10 CFR 20 is the Nuclear Regulatory Commission's revised Title 10 Code of Federal Regulations Part 20, from FR 56 (98) May 21, 1991.

TABLE 5. Recommended Concentration Limits for Unrestricted Areas.

the operator when changing an ion source filament on a cyclotron can vary by a factor of 10 depending on whether or not the design allows easy access to the ion source. A list of the most common repairs and replacements should be made and the procedures needed to put the accelerator back into operation examined to be sure there are no glaring problems.

STABILITY OF OPERATION

The stability of operation is another important factor in the choice of an accelerator. The ideal is an accelerator that can be turned on in the morning with beam instantly available, left running all day with no adjustments, and then shut off at night without a complex shutdown procedure. The reality is somewhat different. The closer one can come to such a machine, the better. The manufacturer should be able to give estimates on start up time, stability during the day and shutdown procedures. It is also wise to speak with another owner of such a machine, if that is possible.

TRACK RECORD OF CURRENT ACCELERATORS

Cyclotrons are by far the most extensively used accelerators for the production of positron emitting radioisotopes. The linacs and Van de Graaff accelerators have been used, but not commonly. It is therefore to be expected that there is a great deal more information concerning the reliability of the cyclotron over the other types of accelerators. In general, cyclotrons are reliable accelerators. Most models and types have proven to be extremely problem free while there are some in need of constant attention. Sometimes these two extremes are seen in cyclotrons from the same manufacturer and depend entirely on the design of the cyclotron. With the advent of crystal oscillators, computer mapping of magnetic fields, and computer control of source position, arc voltage and so forth, the amount of time spent tuning a cyclotron has been greatly reduced. The best indicator of how a machine will perform is how the same model has performed in the past.

Linacs in general are also very stable machines, but the experience in a setting where there is no infrastructure to deal with problems as they arise is limited. As these accelerators are placed, a better idea can be obtained as to how reliable they will be. The lifetime of the rf amplifier tubes and stability of the rf structure await testing.

CONCLUSION

This chapter has attempted to give the reader a very brief outline of how the different accelerators for the production of PET isotopes operate and some indication of the advantages and disadvantages of each. The choice of an accelerator will depend to a great extent on the environment into which the accelerator will be placed. There are however some key questions to be asked when choosing an accelerator for the production of PET isotopes. A short list is:

1. Will this machine produce the amount of radioisotope necessary for our operation?
2. Is this machine reliable?
3. What are the initial costs and how do these compare to the operational costs?
4. What personnel will be needed to maintain the machine and associated hardware?
5. What will be required to shield this accelerator in our environment?
6. What are the power and water requirements and are connections available?
7. Is help from the manufacturer readily available and how much does it cost?
8. How will the radiation safety aspects be implemented and documented?
9. How will radioisotope emissions to the environment be monitored and controlled?
10. Are the target materials required readily available (enriched isotopes)?

This list of questions is a start and the answers should be available for all the accelerators under consideration before a final decision is made. The best course is always to arm yourself with as much information as possible.

ACKNOWLEDGEMENT

This research was carried out at Brookhaven National Laboratory under contract DE-AC02-76CH00016 with the U. S. Department of Energy and supported by its Office of Health and Environmental Research.

REFERENCES

Aglinktlsev K (1965): *Applied Dosimetry*, London: Lliffe Books.

Austin RM, Galousky A and Bortims G (1975): A batch process for the production of ^{13}N-labeled nitrogen gas. *Nucl Inst Methods* 126: 373-379.

Bida GT, Ruth TJ and Wolf AP (1980): Experimentally determined thick target yields for the $^{14}N(p,a)^{11}C$ reaction. *Radiochim Acta* 27: 181-185.

Brodsky A and Beard GV (1965): *A Compendium of Information for use in Controlling Radioation Emergencies, TID-8206 (Rev)*, Washington, D.C.: Atomic Energy Commission.

Casella V, Ido T, Wolf AP, Fowler JS, MacGregor RJ and Ruth TJ (1980): Anhydrous ^{18}F-F_2 radiopharmaceutical preparation. *J Nucl Med* 21: 750-757.

Clark JC and Buckingham PD (1975): Short-Lived Radioactive Gases for Clinical Use. London: Butterworth.

Duggan JL and Morgan IL (1991): Applications of Accelerators in Research and Industry: *Nucl Inst Methods B* Vol 56/57 (entire volume).

Ferrieri RA and Wolf AP (1983): The chemistry of positron emitting nucleogenic (hot) atoms with regard to preparation of labelled compounds of practical utility. *Radiochim Acta* 34: 69-83.

Ido T, Wan CN, Casella V, Fowler JS, Wolf AP, Reivich M and Kuhl, DE (1978): Labeled 2-deoxy-2-glucose analogs. ^{18}F-Labeled 2-deoxy-2-fluoro-D-glucose, 2-deoxy-2-fluoro-D-mannose and ^{14}C-2-deoxy-2-fluoro-D-glucose. *J Label Cmpds Radiopharm* 14: 175-183.

Livingood JJ (1961): Principles of Cyclic Particle Accelerators. New York: D Van Norstrand Co. Inc.

Patterson HW and Thomas RH (1973): Accelerator Health Physics. New York: Academic Press.

Ruth TJ and Wolf AP (1978): Absolute cross sections for the production of ^{18}F via the $^{18}O(p,n)^{18}F$ reaction. *Radiochim Acta* 26: 21-24.

Sajjad M, Lambrecht RM and Wolf AP (1984): Cyclotron isotopes and radiopharmaceuticals. XXXIV. Excitation function for the $^{15}N(p,n)^{15}O$ reaction. *Radiochim Acta* 36: 159-162.

Sajjad M, Lambrecht RM and Wolf AP (1986): Cyclotron isotopes and radiopharmaceuticals XXXVII. Excitation functions for the $^{16}O(p,a)^{13}N$ and $^{14}N(p,pn)^{13}N$ reactions. *Radiochim Acta* 39: 165-168.

Ter-Pogossian MM and Wagner HN Jr (1966): A new look at the cyclotron for making short-lived isotopes. *Nucleonics* 24(10): 50-56.

Vera Ruiz H and Wolf AP (1978): Excitation function for ^{15}O production via the $^{14}N(d,n)^{15}O$ reaction. *Radiochim Acta* 24: 65-67.

Wolf AP and Jones WB (1983): Cyclotrons for biomedical radioisotope production. *Radiochim Acta* 34: 1-7.

Wolf AP (1984): Cyclotrons, radionuclides, precursors and demands for routine vs research compounds. *Ann Neurol* 15(Suppl): S19-S24.

CHEMISTRY OF TRACERS FOR POSITRON EMISSION TOMOGRAPHY

Robert F. Dannals, Hayden T. Ravert and Alan A. Wilson
Department of Nuclear Medicine
The Johns Hopkins Medical Institutions

INTRODUCTION

There are hundreds of naturally occurring and artificial radionuclides; however, only a small number have found use in biomedical research. Of this limited number, a select few have been used to study biochemistry and pharmacology *in vivo* with emission tomography. Positron Emission Tomography (PET) and Single Photon Emission Computed Tomography (SPECT) offer biomedical researchers unique opportunities for studying a wide variety of biochemical and pharmacological interactions in living man. While SPECT offers the advantages of widespread availability of instrumentation and relative low cost of radionuclides (e.g., Technetium-99m and Iodine-123), PET has the distinct advantages of quantitative instrumentation and versatile chemistry of radionuclides (especially, Oxygen-15, Nitrogen-13, Carbon-11, and Fluorine-18). This chapter will focus primarily on these four radioisotopes for PET and highlight certain aspects of the radiochemistry involved in the preparation of tracers for PET studies.

Nuclear Imaging in Drug Discovery, Development, and Approval
H.D. Burns, R. Gibson, R. Dannals, Peter Siegl, Editors

ISOTOPES

Oxygen-15 (^{15}O) is a positron emitting (99.9 %) radionuclide with a half-life of 2.07 minutes. Due to its extremely short half-life, the use of this isotope has the absolute requirement of an on-site accelerator. This radionuclide is commonly produced by deuteron irradiation of Nitrogen-14 gas via the nuclear reaction $^{14}N(d,n)^{15}O$; however, recent interest in the use of single particle accelerators (i.e., accelerating only protons for radionuclide production) has given rise to the production of Oxygen-15 via the $^{15}N(p,n)^{15}O$ nuclear reaction. Oxygen-15 decays to Nitrogen-15 upon emission of a high energy positron (E_{max} = 1.72 MeV).

Nitrogen-13 (^{13}N) is a positron emitting (100 %) radionuclide with a half-life of 9.96 minutes. As with Oxygen-15, the use of Nitrogen-13 requires an on-site accelerator due to its short half-life. This radionuclide is commonly produced by proton irradiation of Oxygen-16 via the nuclear reaction $^{16}O(p,\alpha)^{13}N$, although the $^{13}C(p,n)^{13}N$ nuclear reaction is also used. Nitrogen-13 decays to a stable nuclide, Carbon-13, upon emission of a high energy positron (E_{max} = 1.19 MeV).

Carbon-11 (^{11}C) is a positron emitting (99+%) radionuclide with a half-life of 20.4 minutes. Due to its short half-life, the use of this isotope generally requires an on-site accelerator although it is possible to distribute tracers labeled with Carbon-11 over limited distances with short transportation times. This radionuclide is commonly produced as Carbon-11 labeled carbon dioxide by proton irradiation of Nitrogen-14 gas via the nuclear reaction $^{14}N(p,\alpha)^{11}C$, although other nuclear reactions for its production have been described: $^{11}B(p,n)^{11}C$ and $^{10}B(d,n)^{11}C$. Carbon-11 decays to a stable nuclide, Boron-11, upon emission of a high energy positron (E_{max} = 0.96 MeV).

Fluorine-18 (^{18}F) is a positron emitting (97%) radionuclide with a half-life of 109.7 minutes. Although produced in an on-site accelerator, the longer half-life makes it possible to use either the radionuclide or radiotracer prepared containing Fluroine-18 at sites remote from its production. Fluorine-18 is commonly produced as labeled aqueous fluoride ion by proton irradiation of Oxygen-18 enriched water via the nuclear reaction $^{18}O(p,n)^{18}F$. Other nuclear reactions, however, are used for its production, for example: $^{20}Ne(d,\alpha)^{18}F$. Fluorine-18 decays to a stable nuclide, Oxygen-18, upon emission of a high energy positron (E_{max} = 0.635 MeV).

In addition to the commonly used accelerator produced radioisotopes described above for PET, there are several other isotopes which deserve mention. Two additional radiohalogens have been used: Bromine-75 ($t_{1/2}$ = 101 minutes) which is produced on a cyclotron accelerator (Paans et al., 1980; Stöcklin, 1986), and Iodine-122 ($t_{1/2}$ = 3.5 minutes), which is obtained from a Xenon-122 ($t_{1/2}$ = 20.1 hours) generator system (Stöcklin, 1986; Mathis et al., 1986). Although a radiohalogen with the reactivity of bromine is desirable for synthetic chemistry reasons, the widespread use of Bromine-75 has not been achieved for two reasons: the target material and particle energies used in the isotope production are generally beyond those obtained at small hospital-based biomedical accelerators, and Bromine-75 decays to Selenium-75 which itself is radioactive ($t_{1/2}$ = 118 days) and contributes to the radiation dosimetry of tracers prepared from this isotope. The extremely short half-life of Iodine-122 as well as the relatively short half-life of its parent nuclide have limited the use of this isotope to centers where the parent radionuclide Xenon-122 can be obtained.

There are three other generator systems of PET isotopes that have seen greater utility in PET. Gallium-68 ($t_{1/2}$ = 68 minutes) is obtained from a commercially available Germanium-68 ($t_{1/2}$ = 275 days) generator (Ehrhard and Welch, 1978). Rubidium-82 ($t_{1/2}$ = 1.25 minutes), an analogue of potassium used in PET studies of myocardial blood flow, is obtained from a Strontium-82 ($t_{1/2}$ = 25 days) generator which is also commercially available (Yano et al., 1977). Most recently, Copper-62 ($t_{1/2}$ = 9.7 minutes) has been used in the preparation of metal chelate complexes for studying myocardial and cerebral blood flow (Fujibayashi et al., 1989; Green et al., 1990). Copper-62 is available from a Zinc-62 ($t_{1/2}$ = 9.26 hours) generator.

SIMPLE RADIOCHEMICALS AS TRACER BUILDING BLOCKS

Because of the extremely short half-life of Oxygen-15, the number and complexity of chemical reactions available from a practical standpoint for producing new radiochemicals are limited. As a consequence the number of tracers labeled with Oxygen-15 is limited. Most of the tracers are gases (e.g., molecular oxygen, carbon monoxide, and carbon dioxide) that can be prepared either in the accelerator target used for the production of the radioisotope or via on-line gas processing (Del Fiore et al., 1979). For measurement of blood flow, two Oxygen-15 labeled tracers are prepared: water and butanol. Water is simply prepared on-line from hydrogen and labeled oxygen. The synthesis of Oxygen-15 labeled butanol is quite elegant and far more complex than previously reported chemistry involving this radioisotope. Labeled butanol is prepared by the reaction of molecular oxygen with alumina supported tri-n-butyl-borane, hydrolysis, and rapid purification on reverse phase cartridges (Berridge et al., 1990).

It is not surprising that with increasing half-life, the number and variety of radiochemicals that can be prepared synthetically from a particular radionuclide also increases. Although a half-life of 10 minutes is still rather short, there are a number of Nitrogen-13 labeled tracers reported in the literature. The most prominent is Nitrogen-13 labeled ammonia which it is commonly used in PET to study myocardial blood flow. Nitrogen-13 labeled ammonia can be prepared either by chemical reduction of the species (e.g., nitrous acid) resulting from proton irradiation of water or directly in the accelerator target by proton irradiation of either a slurry of Carbon-13 resin in water or aqueous ethanol mixtures. The ammonia thus produced has been used in classical synthetic organic reactions to produce alkyl amines and with immobilized enzyme catalyzed syntheses to produce several biomolecules of interest (e.g., amino acids) (Gelbard, 1981).

The list of simple radiochemicals labeled with Carbon-11 is quite extensive (Wolf and Fowler, 1986). Generally produced in the accelerator target as $[^{11}C]CO_2$, the carbon dioxide can be rapidly converted on-line to labeled methane, then further converted to labeled hydrogen cyanide, or carbon monoxide. Reduction of labeled carbon dioxide using lithium aluminum hydride rapidly yields methanol which can be oxidized on-line to formaldehyde or further converted in the presence of hydriodic acid to labeled methyl iodide. From methyl iodide, labeled methyl lithium can be prepared by rapid halogen-metal exchange in the presence of an excess of n-butyllithium.

The reaction of $[^{11}C]CO_2$ with Grignard or alkyl lithium reagents yields the corresponding carboxy-labeled carboxylic acids. Subsequent halogenation yields the corresponding acid halides and reduction of the carboxylic acids yields their labeled alcohols which can further be converted in the presence of hydriodic acid to the corresponding iodides. Other simple Carbon-11 labeled radiochemicals include phosgene, nitromethane, and recently methyl triflate.

Much of the chemistry involved in the preparation of these simple Carbon-11 labeled radiochemicals is facilitated by the ability to remotely transfer the large quantities of radioactivity involved in the gas or vapor phase. With larger labeled alkyl and acyl halides that are not easily volitalized, chromatographic procedures such as simple silica-based cartridges or High Performance Liquid Chromatography (HPLC) are often required to purify the labeled intermediates from the reactants.

Although the half-life of Fluorine-18 is nearly two hours, the number of simple radiochemicals used as building blocks in tracer syntheses is limited (Kilbourn, 1990). As mentioned previously, Fluorine-18 is commonly produced as either aqueous fluoride or molecular fluorine - the latter being extremely reactive, the former requiring additional manipulations to prepare reactive fluoride. Labeled molecular fluorine is often used directly in classical synthetic organic reactions, and, as it is mixture of radioactive and stable fluorine, behaves for the most part as predicted. It has been converted on-line to xenon difluoride, diethylaminosulfur trifluoride, and acetyl hypofluorite.

Aqueous radiolabeled fluoride ion itself is relatively unreactive. Generally aqueous solutions of Fluorine-18 labeled fluoride must be dried either in a platinum crucible, in the presence of metal ions (e.g., cesium, rubidium), or in the presence of an azo crown ether (e.g., Kryptofix 2.2.2) and potassium carbonate to prepare "anhydrous" fluoride. Since the fluoride is produced from enriched Oxygen-18 water and this enriched material is not in abundant supply, efforts are generally made to recover the water either by passing the original aqueous fluoride through a ion exchange column to initially trap the water prior to activating the fluoride or by using a cold trap to recover the water during the drying of the fluoride.

Over the past several years, radiolabeled fluoroalkylating agents (e.g., fluoroethyl and fluoropropyl tosylates and mesylates) have been prepared from fluoride ion and the corresponding disubstituted ethanes and propanes. In addition, radiolabeled fluorobenzaldehydes and, after subsequent reduction and halogenation, fluorobenzyl halides have been prepared by nucleophilic displacement of the corresponding nitro or trimethylammonium precursors.

SPECIFIC ACTIVITY CONSIDERATIONS

For the most part, radiochemicals labeled with Oxygen-15 (e.g., molecular oxygen) do not have the requirement for high specific activity of the final product for successful PET studies; for example, Oxygen-15 labeled water, produced by catalytic reduction of carrier added oxygen with hydrogen, is trapped in aqueous sodium chloride for intravenous administration. Only with Oxygen-15 labeled carbon monoxide, for example, which could have serious consequences in PET studies where multiple rapid administrations or a single continuous administration over a lengthy period of time, does concern exist when using a low specific activity Oxygen-15 labeled tracer.

As with Oxygen-15, there is generally little concern with the final specific activity of Nitrogen-13 labeled tracers. To date, almost all of the radiotracers prepared from Nitrogen-13 are bioconstituents (e.g., ammonia, amino acids, etc.) and, at the specific activities attainable with Nitrogen-13, administered in PET studies in small chemical quantities.

Although the theoretical maximum specific activity of Carbon-11 is very high (9.2×10^9 mCi/µmole), in practice this high specific activity is not attainable due to carrier Carbon-12 contamination which enters into the production of the radioisotope and its various radiochemical forms.

Specific activities of 2×10^3 mCi/μmole are more commonly observed at the end of a tracer synthesis. Depending upon the final use of the Carbon-11 labeled tracer, this final specific activity may be adequate; however, there are several examples in the development of tracers for receptor (e.g., Carbon-11 labeled carfentanil for studying opiate μ receptors) and neurotransmitter uptake sites where this final specific activity is the absolute minimum required due to the pharmacological activity of the ligand.

Although the theoretical maximum specific activity of Fluorine-18 is high (1.71×10^9 mCi/μmole), in practice specific activities of 1×10^4 mCi/μmole have been observed at the end of tracer syntheses involving Fluorine-18 labeled fluoride. This is in stark contrast to radiotracers prepared from Fluorine-18 labeled molecular radiofluorine. Fluorine-18 labeled molecular radiofluorine is most commonly produced by the $^{20}Ne(d,\alpha)^{18}F$ reaction using a neon target containing a small amount of carrier fluorine. Although the amount of carrier is small in absolute terms, it is sufficiently high to reduce the specific activity of the tracer by several orders of magnitude compared to Fluorine-18 labeled fluoride. As with Carbon-11, depending upon the final use of the Fluorine-18 labeled tracer, this final specific activity may be more than adequate. Generally speaking, Fluorine-18 tracers designed to interact with molecular recognition sites (e.g., receptors) require preparation from high specific activity labeled fluoride.

TRACER DESIGN AND DEVELOPMENT

During the past decade, the majority of tracers prepared from Carbon-11 and Fluorine-18 have been developed to study some aspect of neurotransmission *in vivo* with PET. The process of neurotransmission has many components. For example, neurotransmission involving biogenic amines includes events which are both pre- and post-synaptic: transmitter synthesis, transmitter storage, transmitter release, transmitter binding to post-synaptic receptors, metabolism of the transmitter, and reuptake of the transmitter. Development of tracers for investigation of these various aspects requires not only an understanding of the underlying pharmacology of the tracers and their interactions within the system, but also an understanding of the strengths and limitations of the radiochemistry involved in their design and preparation. Several articles are suggested for a more comprehensive treatment of this and related topics (Wolf and Redvanly, 1977; Palmer et al., 1977; Fowler, 1978; Fowler and Wolf, 1986; Finn et al., 1984; Wolf, 1984; Wolf and Fowler, 1979; Wolf and Fowler, 1986; Stöcklin, 1986; Kilbourn, 1990).

The process of selecting a particular ligand to be radiolabeled is not simple. It is highly desirable to know the *in vivo* behavior of the compound prior to beginning the tracer synthesis development; however, the first step in the selection process is often satisfying certain *in vitro* criteria for specificity and saturability of binding. An ideal candidate for studying binding sites in the brain should demonstrate high extraction across the blood-brain-barrier with rapid clearance from the blood, a high affinity constant and high specificity for the site of interest, no metabolites, and appropriate kinetics of binding such that the clearance from the sites of interest is slower than from non-specific sites of interaction. From a practical point of view, the compound under consideration should have been available in radiolabeled form (e.g., Tritium, Iodine-125, etc.) and previously studied *in vivo*.

Tracer development for the dopamine D_2 receptor and PET illustrates some of the problems involved in choosing an appropriate compound. In an effort to maximize the specific binding and minimize the nonspecific binding, based on *in vitro* criteria, a compound with a high affinity (K_D) and low dissociation rate (e.g., spiperone with a K_D of 10^{-10} M) is preferred (in contrast to haloperidol, for example, with a $K_D = 10^{-9}$ M). *In vivo* animal distributions appear to support the selection of spiperone as the striatum-to-cerebellum ratio (an index of dopamine D_2 binding) for [^{3}H]spiperone is 6 at one hour, whereas the ratio for [^{3}H]haloperidol at the same time is only 1.5. The lower ratio for haloperidol may be due in part to binding to other receptor sites in the cerebellum.

Ligand specificity is also important to consider. Spiperone with significant interactions at both the serotonin S_2 and adrenergic α_2 receptor sites may be less appropriate than a ligand such as pimozide which binds selectively to the dopamine D_2 receptor. However, when other factors (e.g., lipophilicity, ligand metabolism, etc.) are considered, although spiperone and pimozide have comparable affinities *in vitro*, the higher lipophilicity of pimozide compared to spiperone results in a lower specific-to-nonspecific binding ratio for pimozide due to increased nonspecific binding *in vivo*.

The tracer selection process must also consider constraints imposed by radiochemical synthetic routes and the means for obtaining the non-radioactive precursor to be labeled in addition to *in vitro* and *in vivo* criteria. This is particularly important since many potentially attractive compounds previously studied (e.g., those which were Tritium-labeled) cannot be readily prepared with short-lived radionuclides (e.g., Carbon-11 and Fluorine-18) because of limitations of isotope half-life, specific activity, and radiochemical yields.

In the absence of required information, radiotracer design relies on a more laborious empirical approach. The structure of the molecule and the available synthetic routes for introducing the label into the molecule generally dictate the position of the radiolabel. The most successful radiotracer design will incorporate the radiolabel into the molecule of interest using the smallest possible perturbation to the parent structure. For this reason, tracers derived from simple isotopic substitutions, such as Carbon-11 for Carbon-12 or Fluorine-18 for Fluorine-19, are ideal as it is unlikely that the biological system will recognize the deficiency of neutrons.

A tracer based on a bioisosteric substitution (i.e., the substitution of a radiolabel of similar size, electronic configuration, and lipophilicity for a portion of the parent molecule) may be sufficient when a simple isotopic substitution is not possible due to the molecule's particular chemical structure. In this case, the radiolabeled molecule will no longer be the same as the parent; either a foreign element (e.g., Fluorine-18 for a proton) or a foreign moiety (e.g., a Carbon-11 labeled methyl group or Fluorine-18 labeled fluoroalkyl for a proton, etc.) will be present. The hope is that the "new" molecule will retain the desired characteristics of the parent compound. The best example of a tracer incorporating a bioisosteric substitution is 2-[^{18}F]fluoro-2-deoxy-D-glucose, a tracer which has allowed the extension of the [^{14}C]-2-deoxy-D-glucose autoradiographic method to human imaging using PET.

Several factors influence the decision on which path to follow in the synthesis of a radiolabelled compound. Firstly, the synthesis must be based upon a radioactive precursor which is readily available. Secondly, the total time for the synthetic steps and purification(s) must be sufficiently short to be compatible with the isotope's half-life. Thirdly, large quantities of radioactivity (in excess of 1 Ci) are often required at various stages of the synthesis, and a major effort must be made to eliminate non-essential radiation exposure yet carry out the required manipulations in a rapid, efficient and reproducible manner. Finally, considering these concerns, it is generally best to introduce the radiolabel in a step at or near the end of the synthesis.

The theoretical maximum specific activity of Carbon-11 is extremely high, but a truly carrier-free synthesis of a Carbon-11 labeled radiotracer has never been achieved. Although the syntheses are performed with no carrier intentionally added, there is always contamination from other sources of Carbon-12 that enter into the synthetic procedure and reduce the specific activity. In absolute terms, the amount of carrier is small; however, the reduction in specific activity is profound and often several orders of magnitude. Therefore, it is important to realize that, since the tracer is not free of carrier, the specific activity is changing proportional to the radioactive decay.

It is extremely important to limit carrier (for example, Carbon-12 in a Carbon-11 synthesis) from external sources and maximize the individual radiochemical yields of each step of the synthesis. By limiting carrier contamination, the amount of mass associated with the final product is minimized. By increasing the rate of product formation (i.e., the chemical yield at a particular time for each synthetic step), the radiochemical yield (and the final radioactivity associated with the product) is also increased. Both of these factors directly influence the final specific activity of the tracer. As there will never be less carrier carbon than what one begins with in the target used for producing Carbon-11, it is extremely important to have an isotope production system (target and associated isotope processing) that limits undesirable and unintentional introduction of carrier. For example, in syntheses involving Carbon-11, the use of ultra high purity target gas, trapping impurities chemically, conditioning the production system under inert atmosphere, and appropriate isotope production parameters are extremely important. Additional sources of carrier contamination in Carbon-11 syntheses are all chemical reagents which could potentially introduce carrier carbon contamination, especially, Grignard reagents used in the synthesis of Carbon-11 labeled carboxylic acids and lithium aluminum hydride used to reduce these acids and carbon dioxide itself to their corresponding alcohols. Reagents must be prepared under extremely clean inert atmosphere and only the minimum amounts necessary for the radiochemical synthesis to proceed should be used. It may not be possible to eliminate carrier completely but with appropriate precautions carrier can be minimized and the product's final specific activity can be maximized.

In a synthesis, since carrier is present, the specific activity is always changing and always decreasing; therefore, the synthesis and purification times must also be optimized and kept to a minimum. For Carbon-11, every 3 minutes represents a decrease of 10% of the radioactivity and at best represents only a 10% decrease in the specific activity. Chemists have only limited tools at their disposal to decrease the synthesis times and maintain chemical yields, in particular, changing temperature, reaction solvents, pressures, and reagent concentrations. Perhaps the biggest improvement in synthesis time is achieved by repetition and familiarity with the synthetic steps and manipulations.

The points concerning specific activity for Carbon-11 also apply, in varying degrees, to the production of Fluorine-18 labeled radiotracers. Clearly fluoride is not as ubiquitous as carbon dioxide; however, some of the reagents used in radiofluorinations do contain small absolute quantities of fluoride and, although small, these amounts of fluoride have a profound effect on the specific activity of the tracers synthesized.

Purification of tracers must be accomplished in a rapid and

reproducible fashion. Semipreparative HPLC is often the method of choice for purification of a radiotracer preparation; therefore, its influence on the synthesis time must also be minimized. Changes in stationary and mobile phases and column dimensions along with increases in flow rates achieved by using low viscosity solvent mixtures can increase the speed of chromatographic purification. An appropriate set of HPLC conditions must consider separating the desired radiochemical from all other reactants used in the tracer synthesis.

The precursors used in radiolabeling with Carbon-11 may also be a potential source of carrier. For example, the norbenzyl precursor for [^{11}C]dexetimide (Dannals et al., 1988) is obtained by reductive N-debenzylation of dexetimide itself. Although one milligram of this precursor may be 99% pure, it may contribute as much as 10 μg of stable dexetimide to a radiochemical synthesis. Even this small amount of residual product in the precursor is enough to lower the specific activity. Since many precursors for radiomethylation are obtained from their stable parents via reactions which simply remove the moiety by N-, S-, or O-dealkylations, extreme care must be taken to purify the precursors to remove all traces of the parent starting material.

SOME SPECIFIC EXAMPLES OF RADIOTRACER DEVELOPMENT

Diprenorphine (Fig. 1) is an opiate antagonist which exhibits similar *in vitro* affinities for the mu, delta and kappa opiate receptor subtypes. However, *in vivo* studies suggest that [^{3}H]diprenorphine binds selectively to mu-opioid receptors in mice (Frost et al., 1986). In addition, recent PET studies have demonstrated that [^{11}C]carfentanil and [^{11}C]diprenorphine label different opiate receptor subtypes *in vivo* in humans (Frost et al., 1990). The radiochemical synthesis of [^{11}C]diprenorphine can be accomplished by at least four routes as shown in Fig. 1. These synthetic routes illustrate the potential for labeling a tracer in multiple positions using a simple isotopic substitution. It should be noted that each synthetic route produces authentic diprenorphine, not a nonisotopically substituted derivative of unknown pharmacological behavior.

Route A involves the alkylation of the appropriate precursor with [^{11}C]cyclopropylmethyl iodide (Rimland and Langstrom, 1987) to yield [^{11}C]diprenorphine labeled in the methylene group shown. An alternative route, B, which provides the label in the same position, involves the preparation of [^{11}C]cyclopropyl acid chloride, amide formation, and amide reduction to yield [^{11}C]diprenorphine (Luthra et al., 1985). The third route, C, requires preparation of [^{11}C]methyl lithium *via* halogen-metal exchange using [^{11}C]methyl iodide and

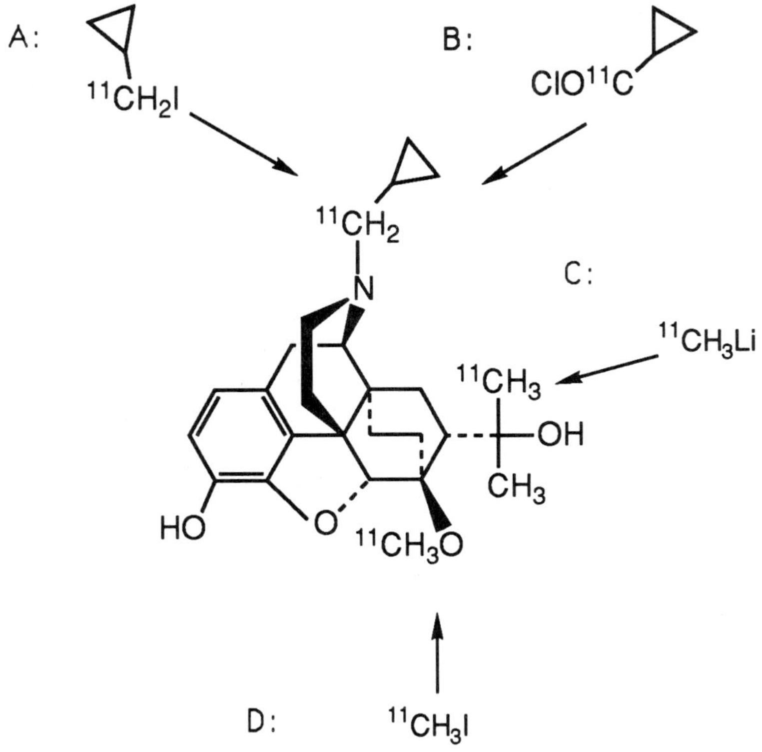

FIGURE 1. Published routes for the synthesis of [^{11}C]Diprenorphine.

n-butyllithium with subsequent reaction with the nor-methyl ketone precursor of diprenorphine (Burns et al., 1984a). The preparative chromatographic purification of [^{11}C]diprenorphine prepared by route C proved difficult, so we developed a selective O-[^{11}C]methylation (route D) on an appropriately protected diprenorphine precursor (Lever et al., 1987). The relative ease of the precursor synthesis, ready availability of high specific activity [^{11}C]CH_3I, and short reaction times makes route C the most attractive synthetic route for the routine production of [^{11}C]diprenorphine at high specific activity and radiochemical purity.

Depending on the purpose of the study, a particular site of metabolism within the molecule can either be avoided or specifically labeled to yield relevant pharmacokinetic data by labeling one of the three positions. In the absence of metabolic information, the ability to label one of a number of positions within a given tracer allows considerable chemical flexibility which increases the likelihood of successfully imaging the receptor population of interest.

Another example of recent radiotracer development is radiolabeled spiperone, and its analogues, for studying dopamine D_2 receptors. Spiperone was developed by Janssen Pharmaceutica and has been the gold standard for *in vitro* and *in vivo* binding studies involving dopamine D_2 receptors. Spiperone has been radiolabeled by isotopic replacement as shown in Figure 2 using multistep syntheses starting with either (A) [^{18}F]fluoride (Kilbourn et al., 1984; Shiue et al., 1985) or (B) hydrogen [^{11}C]cyanide (Fowler et al., 1982). The complex nature of these syntheses led to the search for radiolabeled analogues of spiperone which could be prepared in a more facile manner. In the original spiperone patent, a N-methylated analogue of spiperone was described, but no data were provided regarding its binding characteristics. A radiochemical synthesis (Fig 2C) of the bioisosterically substituted spiperone, 3-N-[^{11}C]methylspiperone ([^{11}C]NMS), was developed (Burns et al., 1984b). The choice was based upon the availability of large quantities of high specific activity [^{11}C]methyl iodide and the commercial availability of the parent compound as the precursor for radiolabeling. N-Methylation decreases the affinity of NMS for the dopamine D_2 receptor, but an increase in overall brain uptake was observed with sufficient specificity to allow determination of dopamine D_2 receptors *in vivo*. Additionally, the binding of the tracer was observed to be essentially irreversible over the time course of a PET study (approximately 90 minutes). Thus the N-alkylation serendipitously resulted in a radiotracer which has received considerable attention in our institution and others (Wagner et al., 1983; Wong et al., 1984; Wong et al., 1986).

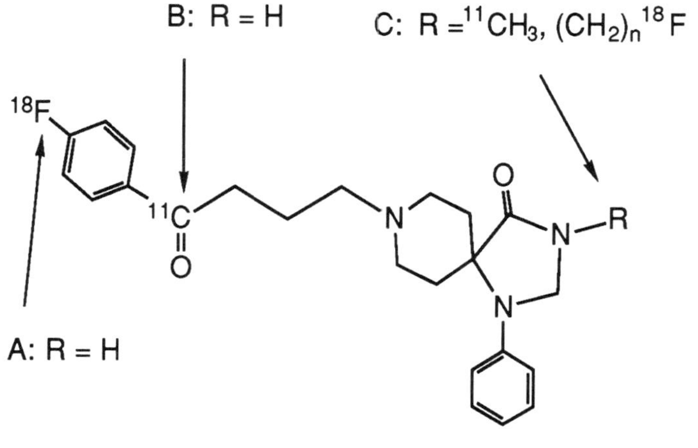

B: R = H

C: R = $^{11}CH_3$, $(CH_2)_n{}^{18}F$

^{18}F

^{11}C

O

N

O

N—R

N

A: R = H

FIGURE 2. Published synthetic routes to radiolabeled spiperone and its analogues.

The demand for some tracers, particularly [^{11}C]NMS, for PET studies is so great that extremely reliable syntheses for their production must be developed. The need for routine production of 3-N-[^{11}C]methylspiperone required that the original synthesis be modified. The original synthesis employed an alkylation with [^{11}C]methyl iodide and derivatization of unreacted starting material (alkylation with n-heptyl iodide) to simplify purification. The process was lengthy and required several manipulations which reduced over-all yield and increased the risk of radiation exposure for the chemists (Burns et al., 1984b). For these reasons, a synthesis was developed which involves fewer manipulations, less time, and lends itself to automation (Dannals et al., 1986).

The volatile solvent originally used for the alkylation reaction, methylene chloride, was replaced with dimethylformamide (DMF). DMF has a higher boiling point which allowed the reaction temperature to be increased from 35° C to 80° C and the reaction time decreased from 5 min to 1 min. A change in the mobile phase used for the preparative HPLC purification to a less viscous mixture (aqueous acetonitrile instead of aqueous methanol) increased the solvent flow through the HPLC column and permitted separation of the product without the derivatization step. By making these changes to the original synthesis, the overall chemical yield was increased and the total time of synthesis was shortened from 30 minutes to less than 17 minutes. This resulted in an increased radiochemical yield (from 8% to 21%, calculated at the end of the synthesis) and, concomitantly, the specific activity of the final product increased.

The perturbations introduced by fluoroalklyation (Fig. 2C) necessitate some knowledge of a structure-activity relationship (SAR) for the class of compound being labeled. For neuroreceptor ligands, the activity measured in the SAR is usually a pharmacological response or affinity (IC_{50}) for a receptor or receptor subtype. Often candidates for labeling with a fluoroalkyl group have been tracers that have been successfully radiomethylated; however, they also include those compounds which already possess a halogen and compounds which display sites of bulk tolerance, i.e., positions on the molecule which, when substituted by large lipophilic groups, maintain or increase their pharmacological activity or affinity for the receptor. Recent efforts to develop a fluoroalkyl analog to substitute for a methyl group have been particularly successful when applied to the spiperone parent structure. Several groups have been working in this area (Coenen et al., 1988; Satyamurthy et al., 1986; Welch et al., 1988) although attempts to extend the analogy to other pharmaceuticals have been less successful (Chesis and Welch, 1989a; Chesis and Welch, 1989b; Moerlein et al., 1989).

The use of [^{11}C]carfentanil with PET to study opiate receptors provides an excellent example of the need for achieving high specific activities in radiochemical syntheses. Carfentanil is an opiate receptor agonist which is approximately 8000 times more potent than morphine *in vivo* (Stahl et al., 1977; Van Daele et al., 1976) and has veterinary use as a tranquilizer for large animals (Haigh et al., 1978; Kock and Berger, 1987). Based on estimated doses scaled for humans, we calculated the minimum specific activity required for PET studies. Using an average dose of 0.05 μg/Kg in a average 70 Kg subject, and a 20 mCi injection, the minimum specific activity would have to be 2280 mCi/μmole at the time of injection to insure that there would be no pharmacological effect. To accomplish this task, a synthesis of [^{11}C]carfentanil was developed in which the nor-methyl carboxylate salt of carfentanil was reacted with [^{11}C]CH_3I (Dannals et al., 1985). Purification by HPLC was facilitated by the polar nature of the precursor which was easily separated from the product. Analytical HPLC was also rapid which allowed the entire synthesis to be completed in less than one half-life.

Many of the PET and SPECT tracers in use today found their beginning as unlabeled chemicals developed within the pharmaceutical industry. Following *in vitro* and *in vivo* screening, the most promising candidates were often labeled with Tritium, Carbon-14, and occasionally Iodine-125, to allow them to be used in autoradiographic studies or *ex vivo* biological distributions. Having passed through the early stages of characterization, a compound may then be chosen to be radiolabeled with an isotope of suitable characteristics for use with PET (e.g., Carbon-11 or Fluorine-18) or SPECT (e.g., Iodine-123). An example of this general scheme in tracer development is seen in the preparation of radiotracers for studying muscarinic cholinergic receptors.

In 1971, Janssen et al., described the synthesis of benzetimide and its enantiomeric resolution into dexetimide (its active form) and levetimide (the inactive form). Within two years, dexetimide was being used as a tool in acetylcholine receptor localization (Soudijn et al., 1973). By the end of the decade, Laduron demonstrated that [^{3}H]dexetimide bound to muscarinic cholinergic receptors *in vitro* (Laduron and Janssen, 1979) and *in vivo* (Laduron et al., 1979) with high affinity, low non-specific binding, and a slow dissociation rate. The accumulation of dexetimide was determined to be stereospecific, saturable, and displaceable while [^{3}H]levetimide did not show preferential uptake or stereospecific displacement in regions known to have high concentrations of muscarinic cholinergic receptors. Based on these results, dexetimide was labeled with Carbon-11 as a PET radiotracer for studying the muscarinic cholinergic receptors, and levetimide was also labeled for assessing non-specific binding associated with the *in vivo* receptor binding studies. The synthesis [^{11}C]dexetimide involved the synthesis of [^{11}C]benzyl iodide and its use in an alkylation reaction of the appropriate nor-benzyl amine precursor (Dannals et al., 1986).

Concomitantly N-benzyl monohalogenated analogs of dexetimide were being synthesized and screened to determine if another derivative might be even better suited for the *in vivo* studies. The results of the analog syntheses led to the development of Iodine-123 labeled 4-iododexetimide, an analog for SPECT studies which binds to the muscarinic cholinergic receptor with an affinity comparable to dexetimide itself (Wilson et al., 1989), and Fluorine-18 labeled 2-fluorodexetimide, a longer half-life analogue for PET studies (Wilson et al, 1990; Wilson et al., 1991).

CONCLUSION

The successful development of radiotracers has relied heavily on cooperation with researchers working in the development of stable pharmaceuticals. The successful utilization of these methodologies, including the design and synthesis of new radiotracers to be used by the ethical pharmaceutical industry, will benefit greatly from the experience acquired by radiopharmaceutical scientists over the past decade. Continued cooperation between researchers working in the development of stable pharmaceuticals and those involved with the development of new radiotracers is likely to be beneficial to both fields. Advances made in one will undoubtedly lead to advances in the other.

ACKNOWLEDGMENTS

This work was supported in part by grant numbers NS-15080 and CA-32845 from the U.S. Public Health Service.

REFERENCES

Berridge MS, Cassidy EH and Terris AH (1990). A routine, automated synthesis of oxygen-15-labeled butanol for positron tomography. *J Nucl Med* 31: 1727-1731.

Burns HD, Lever JR, Dannals RF, Frost JJ, Wilson AA, Ravert HT, Subramanian V, Zemyan SE, Langstrom B and Wagner HN Jr (1984a): Synthesis of ligands for imaging opiate receptors by positron emission tomography: Carbon-11 labeled diprenorphine. *J Label Compd Radiopharm* 21: 1167-1169.

Burns HD, Dannals RF, Langstrom B, Ravert HT, Zemyan SE, Duelfer T, Wong DF, Frost JJ, Kuhar MJ and Wagner HN Jr (1984b):

(3-N-[^{11}C]Methyl)spiperone, a ligand binding to dopamine receptors: Radiochemical synthesis and biodistribution studies in mice. *J Nucl Med* 25: 1222-1227.

Chesis PL and Welch MJ (1989a): Direct analysis of F-18 labeled receptor imaging agents with *in vitro* binding assays. *J Nucl Med* 30: 931.

Chesis PL and Welch MJ (1989b): Synthesis and *in vitro* characterization of fluorinated U-50488 analogs for PET studies. *J Nucl Med* 30: 931.

Coenen H, Wienhard K, Stöcklin G, Laufer P, Hebold I and Pawlik G (1988): PET measurement of D_2 and S_2 receptor binding of 3-N-([2'-^{18}F]fluoroethyl)spiperone in baboon brain. *Eur J Nucl Med* 14: 80-87.

Dannals RF, Ravert HT, Frost JJ, Wilson AA, Burns HD and Wagner HN Jr (1985): Radiosynthesis of an opiate receptor binding radiotracer: ^{11}C-carfentanil. *Int J Appl Radiat Isot* 36: 303-306.

Dannals RF, Ravert HT, Wilson AA and Wagner HN Jr (1986): An improved synthesis of (3-N-[^{11}C]methyl)spiperone. *Int J Appl Radiat Isot* 37: 433-434.

Dannals RF, Langstrom B, Ravert HT, Wilson AA and Wagner HN Jr (1988): Synthesis of radiotracers for studying muscarinic cholinergic receptors in the living human brain using positron emission tomography: [^{11}C]Dexetimide and [^{11}C]levetimide. *Appl Radiat Isot, Int J Radiat Appl Instrum Part A* 39: 291-295.

Del Fiore G, Depresseux J-C, Bartsch P, Quaglia L and Peters J-M (1979): Production of oxygen-15, nitrogen-13 and carbon-11 and of their low molecular weight derivatives for biomedical applications. *Int J Appl Radiat Isot* 30: 543-549.

Ehrhardt GJ and Welch MJ (1978): A new germanium-68/gallium-68 generator. *J Nucl Med* 19: 925-929.

Finn RD, Boothe TE, Vora MM, Hildner JC, Emran AM and Kothari PJ (1984): Syntheses with isotopically labelled carbon. Methyl iodide, formaldehyde, and cyanide. *Int J Appl Radiat Isot* 35: 323-335.

Fowler JS (1978): Positron-emitting nuclides and their synthetic incorporation in radiopharmaceuticals. In: *Chemistry of Radiopharmaceuticals*, Heindel ND, Burns HD, Honda T and Brady LW, eds. New York: Masson Publishing USA, Inc.

Fowler JS, Arnett CD, Wolf AP, MacGregor RR, Norton EF and Findley AM (1982): Carbon-11 labeled spiroperidol. Synthesis, specific activity determination and biodistribution in mice. *J Nucl Med* 23: 437-445.

Fowler JS and Wolf AP (1986): Positron emitter-labeled compounds: Priorities and problems. In: *Positron Emission Tomography and*

Autoradiography: Principles and Applications for the Brain and Heart, Phelps M, Mazziotta J and Schelbert H, eds. New York: Raven Press.

Frost JJ, Smith AC and Wagner HN Jr (1986): ^{3}H-diprenorphine is selective for mu opiate receptors *in vivo*. *Life Sci* 38: 1597-1606.

Frost JJ, Mayberg HS, Sadzot B, Dannals RF, Lever JR, Ravert HT, Wilson AA, Wagner HN Jr and Links JM (1990): Comparison of [^{11}C]diprenorphine and [^{11}C]carfentanil binding to opiate receptors in humans by positron emission tomography. *J Cereb Blood Flow Metab* 10: 484-92.

Fujibayashi Y, Matsumoto K, Yonekura Y, Konishi J and Yokoyama A (1989): A new zinc-62/copper-62 generator as a copper-62 source for PET radiopharmaceuticals. *J Nucl Med* 30: 1838-1842.

Gelbard AS (1981): Biosynthetic methods for incorporating positron-emitting radionuclides into compounds of biomedical interest. *J Label Compds Radiopharm* 18: 933-945.

Green MA, Mathias CJ, Welch MJ, McGuire AH, Perry D, Fernandez-Rubio F, Perlmutter JS, Raichle ME and Bergmann SR (1990): Copper-62-labeled pyruvaldehyde bis(N4-methylthiosemicarbazonato)-copper(II): Synthesis and evaluation as a positron emission tomography tracer for cerebral and myocardial perfusion. *J Nucl Med* 31: 1989-1996.

Haigh JC, Lee LJ and Schweinsburg RE (1983): Immobilization of polar bears with carfentanil. *J Wildl Dis* 19: 140-144.

Janssen PAJ, Niemegeers CJE, Schellekens KHL, Domoen P, Lanaerts FM, van Nueten JM, van Wijngaarden I and Brugmans J (1971): Benzetimide and its optical isomers. *Arzneim Forsch (Drug Res)* 21: 1365-1373.

Kilbourn MR (1990): *Fluorine-18 Labeling of Radiopharmaceuticals* (NAS-NS 3203). Washington, DC: National Academy Press.

Kilbourn MR, Welch MJ, Dence CS, Tewson TJ, Saji H and Maeda M (1984): Carrier-added and no-carrier-added syntheses of fluorine-18 labeled spiroperidol and haloperidol. *Int J Appl Radiat Isot* 35: 591-598.

Kock MD and Berger J (1987): Chemical immobilization of free-ranging North American bison in Badlands National Park, South Dakota. *J Wildl Dis* 23: 625-633.

Laduron PM and Janssen PFM (1979): Characterization and subcellular localization of brain muscarinic receptors labelled *in vivo* by [^{3}H]dexetimide. *J Neurochem* 33: 1223-1231.

Laduron PM, Verwimp M and Leysen JE (1979): Stereospecific *in vitro* binding of [^{3}H]dexetimide to brain muscarinic receptors. *J Neurochem*, 32: 421-427.

Lever JR, Dannals RF, Wilson AA, Ravert HT and Wagner HN Jr (1987): Synthesis of carbon-11 labeled diprenorphine: A radioligand for positron emission tomographic studies of opiate receptors. *Tetrahedron Letters* 28: 4015-4018.

Luthra SK, Pike VW and Brady F (1985): The preparation of carbon-11 labeled diprenorphine. A new radioligand for the study of the opiate receptor system *in vivo*. *J Chem Soc, Chem Comm*: 1423-1425.

Mathis CA, Lagunas-Solar MC, Sargent T III, Yano Y, Vuletich V and Harris LJ (1986): A ^{122}Xe-^{122}I generator for remote radioiodinations. *Appl Radiat Isot Int J Radiat Appl Instrum Part A* 37: 258-260.

Moerlein SM, Lannoye GS and Welch MJ (1989): No-carrier-added synthesis of N w-(F-18)fluoroethyl SCH-23390. *J Nucl Med 30*: 931.

Paans AMJ, J Welleweerd, W Vaalburg, S Reiffers and MG Woldring (1980): Excitation functions for the production of bromine-75: A potential nuclide for the labelling of radiopharmaceuticals. *Int J Appl Radiat Isot* 31: 267-273.

Palmer AJ, Clark J and Goulding RW (1977): The preparation of fluorine-18 labelled radiopharmaceuticals. *Int J Appl Radiat Isot* 28: 53-65.

Rimland A and Langstrom B (1987): Synthesis of carbon-11 iodomethylcyclopropane. An interesting alkylating reagent. *Appl Radiat Isot, Int J Radiat Appl Instrum Part A* 38: 949-952.

Satyamurthy N, Bida G, Barrio J, Luxen A, Mazziotta J, Huang S and Phelps M (1986): No-carrier-added 3-(2'-[18F]fluoroethyl)-spiperone, a new dopamine receptor-binding tracer for positron emission tomography. *Nucl Med Biol, Int J Radiat Appl Instrum Part B* 13: 617-624.

Shiue C-Y, Fowler JS, Wolf AP, Watanabe M and Arnett C (1985): Syntheses and specific activity determinations of no-carrier-added fluorine-18 labeled neuroleptic drugs. *J Nucl Med* 26: 181-186.

Soudijn W, van Wijngaarden I and Ariens EJ (1973): Dexetimide, a useful tool in acetylcholine-receptor localization. *Eur J Pharm* 24: 43-48.

Stahl KD, Van Bever W, Janssen P and Simon EJ (1977): Receptor affinity and pharmacological potency of a series of narcotic analgesic, anti-diarrheal and neuroleptic drugs. *Eur J Pharm* 46: 199-205.

Stöcklin G (1986): Molecules labeled with positron emitting halogens. *Nucl Med Biol Int J Radiat Appl Instrum Part B* 13: 109-118.

Van Daele PGH, De Bruyn MFL, Boey JM, Sanczuk S, Agten JTM and Janssen PAJ (1976): Synthetic analgesics: N-(1-[2-arylethyl]-4-

substituted 4-piperidinyl) N-arylalkanamides. *Arzneim Forsch (Drug Res)* 26: 1521-1531.

Wagner HN Jr, Burns HD, Dannals RF, Wong DF, Langstrom B, Duelfer T, Frost JJ, Ravert HT, Links JM, Rosenbloom S, Lukas SE, Kramer AV and Kuhar MJ (1983): Imaging dopamine receptors in the human brain by positron tomography. *Science* 221: 1264-1266.

Welch M, Katzenellenbogen J, Mathias C, Brodack J, Carlson K, Chi D, Dence C, Kilbourn M, Perlmutter J, Raichle M and Ter-Pogossian M (1988): N-(3-[^{18}F]Fluoropropyl)-spiperone: The preferred ^{18}F labeled spiperone analog for positron emission tomographic studies of the dopamine receptor. *Nucl Med Biol, Int J Radiat Appl Instrum Part B* 15: 83-97.

Wilson AA, Dannals RF, Ravert HT, Frost JJ and Wagner HN Jr (1989): Synthesis and biological evaluation of [^{125}I]- and [^{123}I]-4-iododexetimide, a potent muscarinic cholinergic receptor antagonist. *J Med Chem* 32: 1057-1062.

Wilson AA, Dannals RF, Ravert HT and Wagner HN Jr (1990): Reductive amination of [^{18}F]fluorobenzaldehydes: Radiosyntheses of [2-^{18}F]- and [4-^{18}F]fluoro-dexetimides. *J Label Compd Radiopharm* 28: 1189-1199.

Wilson AA, Scheffel UA, Dannals RF, Stathis M, Ravert HT and Wagner HN Jr (1991): *In vivo* biodistribution of two [^{18}F]-labelled muscarinic cholinergic receptor ligands: 2-[^{18}F]- and 4-[^{18}F]-fluorodexetimide. *Life Sci* 48: 1385-1394.

Wolf AP (1984): Cyclotrons, radionuclides, precursors, and demands for routine versus research compounds. *Ann Neurol* 15: S19-S24.

Wolf AP and Redvanly CS (1977): Carbon-11 and radiopharmaceuticals. *Int J Appl Radiat Isot* 28: 29-48.

Wolf AP and Fowler JS (1979): Organic radiopharmaceuticals - Recent advances. In: *Radiopharmaceuticals II*, New York: Society of Nuclear Medicine.

Wolf AP and Fowler JS (1986): Positron emitter-labeled radiotracers - Chemical considerations. In: *Positron Emission Tomography*, Reivich M and Alavi A, eds. New York: Alan R Liss, Inc.

Wong DF, Wagner HN Jr, Dannals RF, Links JM, Frost JJ, Ravert HT, Wilson AA, Rosenbaum AE, Folstein MF, Petronis JD, Douglass KH, Toung JKT and Kuhar MJ (1984): Effects of age on dopamine and serotonin receptors measured by positron tomography in the living human brain. *Science* 226: 1393-1396.

Wong DF, Wagner HN Jr, Tune LE, Dannals RF, Pearlson GD, Links JM, Tamminga C, Broussolle EP, Ravert HT, Wilson AA, Toung JKT, Malat J, Snyder SH, Kuhar MJ and Gjedde A (1986): Positron emission tomography reveals elevated brain D2 dopamine receptors in schizophrenia. *Science* 234: 1558-1563.

Yano Y, Chu P, Budinger TF, Grant PM, Ogard AE, Barnes JW, O'Brien HA and Hoop B Jr (1977): Rubidium-82 generators for imaging studies. *J Nucl Med* 18: 46-50.

Single Photon, Gamma Emitting Radiotracers for Use in Imaging

H. Donald Burns
Department of Radiopharmacology
Merck Research Laboratories

Kwamena E. Baidoo
Division of Nuclear Medicine
The Johns Hopkins Medical Institutions

Alan A. Wilson
PET Imaging Facility
Clarke Institute of Psychiatry

INTRODUCTION

Non-invasive nuclear imaging techniques are beginning to gain substantial acceptance in pharmaceutical research. Drug candidates can be studied using isotopically labeled compounds (primarily Carbon-11 or Fluorine-18) and Positron Emission Tomography (PET) imaging to obtain information on drug absorption, distribution, metabolism and elimination. Imaging studies in humans with these labeled drugs or drug candidates can be used to compliment information obtained using Tritium and Carbon-14 labeled compounds in animals. Although imaging studies of this type are complicated by a number of factors, such as limitations imposed by the short half-life of Carbon-11 or the need to make corrections for the possible presence of labeled metabolites in

Nuclear Imaging in Drug Discovery, Development, and Approval
H.D. Burns, R. Gibson, R. Dannals, Peter Siegl, Editors

tissues, if designed carefully, this type of study can provide valuable information on the kinetics of a drug in the target tissue or accumulation of the drug in other tissues (suggesting potential toxic side effects in that tissue). They can also be used to evaluate the effect that factors such as formulation and route of administration, have on the biodistribution of a drug. This type of study, however, is limited to the use of short-lived, positron emitting radionuclides at suitably equipped PET centers (see Chapter 4). Other studies using widely available planar or Single Photon Emission Computed Tomographic (SPECT) cameras and gamma emitting radiotracers can also play a valuable role in pharmaceutical research.

Planar and SPECT Imaging

The application of non-invasive, nuclear imaging studies with gamma emitting radiotracers and planar or SPECT cameras is relatively new. These studies can be used in a variety of ways in pharmaceutical research, including studies designed to: i) measure desired and/or adverse effects of drug candidates, ii) provide information on the interaction of an unlabeled drug with a specific binding site (for example, receptors, enzymes, and transport proteins), and iii) provide information on the behavior of a new formulation or drug delivery device *in vivo*.

This new technology provides information which compliments that obtained by more traditional research techniques used by drug companies and has the potential of making the process of drug discovery, development and approval more efficient. A major advantage of nuclear imaging is its non-invasive nature, which makes it possible to conduct studies in humans to obtain information which cannot be obtained by any other means. Many of these studies can be conducted with commercially available radiopharmaceuticals which are currently used in clinical nuclear medicine. Other studies require the design of new radiotracers labeled with suitable gamma emitting radionuclides.

Radionuclides

A variety of factors influence the selection of the radionuclide to be used in an imaging study. Many of these relate to the chemistry of the radionuclide and the particular study design while others depend on the type of imaging instrumentation to be used. Most nuclear imaging is conducted using either planar or SPECT cameras and radionuclides which emit photons with energies between 100 and 250 keV. These cameras are optimized for energies of about 140-150 keV. For lower gamma energies (< 20 keV), a major fraction of the radiation is absorbed by tissue, resulting in decreased sensitivity, difficulty in quantitation and increased radiation dose to the experimental subject (of particular concern in human studies). Higher energy gammas (> 600 keV) are not efficiently detected by these cameras and are difficult to collimate resulting in decreases in sensitivity and resolution. As a result of these considerations and other practical concerns such as cost, half-life and availability, only a few gamma emitting radionuclides are widely used in imaging studies. The most common of these are shown in Table 1. Many radiotracers labeled with these radionuclides are currently used

Radionuclide	Half-Life	Principle Photons (keV)
Gallium-67	78 hr	93 (37)[a] 185 (20) 300 (17)
Indium-111	68 hr	171 (90) 245 (94)
Iodine-123	13 hr	159 (83)
Iodine-131	8.04 day	364 (81)
Selenium-75	120 day	136 (59) 265 (59) 278 (25)
Thallium-201	73 hr	71 (46) 69 (27) 80 (11) 167 (10)
Technetium-99m	6 hr	141 (89)

a) number in parentheses represents % abundance.

TABLE 1. Common radionuclides used in planar and SPECT imaging (data adapted from Stabin et al., 1991).

as diagnostic radiopharmaceuticals in clinical nuclear medicine.

This chapter will review recent advances in the development of radiotracers labeled with the most versatile and widely used gamma emitting radionuclides, Technetium-99m, Iodine-123, Iodine-131 and Indium-111.

TECHNETIUM-99m

Introduction and General Technetium Chemistry

Technetium-99m is the most widely used radionuclide in routine nuclear medicine imaging even though considerable advances have been made in the development of radiopharmaceuticals from other radionuclides. This prominence continues because the radionuclide is inexpensive (< \$1.00/mCi), readily available in generator form, and the practically monoenergetic 140 keV gamma photons are best suited to current planar and SPECT imaging equipment. Furthermore, the high photon yield per disintegration provides good counting statistics during imaging while the lack of particulate emissions coupled to the short half-life of the radionuclide (6 hr) results in relatively low radiation dose to patients.

Chemically, because of its position in the middle of the d-block transition metal series possessing seven electrons beyond the noble gas configuration, technetium (Tc) possesses multiple oxidation states (-1 to +7) and variable coordination numbers and geometries. The element is commercially available in the +7 oxidation state as the pertechnetate ion, $[TcO_4]^-$, the most stable state of the element in aqueous media in the absence of reducing agents. All synthetic transformations with the element must therefore start with this species. Even though pertechnetate is capable of limited complex formation by addition of halides and some nitrogen ligands to give compounds such as $[TcO_3(bipy)Cl]$ (Davison et al., 1981), its major mode of interaction is through electrostatic bonding. Direct preparation of radiopharmaceuticals incorporating the element in this oxidation state is therefore limited. Derivatization of Tc is generally achieved through coordination of ligands to the metal existing in lower oxidation states. Most of the clinically useful or potentially useful Technetium-99m radiopharmaceuticals exist in one of these lower oxidation states.

The element is capable of forming complexes in several different stoichiometries depending on the ligand and the oxidation state. Higher oxidation states ($\geq$ +4) are stabilized by oxotechnetium (e.g., $TcOL_n$, TcO_2L_n, $TcO(OH)L_n$, and Tc_2O_2L) or nitrilotechnetium cores ($TcNL_n$) while lower oxidation states involve ligands complexed to Tc alone (e.g.,

TcL_n, Tc_2L_n) . Multiple complexes are possible from the same reaction using a single ligand. The concentration of Tc in solution can affect both the kinetics and the products formed. The total concentration of Tc in no-carrier-added Technetium-99m from generator eluates is on the order of 10^{-8} - 10^{-6} M. The quantities of Tc available in no-carrier-added Technetium-99m reactions is therefore very small (~1.9 pmol/mCi), and conventional methods of analysis cannot be used in these reactions. Conventional characterization of Tc compounds is performed using Technetium-99 ($T_{1/2}$ = 2.12 x 10^5 yr) which is available in gram quantities. Chromatographic correlations in carrier-added reactions are then performed to characterize the Technetium-99m complexes.

Complexation of ligands to Tc is generally performed by the reduction of $[TcO_4]^-$ in the presence of the coordinating ligand. Unreacted pertechnetate often persists in radiopharmaceutical preparations as a result of incomplete reduction; or in the presence of oxygen, it can be regenerated by decomposition of unstable Tc complexes. The final oxidation state of Tc in a complex is dependent on the type of reducing agent, ligand and the reaction conditions used. In the absence of a strong acidic enviroment and/or a complexing agent, a highly insoluble hydrated oxide, $TcO_2 \cdot xH_2O$, is formed. This species is highly gregarious and forms colloids. Because of its high thermodynamic stability, the formation of this species is often a competing reaction in the preparation of Technetium-99m radiopharmaceuticals. This pathway to the reduction of $[TcO_4]^-$ therefore presents a formidable problem to the synthesis of Tc complexes in aqueous solution.

Several reagents have been used to accomplish the reduction of pertechnetate in aqueous media. It is important that the reducing agent react rapidly, completely, and not interfere with the biodistribution of the resultant Technetium-99m complex. Most organic reducing agents tend to react slowly and at higher pH; conditions which tend to promote the hydrolysis of reduced Tc and the formation of $TcO_2 \cdot xH_2O$. Metal ions tend to act faster but have the disadvantage that they may be incorporated into the complex formed, hydrolyze, or precipitate trapping some Technetium-99m in the process. Even though stannous ion has been shown to be incorporated into some Tc complexes such as $[Tc(DMG)_3SnCl_3(OH)] \cdot 3H_2O$ (DMG = dimethylglyoxime) (Deutsch et al., 1976), stannous compounds have proven to be the most reliable reducing agents for the preparation of Technetium-99m radiopharmaceuticals.

The stability of Tc complexes depends on the effectiveness of a ligand in stabilizing the metal in its lower oxidation state. Inability to stabilize the metal may result in hydrolysis to TcO_2 and/or reoxidation to pertechnetate. In the presence of a more effective stabilizing ligand Tc may be transferred from a weaker complex to the stronger coordinating ligand in the process of transchelation or ligand exchange. Tc complexes having reasonable half-lives in solution can, in this regard, be used as synthetic starting materials to donate reduced Tc during the formation of more stable Tc complexes. Of great utility for this purpose has been several structurally similar Tc complexes derived from polyhydroxy or carboxylate ligands, such as glucoheptonate, citrate, catechol, ethylene glycol and gluconate which can be utilized as donors of prereduced Tc during the synthesis of Tc complexes. The importance of these labile complexes derives from the fact that in certain cases, products could be obtained devoid of contaminants from competing reactions. Conceptually, they could also be used in the synthesis of complexes from ligands that could not withstand the reducing conditions necessary for the reduction of pertechnetate.

The many different oxidation states of the element has allowed formulation into a large variety of radiopharmaceuticals useful for studies of the functions of several organs of the body. These derivatives can be taken advantage of in the drug development effort to non-invasively study the efficacy of drugs designed to affect the overall function or alter the biochemistry or physiology of particular organs. Towards this end, Technetium-99m based radiopharmaceuticals are available for studies of the functions of the brain, heart, kidneys, lungs, spleen, and liver, as well as diseases such as cancer, infection and thrombosis. A further advantage of Technetium-99m is that once radiopharmaceuticals derived from the nuclide are developed, they can be prepared from convenient kits by technologists who have little expertise in chemistry.

Early use of technetium in nuclear medicine involved investigations with $[^{99m}TcO_4]^-$ (Harper et al., 1962). Initial efforts to derivatize Technetium-99m involved adventitious incorporation of the radionuclide in ill-defined formulations of reducing agent, $[^{99m}TcO_4]^-$ and ligands. Products arising out of these early efforts were colloidal preparations, such as $[^{99m}Tc]$sulfur colloid (Harper et al., 1964) and $[^{99m}Tc]$macroaggregated albumin (MAA) (Tow et al., 1966). Later, commonly available compounds such as simple sugars (Subramanian et al., 1971), ethylenediaminetetraacetic acid (EDTA) (Fleary, 1968), and diethylentriaminepentaacetic acid (DTPA) (Richards and Atkins, 1968) were used. Many of the early agents were chemically uncharacterized. Within the last two decades several useful and potentially useful radiopharmaceuticals have been generated. Several efforts have been made within the last decade to clarify the structure and chemistry of these radiopharmaceuticals and to employ the information to design new agents. The areas that have received the most attention recently involve the brain, heart, kidneys and protein labeled agents.

Brain Imaging Agents

Constraints imposed by the blood-brain-barrier demand that radiopharmaceuticals for imaging the uninjured brain be neutral and lipid soluble unless a specific mechanism exists for transport of the particular tracer into the brain. An important breakthrough in Tc chemistry, therefore, was the recent development of neutral lipophilic complexes of the nuclide capable of crossing intact membranes. This has opened the door for the study of brain blood flow imaging in a number of neurological disorders including stroke, dementia, epilepsy and congenital dysphasia using Technetium-99m radiopharmaceuticals.

The first neutral lipohilic Technetium-99m complex to be successfully applied to the imaging of cerebral blood flow in humans was [^{99m}Tc]-*d,l*-HMPAO (**1**) (Ell et al., 1985). This tracer is derived from the propylene amine oxime (PAO) ligand system (Troutner et al., 1984). Complexation of PAO ligands to Tc results in the deprotonation of two amine nitrogens and one oxime hydrogen to compensate for the charges on the $[Tc(V)=O]^{3+}$ core (Jurison et al., 1986). [^{99m}Tc]-*d,l*-HMPAO is commercially available in kit formulation as Ceretec™ (Amersham). As currently formulated [^{99m}Tc]-*d,l*-HMPAO is unstable *in vitro*, decomposing to a hydrophilic complex which does not cross the blood-brain-barrier. This necessitates the use of the radiopharmaceutical within 30 min of preparation (Neirinckx et al., 1987). The long

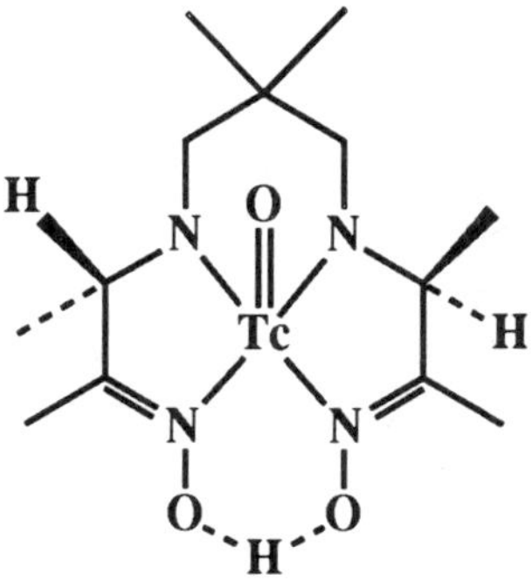

1

retention of the agent *in vivo* may indeed be by virtue of this instability resulting in decomposition in tissue or exchange to intracellular sulfhydryl groups. The agent has a high brain extraction fraction (0.88) and its distribution correlates with microspheres. Prolonged retention in tissue (brain clearance $T_{1/2}$ > 70 hr) permits high resolution SPECT imaging of the brain. The washout of the radiopharmaceutical from other tissues is also slow. The agent has been applied to the study of stroke (Mountz et al., 1990), dementia (Smith et al.,1987), epilepsy (Biersack et al., 1987), congenital dysphasia (Denays et al., 1989) and brain death (Laurin et al., 1989). The limited stability of the radiopharmaceutical however, precludes the use of the tracer in certain acute situations and its preparation in central pharmacies.

A new analogue of the PAO structural class, [^{99m}Tc]-*d,l*-CB-PAO (**2**), has been reported that incorporates a cyclobutane ring on the ligand backbone. This modification imparts rigidity to the Tc complex and thereby results in a structure that is more stable *in vitro*. The *in vivo* characteristics of this analogue are similar to that of [^{99m}Tc]-*d,l*-HMPAO in rats, but in humans, brain uptake may be 30 - 50% less (Messa et al., 1990).

Another class of neutral lipophilic Technetium-99m complexes is the seven coordinate boronic acid adducts of Tc(III)-dioxime complexes (Tc-BATO) (Treher et al., 1989). Derivatives of this class of complexes can be obtained by changing the nature of the dioxime ligands or the substituent on the boron atom. Results obtained with the isobutylboronic acid adduct of the Tc-dimethylglyoxime derivative, [^{99m}Tc]-BATO-2MP (**3**) also called SQ 32097 (Linder et al., 1987), indicate that the agent behaves as a freely diffusible tracer and thus, may require imaging with dedicated multi-headed SPECT cameras within the first 15 min after injection (Drane et al., 1989).

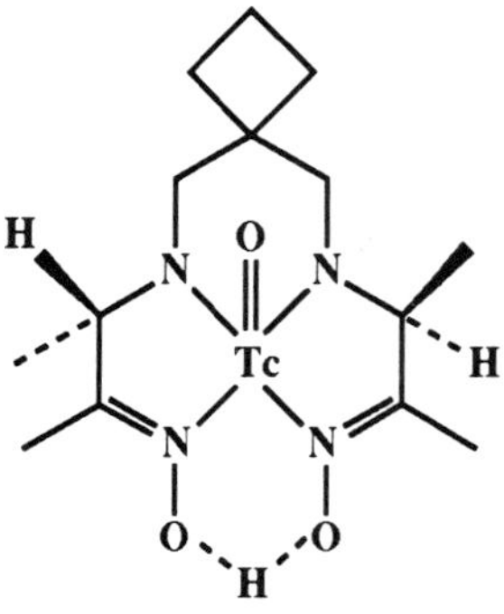

2

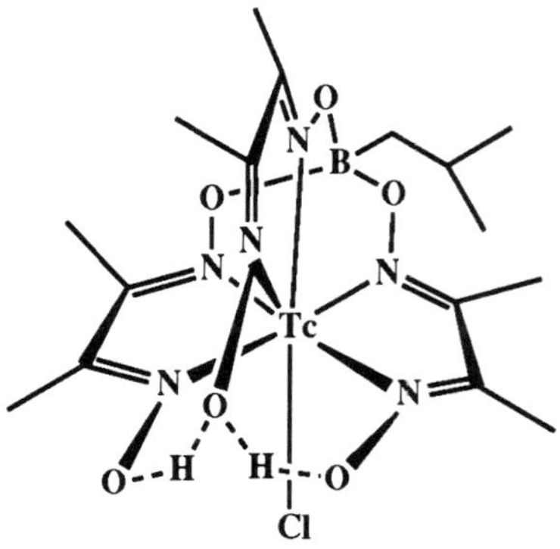

3

At about the same time that the PAO and BATO complexes were being developed, a parallel investigation was undertaken involving neutral lipophilic complexes of the tetradentate diaminedithiol (DADT) ligands (Epps, 1984; Kung et al., 1984). Complexation of these ligands to Tc results in the deprotonation of an amine hydrogen and two sulfur hydrogens to compensate for the charges on the $[Tc(V)=O]^{3+}$ core (Epps, 1984; Epps et al., 1987). Technetium-99m complexes of these ligands are among the most stable complexes of the nuclide known. Derivatives of this class of complexes can be obtained through carbon or nitrogen backbone substitution. The initial complexes developed for studies of blood flow included N-alkyl-amine analogues (**4**) that behaved as freely diffusible tracers (Lever et al., 1985; Scheffel et al., 1988). More recently, a new derivative containing two ester groups, $[^{99m}Tc]$-*l,l*-ethyl cysteinate dimer (**5**) (abbreviated $[^{99m}Tc]$-*l,l*-ECD) was reported (Cheesman et al., 1988). Although $[^{99m}Tc]$-*d,d*-ECD is very stable *in vivo*, the ester groups of $[^{99m}Tc]$-*l,l*-ECD (Neurolite™, Du Pont-Merck Radiopharmaceutical Co.) are hydrolyzed in primate brain to the mono- (**6**) or di-acid (**7**) once the agent is taken up in tissue. The metabolites, which are anionic at physiological pH, are retained in brain tissue with sufficient half-life to permit high resolution SPECT imaging (Leveille et al., 1989; Holman et al., 1989). Because the metabolites are cleared faster from other tissues and blood than from brain tissue, very high

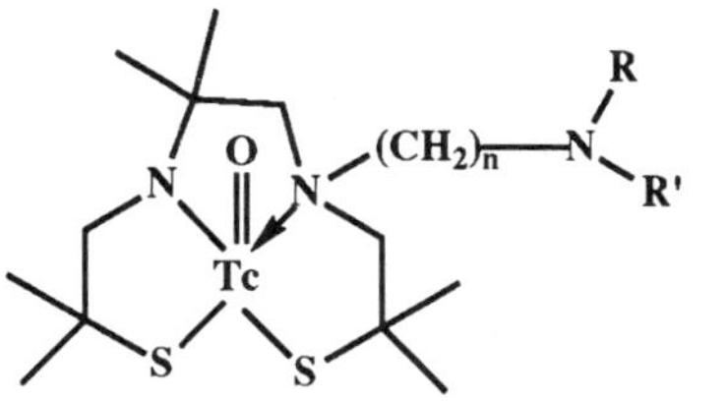

4 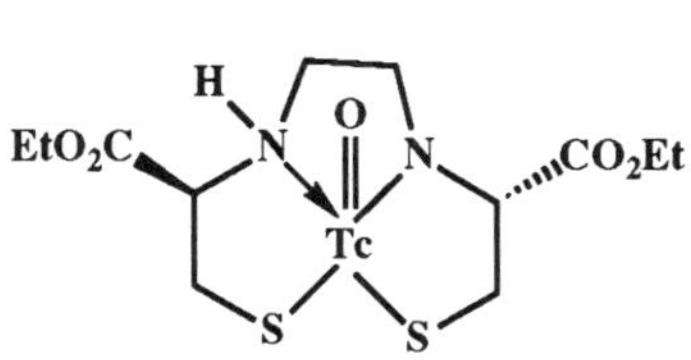 **5**

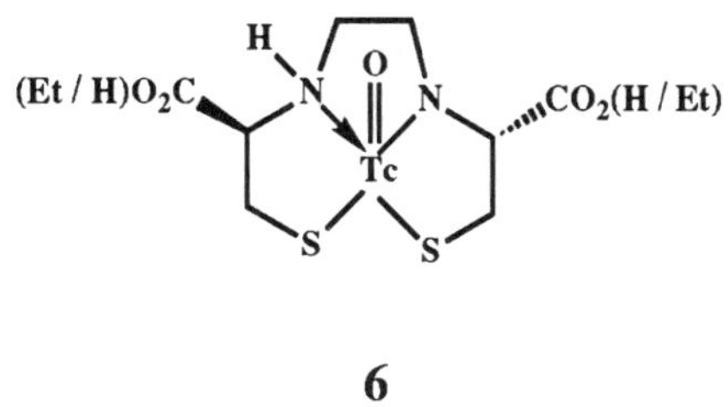

6 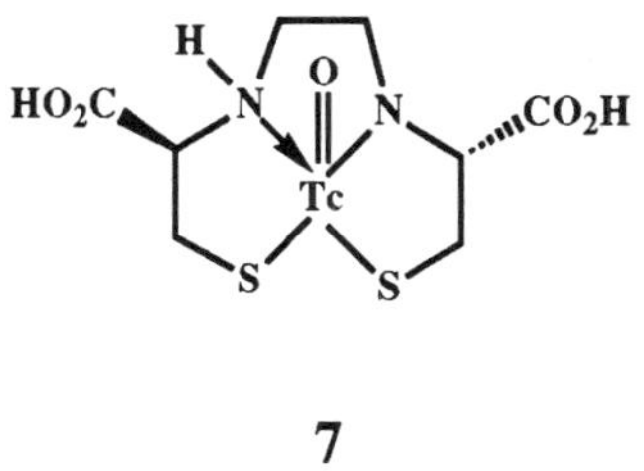 **7**

contrast images are obtained with this agent (Walovitch et al., 1990). [^{99m}Tc]-*l,l*-ECD has great potential for routine studies of cerebral blood flow without the disadvantages of [^{99m}Tc]-*d,l*-HMPAO.

A more recent addition to the list of neutral technetium complexes with the potential for use in brain perfusion studies is [^{99m}Tc]MRP20 (**8**). Preliminary investigations indicate that the agent may have similar *in vivo* properties as [^{99m}Tc]-*d,l*-HMPAO (Morgan et al., 1990).

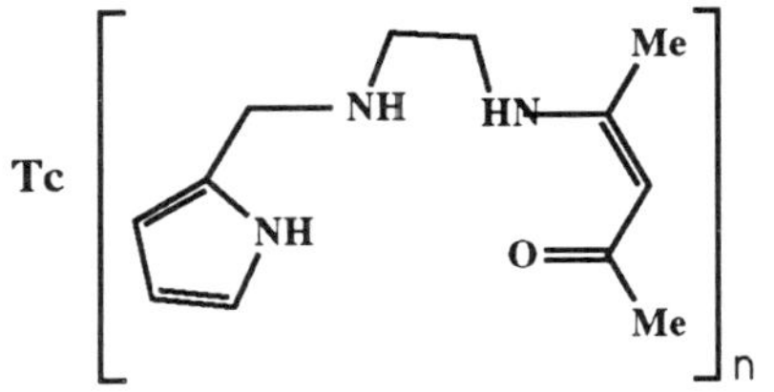

Tc NH HN Me NH O Me n

8

The Technetium-99m radiopharmaceuticals described above adequately provide images of brain blood flow. The next frontier for technetium chemistry, with respect to brain studies, is the preparation of recognition site specific Technetium-99m radiopharmaceuticals for tracing receptors and metabolism to parallel the developments in iodinated and positron radionuclides. Being a Group VIIB transition metal, Tc does not form covalent bonds with carbon. Therefore, it will be difficult to make derivatives of ligands known to bind to specific receptors, for example, by simple utilization of atoms of the putative ligand for coordination to Tc. This approach would involve greater overall transformation of the physico-chemical characteristics of the substrate. A more feasible approach to the development of recognition site specific Tc agents would be to carry the Tc in the form of a suitable chelate. The chelate could then be introduced at a site known to tolerate bulk or built into the molecule as a substituent, e.g., for a lipophilic group in the region of the molecule not essential for biological activity. Success of this approach depends on the use of stable, neutral, lipid soluble multipurpose chelates of small enough size as carriers of Technetium-99m such as the diaminedithiol ligand system.

Significant progress towards the development of receptor-specific radiopharmaceuticals is being made. The latter approach has been used in the development of a [^{99m}Tc]quinuclidinyl complex ([^{99m}Tc]DADT-Qn, **10**) as an analogue of quinuclidinyl benzilate (QNB) (**9**), a muscarinic acetylcholine receptor antagonist (Lever and Wagner, 1990). The Technetium-99m is carried in the multipurpose diaminedithiol

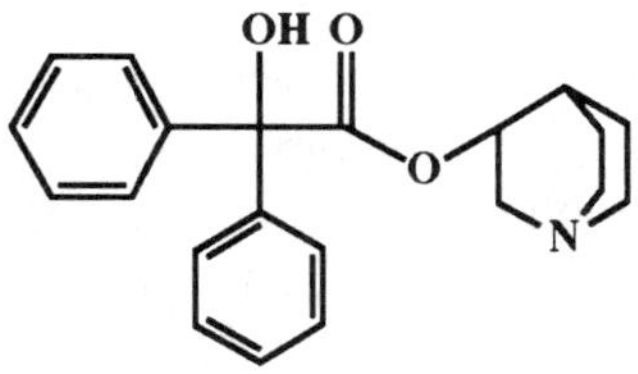

9 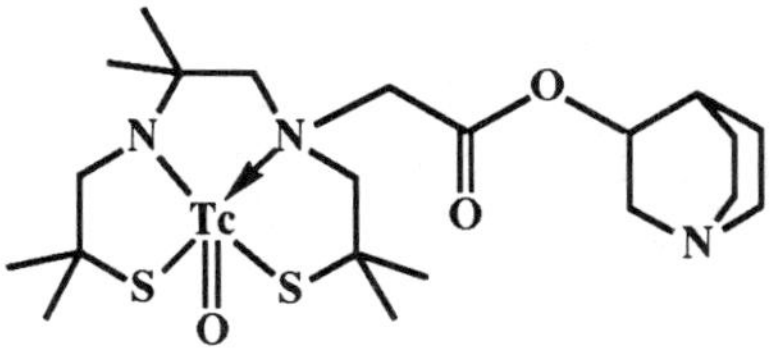 **10**

ligand. The [^{99m}Tc]diaminedithiol complex is small enough to function as a substitute for the two benzene rings of quinuclidinyl benzilate. This region of QNB has been known to have reasonable bulk tolerance. The aminoalcohol head group and the ester functionality which are required in many high affinity muscarinic acetylcholine receptor antagonists have been retained. Two complexes, presumably isomers, were obtained. *In vivo* saturation studies using mice brain homogenates, showed that the K_D of the more lipophilic complex was 1.6 ± 0.9 μM while the K_D of the other complex was 4.0 ± 1.1 μM. In competition studies using [^{125}I]iododexetimide, a muscarinic receptor antagonist of a different chemical class, the IC_{50} of the two complexes were 16.5 ± 3.5 μM and 107.0 ± 7.1 μM, respectively, consistent with the measured affinities. While the affinity of these complexes would not permit their use *in vivo,* they represent the first Technetium-99m complexes with measurable affinity to a central nervous system receptor, and indicate that Technetium-99m, receptor specific complexes of small molecules are possible.

Indeed recently, DiZio et al. (1992) reported the synthesis of a series of Technetium-99m labeled analogues of the antiprogresterone mifepristone (**11**) that retained high binding affinity to the progesterone

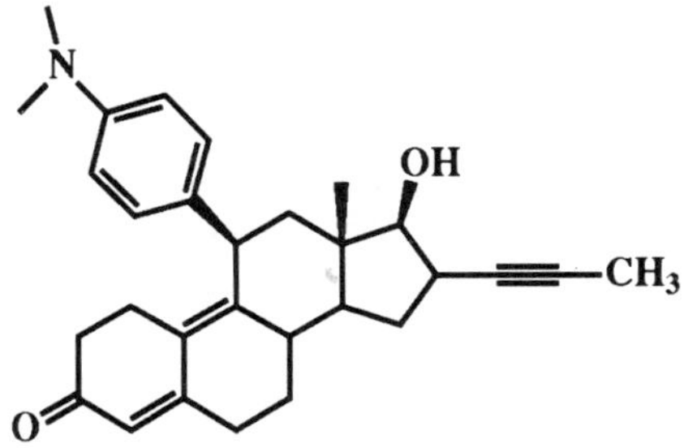

11

receptor. Two syn/anti pairs of the 11-ß-derivatives (**12**) in which the Technetium-99m was carried in the diaminedithiol ligand showed very high affinity for the progesterone receptor. One of these complexes, an anti-derivative, had three times greater affinity for the progesterone receptor than progesterone itself. Even though the complexes were too lipophilic to be useful *in vivo*, they demonstrate the feasiblity of generating high affinity, receptor specific Technetium-99m complexes.

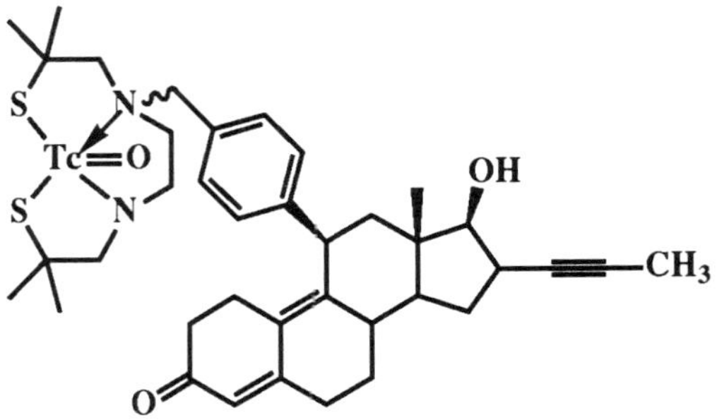

12

The development of neutral recognition site-specific Technetium-99m radiopharmaceuticals can be aided by convenient synthetic intermediates for quick synthesis of the ligands needed for complexation to Technetium-99m. This would make several Technetium-99m complexes available for evaluation in a short period of time. Towards this end, a versatile diaminedithiol bifunctional chelating agent (BCA), Hex-DADT-BCA (**13**), for facile synthesis of diaminedithiol ligands has been designed and synthesized (Baidoo and Lever, 1990). The BCA has a thiolactone reactive functionality for reaction with nucleophiles such as amines. Upon reaction, a previously masked thiol group is released to complete the tetradentate coordinate core of the diaminedithiol ligand system as illustrated below (**14**). The ligand then becomes attached to a substrate through a built-in 2-carbon spacer and a stable amide bond. One of the nitrogen atoms of the BCA has been maintained as a secondary amine so that complexation of its adducts to Technetium-99m would result in deprotonation to achieve a neutral complex core (**15**). It is envisaged that the BCA would be used to generate DADT adducts of metabolic substrates, receptor-specific ligands and drugs.

The metabolism of [^{99m}Tc]-*l,l*-ECD provides evidence that appropriate Technetium-99m complexes could be prepared for studying metabolic processes in the brain *in vivo*. The fact that only the *l,l*-isomer and not the *d,d*-isomer is metabolized indicates that given the

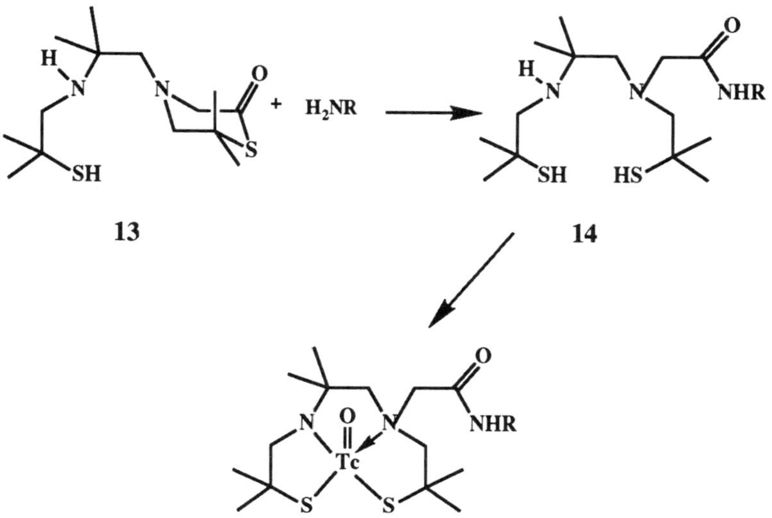

13 + H_2NR → 14 → 15

right configuration and appropriate substrate, specific enzymes could be targeted. This, coupled with the progress made in the development of receptor-specific Technetium-99m labeled tracers, indicates that we stand at the threshold of very exciting developments in [^{99m}Tc]radiopharmaceutical chemistry.

Heart Imaging Agents

An important contribution to cardiac nuclear medicine has been the recent development of two main classes of [^{99m}Tc]radiopharmaceuticals for myocardial blood flow imaging. These are the lipophilic six coordinate Tc(I) isonitrile monocationic complexes (Jones et al., 1984) and derivatives of the neural BATO complexes described above. The most promising of the isonitrile complexes, *hexakis*-(2-methoxy-isobutylisonitrile)-technetium(I) (also called Sestamibi, Hexamibi, MIBI, RP-30, and Cardiolite™ (Dupont-Merck Radiopharmaceutical Co.)), (**16**) shows excellent uptake in the myocardium coupled to rapid washout from background areas such as lungs and liver. Its distribution in the myocardium remains essentially fixed without redistribution of the tracer during the study period (Wackers et al., 1989). The most promising of the BATO complexes for myocardial blood flow studies is [*bis*[1,2-cyclo-hexanedione dioximato-(1-)-O]-[1,2-cyclohexanedione-dioximato-(2-)-O]-methylborato-2N,N',N'',N''',N'''',N''''']-chlorotechnetium(III) (**17**), also called SQ30217, teboroxime, and Cardiotec™ (Squibb Diagnostics).

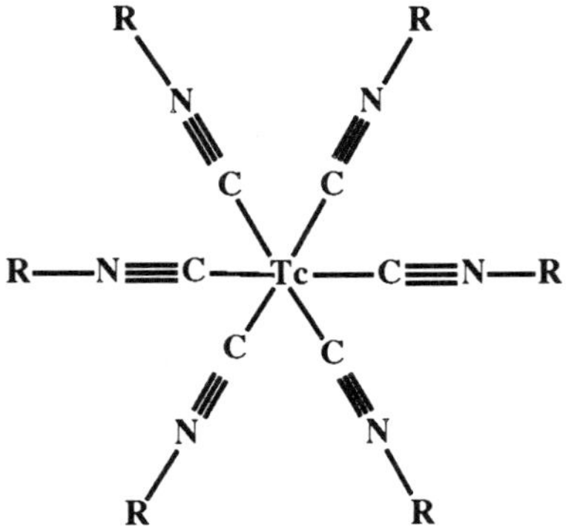

R = $-CH_2C(CH_3)_2OCH_3$

16

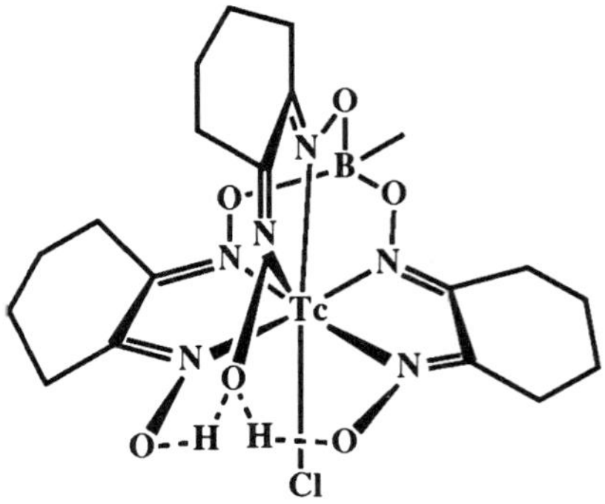

17

SQ30217 shows high intial uptake in the myocardium and lower uptake in background areas, however myocardial uptake clears relatively rapidly compared to Sestamibi (Seldin et al., 1989).

The advantage of these technetium agents is that, perfusion and functional data (right and left ventricular ejection fractions from first pass studies) could be obtained from the same study. These agents could therefore be used in drug research for the evaluation of drugs designed as thrombolytic agents or drugs which affect global cardiac function.

A more recent myocardial perfusion radiopharmaceutical to be described is [^{99m}Tc]tetrofosmin (**18**) also called PPN.1011, [^{99m}Tc]P53,

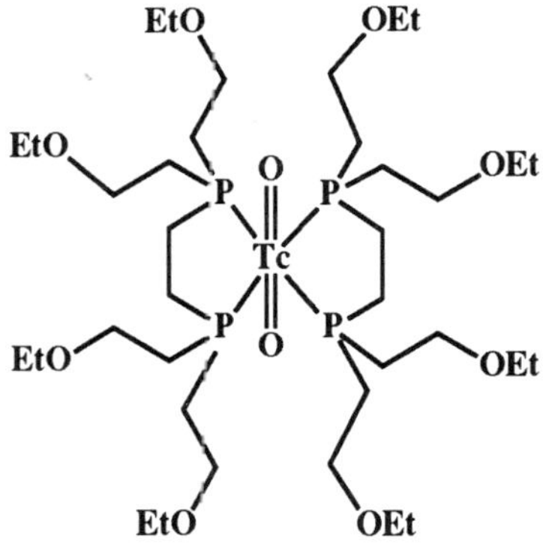

18

Myoview™ (Amersham International) (Kelly et al., 1990; Forster et al., 1992). This agent is a Tc(V) *bis*-diphosphine complex containing a $[TcO_2]^+$ core. The clearance of the agent from the myocardium is slower than SQ30217 and has the added advantage of lower uptake in the liver.

Kidney Imaging Agents

Many technetium based radiopharmaceuticals have been developed for renal function studies. Good renal imaging agents provide fast and homogenous distribution in kidneys followed by accumulation in the ureters and then the bladder. These agents provide valuable detailed information on the integrity of the parenchyma. Many Technetium-99m complexes have been applied to studies of kidney function. These include Technetium-99m complexes of simple sugars, ascobate, glucoheptonate, dimercaptosuccinic acid, EDTA and DTPA. These agents are either glomerular filtration or tubular fixation agents.

Recent efforts to develop Tc-based kidney imaging agents that mimic the tubular secretion agent *o*-iodohippuric acid yielded a new class of agents derived from the diamidedisulfide (DADS) ligand system which are being developed as kidney agents. Several of these Tc(V) complexes have been synthesized, characterized and shown to be promising as renal tubular secretion agents (Jones et al., 1982; Kasina et al., 1986). The best derivative for renal studies was $[^{99m}Tc]$-N,N'-*bis*(mercaptoacetyl)-2,3-diaminopropanoate (**19**) (Fritzberg et al., 1982). The amide nitrogens are deprotonated giving the complex an overall negative charge. In an effort to overcome stereoisomeric problems arising from substituted methylene groups, a triamidemercaptide (N_3S) ligand system has been developed in which one of the sulfurs of the original N_2S_2 system has been replaced by a planar amido nitrogen. Several of these $[^{99m}Tc]N_3S$ complexes show promise as kidney agents with one, $[TcOMAG_3]^{2-}$ (where MAG_3 = mercaptoacetylglycylglycylglycine), presently approved for clinical studies (Fritzberg et al., 1986). $TcOMAG_3$ (**20**) is very rapidly and efficiently extracted by the kidneys

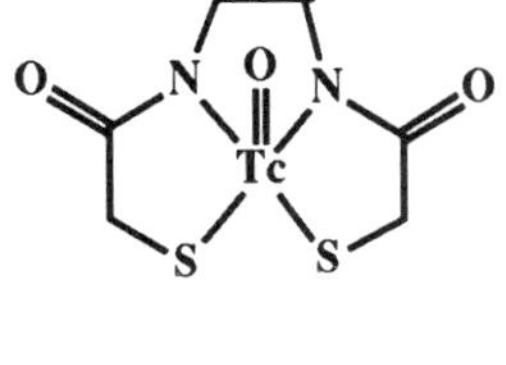

19 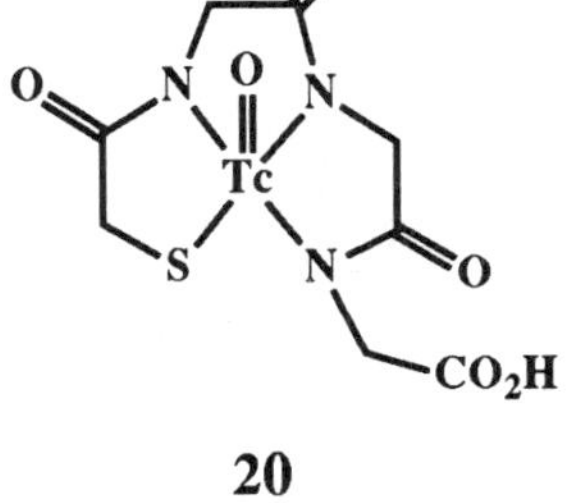 **20**

and excreted in the urine, resulting in renograms similar to those obtained with *o*-iodohippuran.

More recently, the diacid metabolite of Tc-*l*,*l*-ECD (**8**) has also been shown to provide excellent renal tubular imaging properties (Verbruggen et al., 1992).

Technetium-99m Labeled Proteins and Peptides

Proteins and peptides participate in several biochemical and physiological processes *in vivo*. This fact coupled with the availability of monoclonal antibodies which can be directed to specific tissue antigens has driven efforts to label these important biomolecules with Technetium-99m, the ideal radionuclide for SPECT imaging. Labeled proteins have shown utility in studies of cancer, infection and cardiovascular diseases.

Proteins can be labeled with Technetium-99m either directly or via bifunctional chelating agents. Direct labeling methods rely on atoms of the protein for coordination of Tc. The early methods for direct labeling were very unreliable and resulted in products that were unstable *in vivo* and contaminated with colloids. Recently, direct labeling methodology has been greatly improved. Generally, these methods have involved the reduction of disulfide bonds in proteins with reducing agents such as stannous ion, ascorbate and dithiothreotol followed by the Technetium-99m labeling step (Eckelman and Steigmen, 1991; Griffiths et al., 1992).

Several bifunctional chelating agents have also been developed for labeling proteins with Technetium-99m. These include bifunctional chelates derived from the diamidedithiols (DADS, N_2S_2) (Fritzberg et al., 1988), triamidemonothiols (N_3S), diaminedithiols (DADT, N_2S_2) (Baidoo et al., 1990), dithiosemicarbazones (DTS) (Arano et al., 1987), boronic acid adducts of Technetium-99m oximes (BATO) (Linder et al., 1991), diaminetetrathiols (N_2S_4) (Najafi et al., 1991) and hydrazino-nicotinamide derivatives (Abrams et al., 1990).

The development of thrombolytic drugs could benefit from the use of vascular tracers such as [^{99m}Tc]MAA or thrombus imaging agents such as [^{99m}Tc]fibrinogen, Technetium-99m labeled anti-fibrin and anti-platelet antibodies. For example, [^{99m}Tc]fibrinogen could be incorporated into experimental thrombi and the rate of lysis of the thrombus determined after administration of the drug under study. Technetium-99m labeled anti-fibrin and anti-platelet antibodies could be used to evaluate the patency of blocked blood vessels after thrombolytic drug treatment. In the area of cancer imaging, Technetium-99m labeled monoclonal antibodies and fragments directed against several cancer tissue antigens have successfully been used in the diagnosis and treatment evaluation in cancer patients.

RADIOIODINATED TRACERS

The early 1940's saw the introduction of Iodine-131 as the first wide spread radionuclide used in diagnosis of metabolic function by non-invasive imaging techniques. Another first for Iodine-131 is that it was actually a designer radionuclide, created specifically to replace the less suitable Iodine-128 for thyroid studies (Hamilton, 1942). Since then iodine has lost its place as the pre-eminent radionuclide used in nuclear medicine to the convenience and economy of generator produced radionuclides, most notably Technetium-99m. However a number of factors has resulted in a resurgence of interest in the use of radiotracers containing isotopes of iodine in nuclear medicine imaging.

The early dominance of Iodine-131 was due to its cheapness and availability from nuclear reactors. However, Iodine-123 has long been recognized as a far superior imaging isotope (Fly, 1977). The properties of the two radionuclides are summarized in Table 1. While Iodine-123 is more expensive than Iodine-131, in all other respects it is clearly a more useful imaging radionuclide. Its principal emission is of an ideal energy for imaging while the shorter half-life poses fewer contamination problems. In addition, the absorbed radiation dose to the patient is only a few percent of that incurred using Iodine-131. Three nuclear reactions are significant for the commercial production of Iodine-123. The $^{123}Te(p,2n)^{123}I$ reaction produces large amounts of Iodine-124 which is a major problem given its high energy gamma emissions and half-life of 4.2 days. The second process, the $^{127}I(p,5n)^{123}Xe \rightarrow {}^{123}I$ reaction, has small amounts of Iodine-125 as its only radionuclidic impurity but requires high energy protons (and consequently a large cyclotron) to effect the nuclear reaction. Of late, another process has entered commercial production namely, $^{124}Xe(p,2n)^{123}Cs \rightarrow {}^{123}I$. While this method results in high yields of high purity Iodine-123 from moderate energy accelerators (25-40 MeV), the >99% enriched Xenon-124 gas required as target fodder is a significant capital outlay. The target gas is recovered and recycled but accidents do happen. Nevertheless, the cost of high quality Iodine-123 has remained steady over the past five years, an actual reduction in price when inflation is taken into account.

Chemistry

For the most part, iodine is monovalent in organic chemistry. This enables us to envisage, superficially, analogues of biologically interesting molecules which are derived from their parent(s) by direct substitution of the iodine radical for other moeities, e.g., for hydrogen, methyl and phenyl groups, or other halogens. Of course, information from structure-activity relationships both *in vivo* and *in vitro* must be applied whenever possible. Unfortunately, when a series of related compounds are prepared for biological evaluation, iodine substituents are usually conspicuous by their absence. The medicinal chemist must appreciate, and allow for the effects of the steric bulk and lipophilicity of the iodine moiety on the prototype compound as well as the reactivity of the carbon-iodine bond *in vivo*. Despite the crudity of this approach, it continues to be the *modus operandi* by which potential radiopharmaceuticals are designed. Most of the examples of SPECT ligands listed in Tables 2 and 3 are direct iodine analogues of lead compounds with iodine replacing hydrogen, methyl, chlorine etc. at a site on the molecule dictated by synthetic convenience and/or by knowledge of site-specific bulk tolerance.

Having identified a potential radiopharmaceutical, there is now a large and increasing variety of techniques available for its radiosynthesis. The chemistry of radioiodine and methods for its introduction into complex and simple molecules has been thoroughly and excellently reviewed recently (Seevers and Counsell, 1982; Coenen et al., 1983; Baldwin, 1986), thus only a few trends will be presented here. Increasingly popular in the field of radioiodination is the use of radioiododemetallations to synthesize Iodine-125 and Iodine-123 radiotracers, resulting in reactions which proceed in a highly regiospecific manner at high yields and high specific activities. This "new" strategy has been encouraged by improved methods of preparing the neccessary metallated precursors (Azizaian et al., 1981; Hatanaka and Hiyama, 1987) and by detailed systematic examinations of the parameters (substituent effects, solvent, temperature, oxidizing agent etc.) which influence the course of these reactions (Eaborn, 1975; Moerlein, 1985; Moerlein, 1987; Moerlein et al., 1987). Many of the compounds in Tables 2 and 3 have been radiolabeled by the radioiododemetallation method.

The success of the demetallation approach to Iodine-123 radiotracer synthesis is exemplified in the area of Iodine-123 labeled benzamides. This well-studied class of compounds has been scrutinized over the past few years as potential SPECT agents for imaging dopamine D_2 post-synaptic receptors (Table 3). As exemplified by $[^{123}I]$IBZM (**21**) in Scheme 1, the original approach to labeling was the "classical" aromatic electrophilic substitution on an activated aromatic ring (Kung et al., 1988a). However the next generation of ^{123}I-labeled dopamine D_2 receptor ligands have for the most part been prepared by radioiododestannylation (Clanton et al., 1991) as depicted for

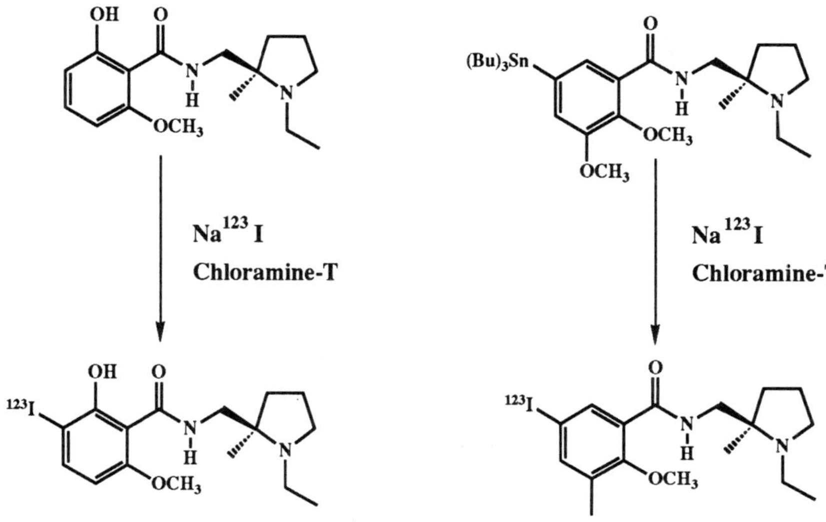

$Na^{123}I$
Chloramine-T

$Na^{123}I$
Chloramine-T

21 **22**

Scheme 1

[^{123}I]epidepride (**22**). While the second approach requires a less accessible precursor than the first, the advantages of short reaction times, radiochemical yield, reproducibility, and ease of purification justify the extra effort required. A much different approach to labeling with iodine is the Cu(I) catalyzed exchange of aryl halogen with radioactive iodide. This method, shown in Scheme 2 for [^{123}I]iodoketanserin (**23**) (Mertens et al., 1990), has the advantage of a more readily available precursor, but careful separation of product from precursor may be required. However this technique has been shown to be possible using very small quantities of precursor (a few micrograms) and thus exchange reactions of hot for cold iodine result in products with a moderately high specific activity (Mertens et al., 1987) . Using such conditions, purification problems are minimized, and allows a "kit" approach to the preparation of the radiopharmaceutical. Recently, the radiosynthesis of high specific activity [^{123}I]QNB by this exchange procedure has been reported (Owens et al., 1992).

It is unlikely that Iodine-123 will ever compete with generator-produced radionuclides in terms of convenience or price. However, despite the stress that labeling with iodine places on the radiotracer principle, strength lies in its chemistry. Despite some success in the area of blood flow studies, the future of Iodine-123 radiopharmaceuticals would appear to lie in imaging of receptors. It is possible that an agent

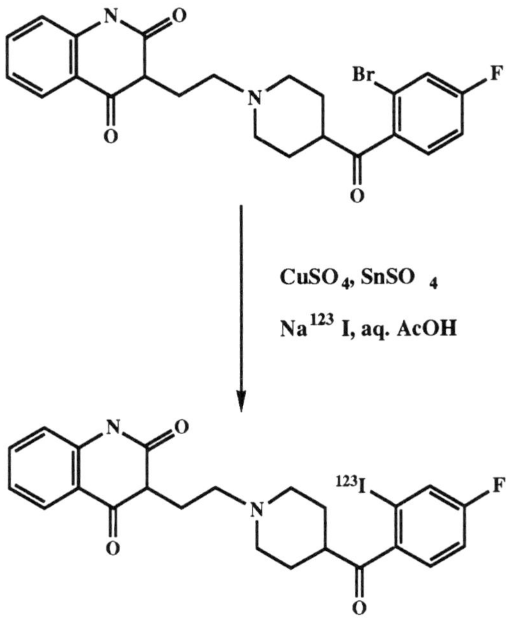

$CuSO_4$, $SnSO_4$

$Na^{123}I$, aq. AcOH

23

Scheme 2

for imaging brain tumors will be developed (Bomanji et al., 1991) but receptor binding will probably be involved, i.e., detection and characterization will be based on changed affinity or concentration of receptors in or surrounding the tumor, rather than a change in brain tissue metabolism. Many ligands have been developed whose distribution has been shown by SPECT imaging to represent neuroreceptor binding in human or primate brain (Tables 2 and 3) and many more are being pursued (Kung, 1990).

The synergistic effect of many factors are slowly manifesting in a new era of radioiodine imaging in nuclear medicine. These include new and improved imaging techniques, radiopharmaceuticals, production methods of Iodine-123, synthetic methods, and increased emphasis on reducing radiation dose to patients. In addition the techniques and tracers under active refinement for PET imaging and quantification of receptors can be transplanted to the field of SPECT where the advantages of longer half-life, (clearance of non-specific binding), lower costs (no cyclotron required on site), and potentially, commercial development of radiopharmaceuticals will ensure that Iodine-123 SPECT imaging of neuroreceptors remains a rich field.

N-isopropyl-4-iodoampetamine (IMP)
Blood Flow [1]

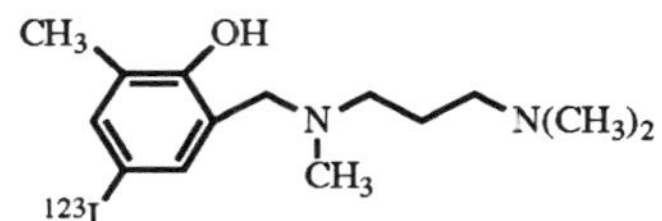

HIPDM
Blood Flow [2]

4-IodoQNB
Muscarinic Acetylcholine Receptors [3]

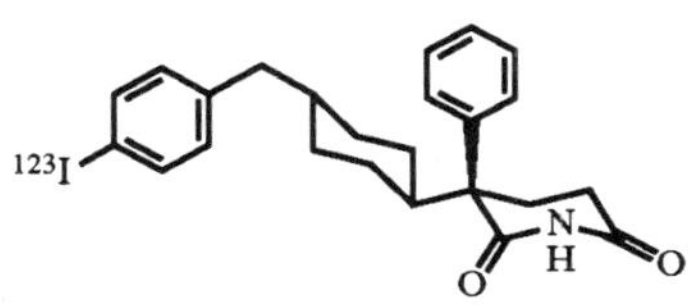

4-Iododexetimide
Muscarinic Acetylcholine Receptors [4]

2-Iodoketanserin
$5HT_2$ Receptors [5]

IodoethylLSD (IEL)
$5HT_2$ Receptors [6]

RTI-55, CIT
Dopamine Transporter [7]

L-3-Iodo- α-methyl tyrosine
Amino Acid Transporter [8]

1) Winchell et al., 1980; Kuhl et al., 1982; 2) Kung et al., 1983; 3) Eckelman et al., 1984; Weinberger et al., 1989; 4) Mueller-Gaertner et al., 1992; 5) Mertens et al., 1990; 6) Lever et al., 1991; 7) Innis et al., 1991; Shaya et al., 1992; 8) Langen et al., 1990.

TABLE 2. [^{123}I]Labeled SPECT Radiotracers.

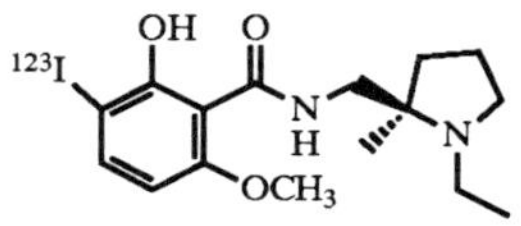

IBZM
Dopamine D-2 Receptors [1]

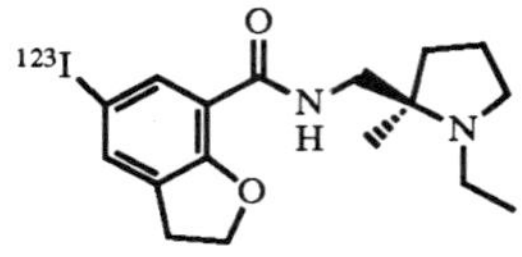

IBF
Dopamine D-2 Receptors [2]

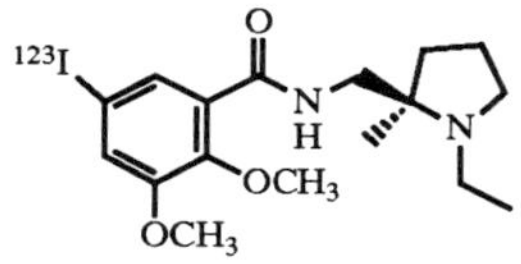

Epidepride
Dopamine D-2 Receptors [3]

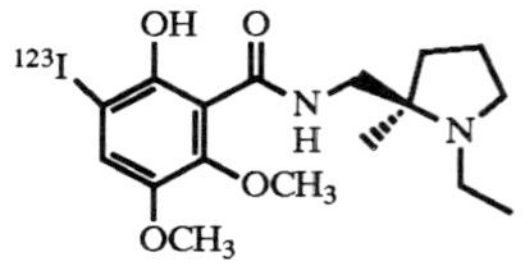

NCQ 298 (Ioxipride)
Dopamine D-2 Receptors [4]

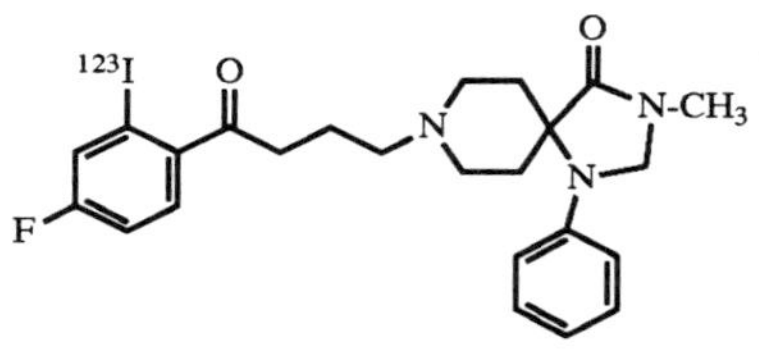

2'-Iodospiperone
Dopamine D-2 Receptors [5]

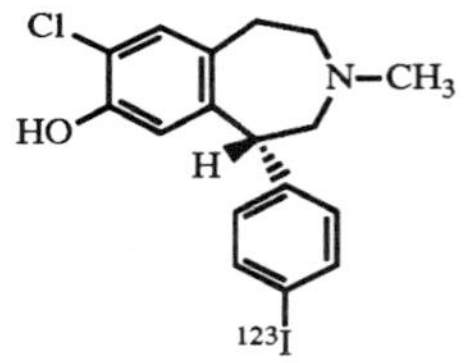

FISCH
Dopamine D-1 Receptors [6]

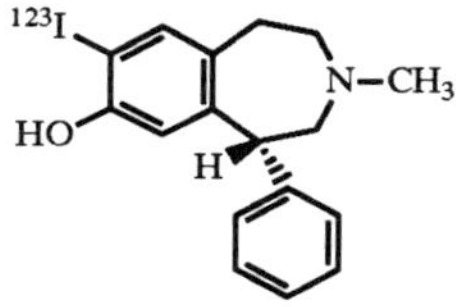

IBZP
Dopamine D-1 Receptors [7]

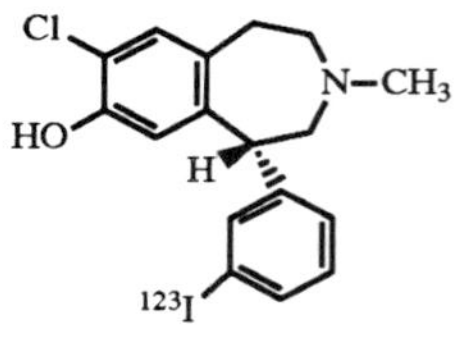

TISCH
Dopamine D-1 Receptors [8]

1) Kung et al., 1990; 2) Kung et al., 1991; 3) Kessler et al., 1991; 4) Hall et al., 1991; 5) Saji et al., 1987; Mertens et al., 1989; 6) Chumpradit et al., 1989; 7) Kung et al., 1988b; 8) Billings et al., 1992.

TABLE 3. [123]Labeled Radiotacers for Dopamine Receptor Imaging.

INDIUM LABELED RADIOTRACERS

Indium is a Group III metal ion which exists as a cation in aqueous solution. Two isotopes of indium are of interest in nuclear imaging studies, Indium-111 and Indium-113m. Indium-113m is a short-lived, generator produced, gamma emitting radionuclide (390 keV, $t_{1/2}$ = 1.7 hr) which received considerable attention in the late 1960s and early 1970s. With the widespread acceptance of Technetium-99m as the most practical, generator produced radionuclide for use in nuclear medicine, interest in In-113m has decreased. Interest in Indium-111, on the other hand, has increased over the past 20 years, primarily due to its ever increasing application to the labeling of macromolecules and cells, particularly monoclonal antibodies, leukocytes and platelets. Indium-111 is cyclotron produced and, as a result, considerably more expensive than Technetium-99m. Indium-111 has a half-life of 67 hours and emits principal photons of 171 keV (90%) and 245 keV (94%). Because of its longer half-life (67 hr compared to 6 hr for Technetium-99m), Indium-111 can be used in studies which require longer times after injection than is practical with Technetium-99m.

The chemistry of indium in aqueous solution is quite similar to that of Iron except that only the +3 oxidation state of indium is stable in water. In this regard, the chemistry of indium is much simpler than that of technetium and, as a result, much more limited. Generally, the radiopharmaceuticals labeled with Indium-111 which are currently of interest can be divided into two types. The first are the labile complexes which are stable in aqueous solution but which readily undergo exchange reactions transferring indium to ligands which form more thermodynamically stable complexes. The second type are those which possess sufficient thermodynamic and kinetic stability to prevent dissociation in aqueous solution and which will not release indium to transferrin via ligand exchange. These stable complexes are primarily used for labeling proteins (e.g., monoclonal antibodies) via conjugation reactions discussed below.

Indium-111 is commercially available as indium chloride in dilute aqueous HCl and it is this form of Indium which is used for most labeling experiments. In acidic solutions, In^{+3} is hydrated, while at neutral pH, trivalent Indium forms a hydroxide ($In(OH)_3$) which is insoluble at pH > 3.5. Formation of this insoluble complex can be prevented by the coordination with weak ligands such as lactate, citrate and acetate. These complexes can be prepared in neutral solution and used to label other ligands via exchange reactions. As a result of its similarity to Iron, indium forms a very stable complex with transferrin in plasma. This leads to significant restrictions on the type of indium complexes which can be used *in vivo.* Only ligands which form very stable complexes with indium such as the polyamincarboxylates, EDTA (**24**) and DTPA (**25**), are useful for coupling Indium-111 to other molecules.

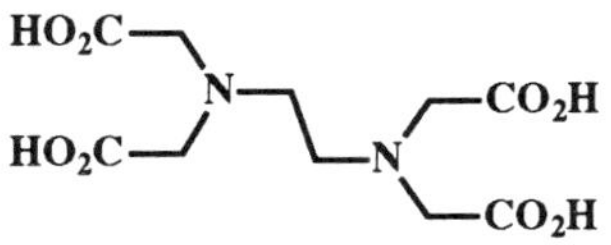

HO_2C HO_2C N N CO_2H CO_2H

24

HO_2C HO_2C N N N CO_2H CO_2H CO_2H

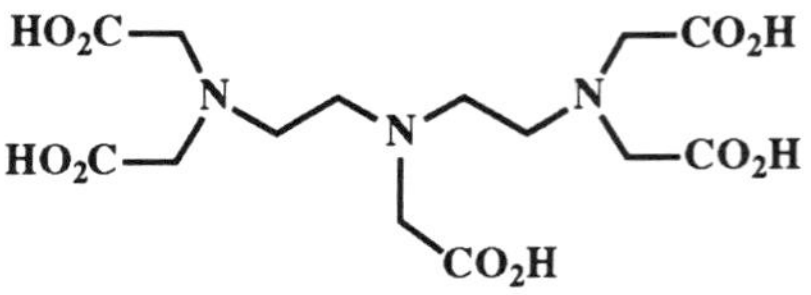

25

Indium Labeled Blood Cells

Much of the recent work with Indium-111 has involved labeling of blood cells (platelets and leukocytes) or proteins (primarily monoclonal antibodies). Platelets and leukocytes can be labeled with any of a variety of neutral, lipophilic Indium-111 complexes such as Indium-111 oxine (8-hydroxyquinoline) (McAfee and Thakur, 1976; Thakur et al., 1976), Indium-111 tropolonate (Dewanjee et al., 1981; Danpure et al., 1982) and Indium-111 2-mercaptopyridine-N-oxide (Danpure and Osman, 1986). All of these agents are non-selective and will label all cells if the lipophilic complex is added to whole blood (primarily red cells since they far outnumber others). Thus, it is necessary to isolate the cells to be labeled prior to treating them with the labeling agent. These lipophilic complexes all label cells by a similar mechanism. The complex enters the cell by passive diffusion across the cell membrane and the indium becomes trapped within the cell (presumably via exchange of indium from the labile complex to another ligand forming a stable, hydrophilic complex that cannot diffuse from the cell). Liposomes have been labeled by the same mechanism for use in drug delivery studies (Gabizon et al. 1991).

Indium-111 oxine is commercially available and is the most widely used agent for labeling blood cells with indium. Although there have been reports that the tropolone complex is superior (Peters et al., 1986; Peters, 1988; Dewanjee et al., 1981), a recent report claims that, in humans, there is no significant difference between the kinetic behavior of platelets labeled by either of these methods (Kotze et al., 1991). The various methods for labeling platelets and leukocytes have recently been reviewed (Danpure and Osman, 1989; Thakur, 1986).

Protein and Peptide Labeling

A variety of bifunctional chelating agents (BCAs) have been developed for use in labeling peptides with Indium-111 (as well as other metal ions). These BCAs all possess a reactive functional group that forms stable covalent bonds with peptides under mild conditions. They also possess a chelating group that forms a stable complex with indium. Most of the BCAs developed to date were designed to couple to either the ε-amino group of lysine or to the free thiol of cysteine. These BCAs are first reacted with the protein and the conjugate is purified to remove any chelating agent which is not covalently bound to the protein. The protein-conjugate is then treated with Indium-111 (either as indium chloride or some other labile complex) to provide the indium labeled protein. The resulting indium-complex must be sufficiently stable to avoid exchange of indium from the labeled peptide to transferrin. Since a rather bulky, charged group is coupled to the protein by this process, care must be taken to demonstrate that the indium labeled protein retains the biological properties of interest.

Many of the BCAs used for labeling peptides with Indium-111 are analogs of the well known chelating agents EDTA and DTPA since both of these ligands form complexes which are stable in the presence of transferrin. A series of analogues of EDTA have been reported which incorporate a phenyl substituent attached to the EDTA backbone (Meares et al., 1984). Each of these EDTA analogues also possesses a reactive functional group which can be used to couple the EDTA chelating group to proteins. Two of these couple to the amino group of lysine to form either a thiourea (**26**) or carboxamido (**27**) linkage. The bromoacetamido compound (**28**) couples to the thiol group of cysteine and the diazo group of compound **29** can form a stable bond to either lysine or tyrosine.

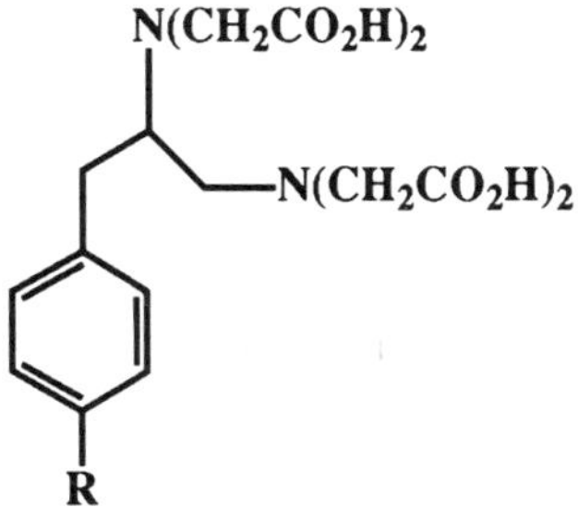

$N(CH_2CO_2H)_2$

$N(CH_2CO_2H)_2$

R

26 R = SCN- **27** R = HO_2CCH_2O-

28 R = $BrCH_2CONH$- **29** R = $^+N_2$-

A variety of DTPA based BCAs has also been developed for labeling proteins (primarily monoclonal antibodies) with Indium-111. These agents are also designed to couple to a lysine amino group. In three of these compounds (**31, 32, 33**) one of the carboxyl groups of DTPA is activated so it will react readily with the amino group resulting in formation of a stable amide bond. Compound **31** is a symmetrical, dianhydride which is commercially available and widely used (Hnatowich et al., 1983). One disadvantage of **31** is the fact that two anhydrides from the same molecule can react with amine containing residues on the protein resulting in inter- and intra- molecular crosslinking. The mixed anhydride (**32**) and the N-hydroxysuccinate

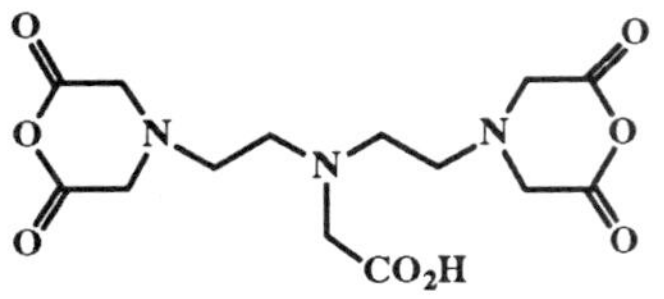

31

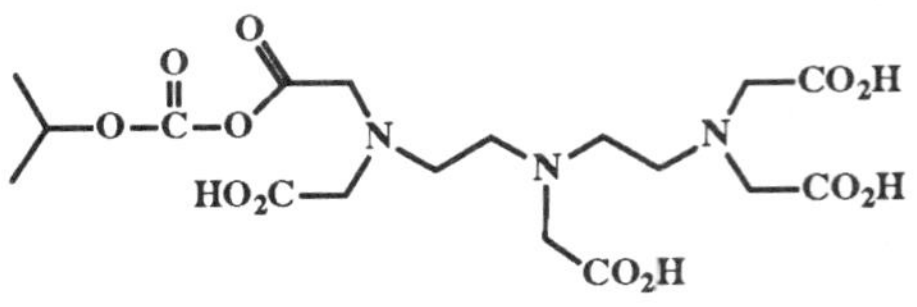

32

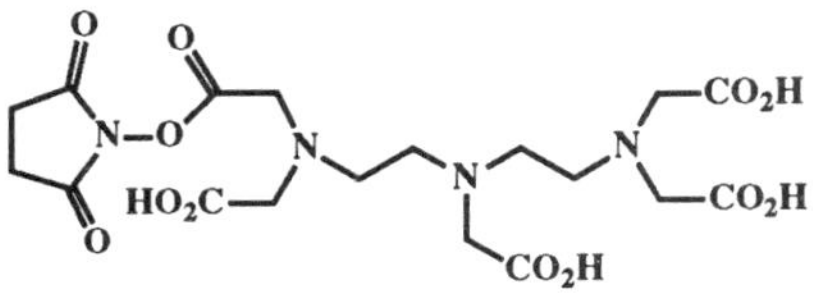

33

ester (**33**) each react to form derivatives of proteins with the same structure as that formed with the dianhydride. The isothiocyanto substituted compound (**34**) reacts with amino groups to form a thiourea linkage leaving all five of the carboxyl groups of DTPA to coordinate to Indium (Brechbiel et al., 1986).

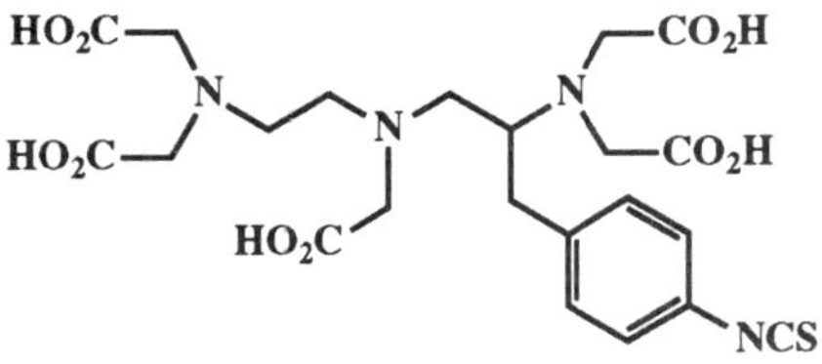

34

The stability constant for indium EDTA is 2-3 orders of magnitude lower than that of indium DTPA and the thermodynamic stabilities of both these complexes are significantly lower than that of indium transferrin (Welch et al., 1975). Nevertheless, exchange of indium from these complexes to transferrin is slow due to the kinetics of the reaction (Cole et al., 1987; Yeh et al., 1979). When DTPA anhydride has been used to couple Indium-111 to monoclonal antibodies, clearance of radioactivity from the liver and kidney is sometimes slow (Mathias, et al., 1990). This has prompted the development of new BCAs which form even more stable complexes. Mathias has reported on a new class of Indium complexes which clear quickly from the plasma and liver (Mathias et al., 1988). Based on these results, a new BCA was developed (BrHBED, **35**) that possesses a bromoacetamido substituent that can couple with proteins by alkylation of either a lysine amino group or the thiol of cysteine (Mathias, et al., 1990). The ndium complex of this ligand is significantly more stable than the indium complex of DTPA. Indium-111 labeled antibodies prepared by this method showed high uptake in tumors and plasma clearance of radioactivity was rapid. Development of new BCAs for use in labeling proteins with Indium-111 continues as an active area of research and a number a reviews on this subject have been published (Britton et al., 1991; Brandt and Johnson, 1992; Gansow, 1991; Hnatowich, 1990; Mather, 1987; Otsuka and Welch, 1987; Quadri et al., 1990; Washburn, 1991).

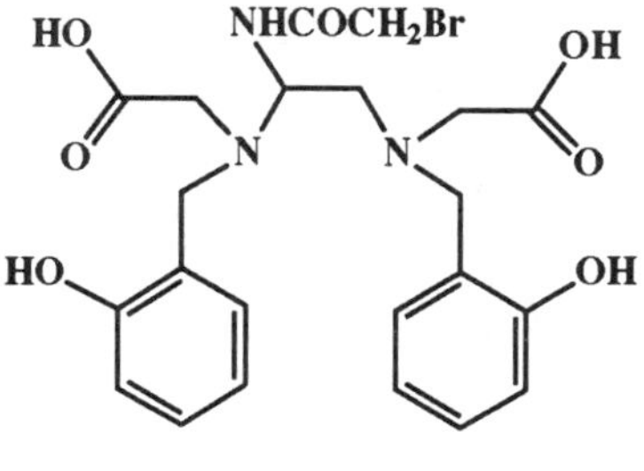

35

CONCLUSION

Non-invasive nuclear imaging techniques are beginning to play an important role in pharmaceutical research. These studies can be performed in a variety of experimental subjects, including valuable non-human primates and man. Although some of these studies require use of short-lived, positron emitting radionuclides and can only be carried out at a cyclotron/PET facility, a great deal of information can be obtained using readily available, single photon, gamma emitting radionuclides. We have reviewed recent advances in the development of new radiotracers labeled with the most widely used gamm-emitting radionuclides, Technetium-99m, Iodine-123 and Indium-111. With these radionuclides, radiotracers can be made which can be used in a wide variety of imaging studies which provide information on the biochemistry or physiology of the experimental subject. Information can be obtained concerning the interaction of candidate drugs with specific binding sites *in vivo* and insight into the *in vivo* behavior of new drug formulations or delivery devices. The chemical and physical properties of each of these radionuclides are sufficiently versatile to permit researchers to design a wide variety of experiments which can contribute significantly to improving the efficiency of the drug discovery and development process.

REFERENCES

Abrams MJ, Juweid M, Tarkate CI, Schwartz DA, Hauser MM, Gaul FE, Fuccello AJ, Rubin RH, Strauss HW and Fischman AJ (1990): Technetium-99m-human polyclonal antibody labeled via the hydrazino nicotinamide derivative for imaging focal sites of infection in rats. *J Nucl Med* 31: 2022-2028.

Arano Y, Yokoyama A, Furukana T, Horiuchi K, Yahata T, Saji H., Sakahara H, Nakashima T, Koizumi M, Endo K and Torizuka K (1987): Technetium-99m labeled monoclonal antibody with preserved immunoreactivity and high *in vivo* stability. *J Nucl Med* 28: 1027-1033.

Azizaian H, Eaborn C and Pidcock A (1981): Synthesis of organotrialkylstannanes. The reaction between organic halides and hexaalkyldistannes in the presence of palladium complexes. *J Organomet Chem* 215: 49-58.

Baidoo KE and Lever SZ (1990): Synthesis of a versatile presursor to technetium and rhenium complexes. *Tetrahedron Lett* 31: 5701-5704.

Baidoo KE, Scheffel U and Lever SZ (1990): Technetium-99m labeling of proteins: initial evaluation of a novel diaminedithiol bifunctional chelating agent. *Cancer Res* 50 (Suppl.): 799s-803s.

Baldwin R (1986): Chemistry of radioiodine. *Appl Radiat Isot Int J Radiat Appl Instrum (Part A)* 37: 817-821.

Biersack HJ, Linke D, Brassel F, Reichmann K, Kurthen M, Durwen HF, Reuter BM, Wappenschmidt J and Stefan H (1987): Technetium-

99m HM-PAO brain SPECT in epileptic patients before and during unilateral hemisphere anesthesia (Wada test): report of three cases. *J Nucl Med* 28: 1763-1767.

Billings JJ, Kung M-P, Chumpradit S, Mozley D, Alavi A and Kung HF (1992): Characterization of radioiodinated TISCH: A high affinity and selective ligand for mapping CNS D1 dopamine receptor. *J Neurochem* 58: 227-236.

Bomanji J, Moyes J, Huneida AH, Solanki K, Hawkins L, Nimmon CC, Kingston JE and Britton KE (1991): Does cerebral uptake of radiolabelled MIBG represent adrenergic receptors. *J Nucl Med* 32: 992.

Brandt KD and Johnson DK (1992): Strudture-function relationships in indium-111 radioimmunoconjugates. *Bioconjugate Chem* 3: 118-125.

Brechbiel MW, Gansow OA, Atcher RW, Schlom J, Estaban J, Simpson D and Colcher D (1986): Synthesis of 1-p-isothiocyanatobenzyl) derivatives of DTPA and EDTA. Antibody labeling and tumor imaging studies. *Inorg Chem* 25: 2772-2781.

Britton KE, Granowski M and Mather SJ (1991): Radiolabelled antibodies in oncology. I. Technical aspects. *Nucl Med Commun* 12: 65-76.

Cheesman EH, Blanchette MA, Ganey MV, Maheu EJ, Miller SJ and Watson AD (1988): Technetium-99m ECD: Ester derivatized diamine-dithiol Tc complexes for imaging brain perfusion. *J Nucl Med* 29: 788.

Chumpradit S, Billings J, Kung M, Pan S and Kung H (1989): An improved CNS D-1 dopamine receptor imaging ligand: [^{123}I](±)FISCH. *J Nucl Med* 30: 803.

Clanton JA, Depaulis T, Schmidt DE, Ansari MS, Manning RG, Baldwin RM and Kessler RM (1991): Preparation of [I-123]-Epidepride and [I-125]Epidepride - a dopamine D-2 receptor antagonist radioligand. *J Labelled Compd Radiopharm* 29: 745-751.

Coenen H, Moerlein S and Stöcklin G (1983): No carrier added radiohalogenation methods with heavy halogens. *Radiochim Acta* 34: 47-68.

Cole WC, DeNardo SJ, Meares CF, McCall MJ, DeNardo GL, Epstein AL, O'Brien HA and Moi MK (1987): Comparative serum stability of radiochelates for antibody radiopharmaceuticals. *J Nucl Med* 28: 83-90.

Danpure HJ, Osman S and Brady F (1982): The labeling of blood cells in plasma with Indium-111 tropolonate. *Br J Radiol* 55: 247-249.

Danpure HJ and Osman S (1986): The importance of radiolabeling human granulocytes with [^{111}In]tropolonate or [^{111}In]2-

mercaptopyridine-N-oxide in plasma containing acid-citrate-dextrose. *Br J Radiol* 59: 907-910.

Danpure HJ and Osman S (1989): Radiolabeling human leukocytes and platelets for clinical studies. In: *Radiopharmaceuticals: Using radioactive compounds in pharmaceutics and medicine*, Theobald AE, ed. Chichester: Ellis Horwood Ltd.

Davison A, Jones A and Abrams MJ (1981): New 2,2-bipyridine and 1,10-phenanthroline oxohalide complexes of Tc(VII) and Tc(V). *Inorg Chem* 20: 4300-4302.

Denays R, Tondeur M, Foulon M, Verstraeten F, Ham H, Piepsz A and Noel P (1989): Regional brain blood flow in congenital dysphasia: Studies with technetium-99m HM-PAO SPECT. *J Nucl Med* 30: 1825-1829.

Deutsch E, Elder RC, LAnge BA, Vaal MJ and Lay DG (1976): Structural characterization of a bridged ^{99m}Tc-Sn-dimethylglyoxime complex: Implications for the chemistry of ^{99m}Tc radiopharmaceuticals prepared by the Sn(II) reduction of pertechnetate. *Proc Natl Acad Sci* 73: 4287-4289.

Dewanjee MK, Rao SA and Didisheim P (1981): Indium-111 tropolone, a new, high affinity label. Preparation and evaluation of labeling parameters. J Nucl Med 22: 981-987.

DiZio JP, Anderson CJ, Davison A, Ehrhardt GJ, Carlson KE, Welch MJ and Katzenellenbogen JA (1992): Technetium and rhenium-labeled progestins: Synthesis, receptor binding and *in vivo* distribution of an 11-ß-substituted progestin labeled with technetium-99 and rhenium-186. *J Nucl Med* 33: 558-569.

Drane WE, Strickland P and Zmuda S (1989): Regional cerebral localization of SQ 32097 (BATO agent) in normal volunteers: Evaluation with dynamic SPECT using a dedicated, multi-headed SPECT device. *J Nucl Med* 30: 1Ab.

Eaborn C (1975): Cleavage of aryl-silicon and related bonds by electrophiles. *J Organomet Chem* 100: 43-57.

Eckelman WC, Reba RC, Rzeszotarski WJ, Gibson RE, Hill T, Holman BL, Budinger T, Conklin JJ, Eng R and Grissom MP (1984): External imaging of cerebral muscarinic acetylcholine receptors. *Science* 223: 291-293.

Eckelman WC and Steigman J (1991): Direct labeling with ^{99m}Tc. *Nucl Med Biol* 18: 3-7.

Ell PJ, Hocknell JML, Jarrit PH, Cullum I, Lui D, Campos-Costa D, Nowotnik DP, Pickett RD, Canning LR and Neirinckx RD (1985): A ^{99m}Tc-labelled radiotracer for the investigation of cerebral vascular disease. *Nucl Med Commun* 6: 437-441.

Epps LA, Burns HD, Lever SZ, Goldfarb H and Wagner HN Jr (1987):

Brain imaging agents: Synthesis and characterization of (N-piperidinylethyl)-hexamethyl diaminodithiolate oxo technetium(V) complexes. *Appl Radiat Isot Int. J. Radiat. Appl. Instrum. (Part A)* 38: 361-36.

Epps LA (1984): The chemistry of neutral, lipid soluble technetium(V) complexes of aminoalcohols and aminothiols. Ph.D. Thesis, The Johns Hopkins University, Baltimore.

Fleary RF (1968): ^{99m}Tc labeled EDTA for renal scanning. *Aust Radiol* 12: 265-270.

Fly R (1977): I-123: Less radiation, better results. *Applied Radiol* 6: 177-179.

Forster AM, Storey AE, Archer CM, Nagle KR, Booker FS, EdwardsPG, Gill HK, Kelly JD and McPartlin M (1992): Structural characterization of the new myocardial imaging agent technetium-99m tetrofosmin (^{99m}Tc-P53). *J Nucl Med* 33: 850-851.

Fritzberg AR, Kuni CC, Klingensmith WC III and Stevens J (1982): Synthesis and biological evaluation of Tc-99m N,N'-*bis*(mercaptoacetyl)-2,3-diaminopropanoate: a potential replacement for [^{131}I]o-iodohippurate. *J Nucl Med* 23: 592-598.

Fritzberg AR, Kasina S, Eshima D and Johnson DL (1986): Synthesis and biological evaluation of technetium-99m-MAG_3. *J Nucl Med* 27: 111-116.

Fritzberg AR, Abrams PG, Baumier PL, Kasina S, Morgan AC Jr, Rao TN, Reno JM, Sanderson JA, Srinivasan A, Wilbur DS and Vandrrheyden J-L (1988): Specific and stable labeling of antibodies with technetium-99m with a diamide dithiolate chelating agent. *Proc Natl Acad Sci* 85: 4025-4029.

Gabizon A, Chisin R, Amselem S, Druckmann S, Cohen R, Goren D, Fromer I, Peretz T, Sulkes A, and Barenholz Y (1991): Pharmacokinetic and imaging studies in patients receiving a formulation of liposome-associated adriamycin. *Br J Cancer* 64: 1125-1132.

Ganzow OA (1991): Newer approaches to the radiolabeling of monoclonal antibodies by use of metal chelates. *Nucl Med and Biol* 18: 369-81.

Griffiths GL, Goldenberg DM, Jones AL and Hansed HJ (1992): Radiolabeling of monoclonal antibodies and fragments with technetium and rhenium. *Bioconj Chem* 3: 91-99.

Hall H, Högberg T, Halldin C, Köhler C, Ström P, Ross SB, Larsson SA and Farde L (1991): NCQ 298, a new selective iodinated salicylamide ligand for the labelling of dopamine D2 receptors. *Psychopharmacol* 103: 6-18.

Hamilton JG (1942): The use of radoactive tracers in biology and medicine. *Radiology* 39: 541-551.

Harper PV, Andros G and Lathrop KA (1962): Progress report. In: *Semiannual report to the Atomic Energy Commission*, Jacobson LO, ed. ACRH-18: 76-87.

Harper PV, Lathrop KA and Richards P (1964): A radiocolloid of ^{99m}Tc. *J Nucl Med* 5: 382.

Hatanaka Y and Hiyama T (1987): Palladium-mediated silylation of organic halides with dilsilane/F^- reagent. *Tetrahedron Lett* 28: 4715-4718.

Hnatowich DJ (1990): Recent developments in the radiolabeling of antibodies with iodine, indium, and technetium. *Semin Nucl Med* 20: 80-91.

Hnatowich DJ, Layne WW, Childs RL, Lanteigne D, Davis MA, Griffin TW and Doherty PW (1983): Radioactive labeling of antibody: A simple and efficient method. *Science* 220: 613-615.

Holman BL, Hellman RS, Goldsmith SJ, Mena I, Leveille L, Gherardi PG, Moretti J-L, Bischof-Delaloye A, Hill T, Rigo PM, Van Heertum RL, Ell PJ, Buell U, De Roo MC and Morgan RA (1989): Biodistribution, dosimetry, and clinical evaluation of technetium-99m ethyl cysteinate dimer in normal subjects and in patients with chronic cerebral infarction. *J Nucl Med* 30: 1018-1024.

Innis R, Baldwin R, Sybirska E, Zea Y, Laruelle M, Al-Trkriti M, Charney D, Zoghbi S, Smith E, Wisniewski G, Hoffer P, Wang A, Milius R and Neumeyer J (1991): Single photon emission computed tomography imaging of monamine reuptake sites in primate brain with [^{123}I]CIT. *Eur J Pharmacol* 200: 369-370.

Jones AG, Davison A, Lategola MR, Brodack JW, Orvig C, Sohn M, Toothaker AK, Lock CJL, Franklin KJ, Costello CE, Carr SA, Biemann K and Kaplan ML (1982): Chemical and *in vivo* studies of the anion oxo [N,N'-ethylene-*bis*-(2-mercaptoacetamido)]-technetate(V). *J Nucl Med* 23: 801-809.

Jones AG, Abrams MJ, Davison A, Brodack JW, Toothaker AK, Aldelstein SJ and Kassis AI (1984): Biological studies of a new class of technetium complexes: The hexakis-(alkylisonitrile)-technetium(I) cations. *Int J Nucl Med Biol* 11: 225-234.

Jurison S, Schlemper E, Troutner DE, Canning LR and Nowotnik DP (1986): Synthesis, characterization, and x-ray structural determinations of neutral Tc(V) oxotetradentate amine oxime complexes. *Inorg Chem* 25: 543-549.

Kasina S, Fritzberg AR, Johnson DL and Eshima D (1986): Tissue distribution properties of technetium-99m-diamide-dimercaptide complexes and potential use as renal radiopharmaceuticals. *J Med Chem.* 29: 1933-1940.

Kelly JD, Higley B, Archer CM, Canning LR, Chiu KW, Edwards B, Forster AM, Gill HK, Latham IA, Pickett RD, Webbon P, Edwards PG, Imran A, Griffiths DV, York DC, Mahoney PM, Tonkinson DJ, Dilworth JR and Lahiri A (1990): Technetium-99m complexes of functionalized diphosphines for myocardial imaging. In: *Technetium and Rhenium in Chemistry and Nuclear Medicine, Vol. 3*, Nicolini M, Bandoli G and Mazzi U, eds. New York: Raven Press.

Kessler RM, Ansari MS, Schmidt DE, Depaulis T, Clanton JA, Innis R, Altikriti M, Manning RG and Gillespie D (1991): High affinity dopamine-D2 receptor radioligands 2. [I-125]Epidepride, a potent and specific radioligand for the characterization of striatal and extrastriatal dopamine-D2 receptors. *Life Sci* 49: 617-628.

Kotze HF, Heyns AdP, Lotter MG, Pieters H, Roodt JP, Sweetlove MA, Badenhorst PN (1991): Comparison of oxine and tropolone methods for labeling human plateltes with Indium-111. J. Nucl Med 32: 62-66.

Kuhl DE, Barrio JR, Huang H-C, Selin C, Ackerman RF, Lear JL, Wu JL, Lin TH and Phelps ME (1982): Quantifying local cerebral blood flow by N-isopropyl-p-[^{123}I]iodoamphetamine (IMP) tomography. *J Nucl Med* 23: 196-203.

Kung HF, Tramposch KM and Blau M (1983): A new brain perfusion imaging agent: [I-123]HIPDM: N,N,N'-Trimethyl-,[2'-hydroxy-3-methyl-5-iodobe nzyl]-1,3-propanediamine. *J Nucl Med* 24: 66-72.

Kung HF, Molnar M, Billings J, Wicks R and Blau M (1984): Synthesis and biodistribution of neutral, lipid soluble Tc-99m complexes that cross the blood-brain barrier. *J Nucl Med* 25: 326-332.

Kung HF, Guo YZ, Billings J, Xu X, Mach RA, Blau M and Ackerhalt RE (1988a): Preparation and biodistribution of [^{125}I]IBZM: a potential CNS D-2 dopamine receptor imaging agent. *Nucl Med Biol* 15: 195-201.

Kung H, Alavi A, Billings J, Kung M, Pan S and Reilley J (1988b): [^{123}I]IBZP: A potential CNS D-1 dopamine receptor imaging agent: *In vivo* biodistribution in a monkey. *J Nucl Med* 29: 758.

Kung HF (1990): Radiopharmaceuticals for CNS receptor imaging with SPECT. *Nucl Med Biol* 17: 85-92.

Kung HF, Alavi A, Chang W, Kunf M-P, Keyes HW Jr, Velchik MG, Billings J, Pan S, Noto R, Rausch A and Reilley J (1990): *In vivo* SPECT imaging of CNS D-2 dopamine receptors : Initial studies with Iodine-123 IBZM in humans. *J Nucl Med* 31: 573-579.

Kung HF, Kung MP, Billings J, Shao L, Mozley D, Broich K, Pfeiffer N, McElgin W, Ranger N and Alavi A (1991): *In vivo* study of I-123 IBF: a new D2 dopamine receptor imaging agent. *J Nucl Med* 32: 1008.

Langen K-J, Coenen HH, Roosen N, Kling P, Muzik O, Herzog H, Kuwert T, Stöcklin G and Feinendegen LE (1990): SPECT studies of brain tumors with L-3-[^{123}I]Iodo-α-methyl tyrosine: Comparison with PET, ^{124}IMT and first clinical results. *J Nucl Med* 31: 281-286.

Laurin NR, Dreidger AA, Hurwitz GA, Mattar AG, Powe JE, Chamberlein MJ, Zabel PL and Pavlosky WF (1989): Cerebral perfusion imaging with technetium-99m HM-PAO in brain death and severe central nevous system injury. *J Nucl Med* 30: 1627-1635.

Leveille L, Demonceau G, De Roo M, Rigo P, Taillefer R, Morgan RA, Kupranick D and Walowitch RC (1989): Characterization of technetium-99m-L,L-ECD for brain perfusion imaging, Part 2: Biodistribution and brain imaging in humans. *J Nucl Med* 30: 1902-1910.

Lever SZ and Wagner HN (1990): The status and future of technetium–99m radiopharmaceuticals. In: *Technetium and Rhenium in Chemistry and Nuclear Medicine, Vol. 3*, Nicolini M, Bandoli G and Mazzi U, eds. New York: Raven Press.

Lever SZ, Burns HD, Kervitsky TM, Goldfarb HW, Woo DV, Wong DF, Epps LA, Kramer AV and Wagner HN Jr (1985): Design, preparation, and biodistribution of a technetium-99m triaminedithiol complex to assess regional cerebral blood flow. *J Nucl Med* 26: 1287-1294.

Lever JR, Scheffel UA, Musachio JL, Stathis M and Wagner HN Jr (1991): Radioiodinated D-(+)-N1-Ethyl-2-iodolysergic acid diethylamide: a ligand for *in vitro* and *vivo* studies of serotonin receptors. *Life Sci* 48: PL73-78.

Linder K, Feld T, Juri PN, Nunn AD and Treher EN (1987): The synthesis of a new technetium agent, SQ 32097, which can be used to assess regional cerebral blood flow. *J Nucl Med* 28: 592-593.

Linder KE, Wen MD, Nowotnic DP, Malley MF, Gougoutas L-L, Nunn AD and Eckelman WC (1991): Technetium labeling of monoclonal antibodies with functionalized BATOs. 1. $TcCl(DGM)_3PITC$. *Bioconjugate Chem* 2: 160-170.

Mather SJ (1987): Labeling with Indium-111. In: *Safety and Efficacy of Radiopharmaceutical 1987*. Kristensen K and Norbygaard E eds. Dordrecht: Martinus Nijhoff Publishers.

Mathias CJ, Sun Y, Welch MJ, Connett JM, Philpott GW and Martell AE (1990): N,N'-Bis(20hydroxybenzyl)-1-(4-bromoacetamido-benzyl)-1,2-ethylenediamine-N,N'-diacetic acid: A new bifunctional chelate for radiolabeling antibodies. *Bioconjugate Chem* 1: 204-211.

Mathias CJ, Sun Y, Welch MJ, Green MA, Thomas JA, Wade KR and Martell AE (1988): Targeting radiopharmaceuticals: Comparative biodistribution biodistribution studies of gallium and indium complexes of multidentate ligands. *Nucl Med Biol* 15: 69-81.

McAfee JG and Thakur ML (1976): Survey of radioactive agents for *in vitro* labelling of phagocytic leucocytes. *J Nucl Med* 17: 480-487.

Meares CF, McCall MJ, Reardon DJ, Goodwin DA, Diamanti CJ and McGuire M (1984): Conjugation of antibodies with bifunctional chelating agents: isothiocyanato and bromoacetamido reagents, methods of analysis and subsequent addition of metal ions. *Anal Biochem* 142: 68-78.

Mertens J, Vanryckeghem W, Gysemens M, Eersels J, Finda-Panek E and Carlsen L (1987): New fast preparation of ^{123}I labelled radiopharmaceuticals. *Eur J Nucl Med* 13: 380-381.

Mertens J, Bossuyt-Piron C, De Geeter F, Christiaens L, Cantineau R, Guillaume M and Leyssen J (1989): Evaluation of pure 2'-I-123-spiperone as a promising tracer for *in vivo* receptor studies with SPECT. *J Nucl Med* 30: 926.

Mertens L, Gysemans M, Bossuyt-Piron C and Thomas M (1990): High yield preparation of pure 2-radioiodo-ketanserin of high specific activity, a serotonin S2 receptor tracer for SPECT. *J Labelled Compd Radiopharm* 28: 732-737.

Messa C, Zito F, Colombo F, Rossetti C, Taddei G, Lucignani G, Fazio F and Deutsch E (1990): Preliminary evaluation of the rCBF tracer [^{99m}Tc]d,l-CB-PAO in human volunteers. In: *Technetium and Rhenium in Chemistry and Nuclear Medicine, Vol. 3*, New York: Raven Press.

Moerlein S (1985): Regiospecific incorporation of no-carrier-added radiobromine and radioiodine into aromatic rings via halogenodegermylation. *J Chem Soc Perkin Trans I*: 1687-1692.

Moerlein S (1987): Use of aryltrimethylgermanium substrates for facile aromatic chlorination, bromination, and iodination. *J Org Chem* 52: 664-667.

Moerlein S, Mathis C and Yano Y (1987): Comparative evaluation of electrophilic aromatic iododemetallation techniques for labeling radiopharmaceuticals with Iodine-122. *Appl Radiat Isot Int J Radiat Appl Instrum (Part A)* 38: 85-90.

Morgan GF, Deblaton M, Clements P, Van den Broeck P, Huddain W, Bossuyt A and Thornback JR (1990): [^{99m}Tc]TcMRP20 - An example of a new class of neutral complexes with implications for brain scintigraphy. In: *Technetium and Rhenium in Chemistry and Nuclear Medicine, Vol. 3*, Nicolini M, Bandoli G and Mazzi U, eds. New York: Raven Press.

Mountz JM, Modell JG, Foster NL, DuPree ES, Ackermann RJ, Petry NA, Bluemlein LE and Kuhl DE (1990): Prognostication of recovery following stroke using comparison of CT and technetium-

99m HM-PAO SPECT. *J Nucl Med* 31: 61-66.

Mueller-Gaertner H, Wilson AA, Dannals RF, Wagner HN Jr and Frost JJ (1992): Imaging of muscarinic cholinergic receptors in human brain *in vivo* with SPECT, [^{123}I]-4-Iododexetimide, and [^{123}I]-4-Iodolevetimide. *J Cereb Blood Flow Met*, in press.

Najafi A, Alauddin MM, Siegel ME and Epstein AL (1991): Synthesis and preliminary evaluation of a new chelate N_2S_4 for use in labeling proteins with metallic radionuclides. *Nucl Med Biol* 18: 179-185.

Neirinckx RD, Canning LR, Piper IM, Nowotnik DP, Pickett RD, Holmes RA, Volkert WA, Forster AM, Weisner PS, Marriott JA and Chaplin SB (1987): Technetium-99m-*d,l*-HM-PAO. A new radiopharmaceutical for SPECT imaging of regional cerebral blood perfusion. *J Nucl Med* 28: 191-202.

Otsuka FL and Welch MJ (1987): Methods to label antibodies for use in tumor imaging. *Nucl Med and Biol* 14: 243-249.

Owens J, Murray T, McCulloch J and Wyper D (1992): Synthesis of (R,R)^{123}I-QNB, a SPECT Imaging agent for cerebral muscarinic acetylcholine receptors *in vivo. J Labelled Compd Radiopharm* 31: 45-60.

Peters AM, Saverymutti SH, Danpure HJ and Osman S (1986): Cell labeling. In: *Methods in Hematology*, Lewis SM and Bayly RJ, eds. New York: Churchill Livingston.

Peters AM (1988): Review on platelet labelling and kinetics. *Nucl Med Comm*, 9: 803-808.

Quadri SM, Paik CH, Reba RC and Hong WP (1990): Optimization of biodistribution by introducing different chemical linkages between antibody and an indium-111 chelate. *Cancer Treat Res* 51: 201-13.

Richards P and Atkins HL (1968): ^{99m}Tc labeled compounds. *Jpn Nucl Med* 7:165.

Saji H, Nakatsuka I, Shiba K, Tokui T, Horiuchi K, Yoshitake A, Torizuka K and A(?) Y (1987): Radioiodinated 2'-Iodospiperone: a new radioligand for *in vivo* dopamine receptor studies. *Life Sci* 41: 1999-2006.

Scheffel U, Goldfarb HW, Lever SZ, Gungon RL, Burns HD and Wagner HN Jr (1988): Comparison of technetium-99m aminoalkyl diaminodithiol analogs as potential brain blood flow imaging agents. *J Nucl Med* 29: 73-82.

Seevers R and Counsell R (1982): Radioiodination techniques for small organic molecules. *Chem Rev* 82: 575-582.

Seldin DW, Johnson LL, Blood DK, Muschel MJ, Smith KF, Wall RM and Cannon PJ (1989): Myocardial perfusion imaging with technetium-99m SQ30217: Comparison with thallium-201 and coronary anatomy. *J Nucl Med* 30: 312-319.

Shaya ES, Scheffel U, Dannals RF, Ricaurte GA, Carrol FI, Wagner HN Jr, Kuhar MJ and Wong DF (1992): *In vivo* imaging of dopamine reuptake sites in the primate brain using single photon emission computed tomography (SPECT) and Iodine-123 labeled RTI-55. *Synapse* 10: 169-172.

Smith FW, Gemmell HG and Sharp PF (1987): The use of 99m-Tc-HM-PAO for the diagnosis of dementia. *Nucl Med Commun* 8: 525-533.

Stabin M and Schlafke-Stelson A (1991): A list of nuclear medicine radionuclides and potential contaminants for operators of *in vivo* counters. *Health Physics* 61: 427-30.

Subramanian G, McAfee JG, Bell EG, Sandler M, Keesee RE and Henry C (1971): New ^{99m}Tc labeled radiopharmaceuticals for renal imaging. *J Nucl Med* 12: 399.

Thakur ML (1986): Radiolabeled Leukocytes. In: *Radiopharmaceuticals: progress and clinical perspectives*, Fritzberg AR, ed. Boca Raton: CRC Press.

Thakur ML, Welch MJ, Joist JH and Coleman RE (1976): Indium-111 labelled platelets: studies on preparation and evaluation of *in vivo* and *in vitro* functions. *Thrombosis Res* 9: 345-357.

Tow DE, Wagner HN, Lopex-Majano V, Smith EM and Magita T (1966): Validity of measuring regional pulmonary arterial blood flow with macroaggregates of human derum albumin. *Am J Roentgen* 96: 664-676.

Treher EN, Francesconi LC, Gougoutas JZ and Malley MF (1989): Monocapped tris(dioxime) complexes of technetium(III): Synthesis and structural characterization of TcX(dioxime)$_3$B-R (X = Cl, Br; dioxime = dimethylglyoxime, cyclohexanedione dioxime; R = CH_3, C_4H_9). *Inorg Chem* 28: 3411-3416.

Troutner DE, Volkert WA, Hoffman TJ and Holmes RA (1984): A neutral lipophilic complex of ^{99m}Tc with a multidentate amine oxime. *Int J Appl Radiat Isot* 35: 467-470.

Verbruggen AM, Nosco DL, Van Nerom CG, Bormans GM, Adriaens PJ and De Roo MJ (1992): Technetium-99m-L,L-ethylenedicysteine: a renal imaging agent. I. Labeling and evaluation in animals. *J Nucl Med* 33: 551-557.

Wackers FJT, Berman DS, Maddahi J, Watson DD, Beller GA, Strauss HW, Boucher CA, Picard M, Holman BL, Fridrich R, Inglese E, Delaloye B, Bischof-Delaloye A, Camin L and McKusick K (1989): Technetium-99m hexakis 2-methoxyisobutyl isonitrile: Human biodistribution, dosimetry, safety, and preliminary comparison to thallium-201 for myocardial perfusion imaging. *J Nucl Med* 30: 301-311.

Walovitch RC, Williams SJ and Lafrance ND (1990): Radiolabeled agents for SPECT imaging of brain perfusion. *Nucl Med Biol* 17: 77-83.

Washburn LC (1991): Radiochemistry of monoclonal antibody labeling. In: *New Trends in Radiopharmaceutical Synthesis, Quality Assurance, and Regulatory Control*, Emran AM, ed. New York: Plenum Press.

Weinberger DR, Gibson RE, Coppola R, Jones DW, Braun AR, Mann U, Berman KF, Sunderland T, Chase TN and Reba RC (1989): Distribution of muscarinic receptors in patients with dementia: A controlled study of ^{123}I QNB and SPECT. *J Cereb Blood Flow Metab* 9: S537.

Welch MJ and Welch TJ (1975): Solution chemistry of carrier-free indium. In: *Radiopharmaceuticals*, Subramanian G, Rhodes BA, Cooper MD and Sodd VJ, eds. New York: Society of Nuclear Medicine.

Winchell HS, Horst WD, Braun L, Oldendorf WH, Hatter R and Parker H (1980): N-Isopropyl-(^{123}I)-p-iodoamphetamine: Single pass uptake and washout, binding to brain synaptosomes and localization in dog and monkey brain. *J Nucl Med* 21: 947-952.

Yeh SM, Meares CF, and Goodwin DA (1979): Decomposition rates of radiopharmaceutical indium chelates in serum. *J Radioanal Chem* 53: 327-336.

THE DESIGN OF SITE-DIRECTED RADIOPHARMACEUTICALS FOR USE IN DRUG DISCOVERY

William C. Eckelman
PET Department
National Institutes of Health

Raymond E. Gibson
Radiopharmacology Department
Merck Research Laboratories

INTRODUCTION

The application of Nuclear Medicine technologies to drug discovery can be divided into two approaches: 1) the use of currently available radiopharmaceuticals to determine the presence of either a desired or unwanted physiological effect or 2) the ability to determine the interaction of the drug with a desired target (e.g., receptor or enzyme) in a specific organ. The first of these is described in Chapters 9, 15 and 16 of this book. Although drugs can be labeled with either positron emitting or single photon emitting nuclides, the selection of the drug to be labeled and the validation of the radiolabeled drug as one providing the desired information is still problematic. An entire field of basic research in Nuclear Medicine involving receptor binding radiotracers has grown since 1970, but many aspects of radioligand design remain empirical.

We have chosen the designation "site directed" as compared to receptor-binding radiotracer since the site-directed radiotracer includes

Nuclear Imaging in Drug Discovery, Development, and Approval
H.D. Burns, R. Gibson, R. Dannals, Peter Siegl, Editors

enzyme substrates and inhibitors, and ligands that interact with transport proteins as well as membrane bound or cytosolic receptors. The phenomenology which leads to a suitable radiotracer *in vivo* for receptors applies equally well for these additional targets. We will use the term receptor to represent the drug binding site for each of these target macromolecules.

EVOLUTION OF THE RECEPTOR CONCEPT

The evolution of the receptor concept is a popular topic (Aronstam, 1982; Parascandola, 1986). The concept that a pharmaceutical must act through a specific site was suggested as early as 1685 by Robert Boyle who proposed a mechanical picture of a drug being more fit to be detained by one organ than another according to its size, shape and motion. Through the years, there were numerous other hypotheses leading to the elaboration of the receptor theory by Ehrlich and Langley. Based on his observations of the physiological antagonism of drug pairs such as opium and belladonna or atropine and pilocarpine, Langley postulated a single substance that is capable of forming an association with both compounds. Ehrlich came to similar conclusions using the "specific" localization of dyes in various organs. This theory was further refined by studying the immunologic binding of toxin with antibody. However, Ehrlich did not apply this concept to drugs for another 10 years because of the ease of washing drugs from the target tissue. The higher affinity of the toxin-antibody bond compared to the lower affinity of most drugs was misinterpreted as the presence or absence of specific binding sites. Ehrlich later adapted the receptor theory of drugs based on earlier publications by Langley and his own work on drug resistance in the treatment of trypanosomes. As a result, he added the word "chemoreceptor" to his previously coined word "nutrireceptor" to describe the toxin-antibody interaction. The mathematical approach to these binding phenomena was first developed by A.V. Hill and then A.J. Clark.

The approach to understanding the receptor binding process was enhanced by the development of radiolabeled probes first used by Jensen and Jacobson (1962) and Paton and Rang (1966). Farrow and O'Brien later observed high affinity binding in rat brain fractions using [^{3}H]atropine (Farrow and O'Brien, 1973). Many antagonists to the muscarinic receptor were synthesized based on the structure of acetylcholine and atropine. However, higher affinity ligands such as dexetimide were still being prepared at low specific activity (< 600 Ci/mol) and probably would not reflect receptor concentration when injected intravenously. Not until 3-quinuclidinyl benzilate (QNB) was tritiated at about 4 Ci/mmol, was a distribution reflective of muscarinic receptor concentration obtained. Yamamura and Synder (1974) used the higher affinity and higher specific activity ligand [^{3}H]QNB to define the affinity constant and the receptor concentration necessary to provide receptor-mediated localization. With the injection of 60 μCi in rats, the corpus striatum contained the highest concentration of radioactivity, the hippocampus was intermediate, and the cerebellum was lowest which is in line with the receptor concentrations obtained *in vitro*. It is interesting to note that when this experiment was repeated with 5 μCi of [^{3}H]QNB at 60 Ci/mmol, the radioligand concentration in the three tissues was similar (Gibson et al., 1984).

USES OF RADIOLIGANDS *IN VITRO* AND *IN VIVO*

Perhaps no other factor has influenced the study of receptors and enzyme binding sites than the development of tritiated and radioiodinated ligands with sufficient affinity and specific activity to permit the study of these species *in vitro*. Attempts to use radiolabeled drugs *in vivo* are, however, far more difficult than *in vitro* techniques such as receptor assays using tissue homogenates. For example, *in vitro* assays permit washing to reduce nonspecific interactions, but *in vivo* the site-specific signal is dependent on the inherent properties of affinity, lipophilicity, and structural features which lead to interactions with other binding sites (others receptors, enzymes, drug-acceptor sites, etc.). In addition, the radioligand is metabolized *in vivo*. Non-receptor interactions may be increased by metabolites which persist in plasma but do not interact with the desired receptor or the radiolabel may be rapidly removed thus obviating any specific signal. While *in vitro* assay conditions can often be modified to allow the use of a "less-than-optimal" radioligand, the only solution for a radioligand which does not permit visualization of a specific receptor or enzyme *in vivo* is to seek a different drug with better properties. Sometimes derivatives of the unsuccessful drug will overcome the difficulties, but frequently an entirely different structure is necessary. Thus, the design of a site-directed radiotracer for *in vivo* imaging is not particularly easy.

To develop a drug with the necessary pharmacological properties to be successfully marketed may require synthesis of 10,000 compounds. Although the development of *in vivo* radiotracers can rely heavily upon the knowledge gained from medicinal chemistry efforts in developing a new drug, a successful imaging agent may require synthesis of tens to hundreds of compounds. And, like any potential pharmaceutical, a seemingly ideal radiotracer may founder far into development because of unexpected complications. For example, [^{11}C]scopolamine was developed to image and quantify the muscarinic receptor in brain (Frey et al., 1992). However, poor sensitivity of the radiotracer to the concentration of receptor in the target tissue, similar to that observed for [^{3}H]QNB cited above, and the rapid appearance of extensive metabolites in serum reduce the usefulness of [^{11}C]scopolamine, but led to the design of Carbon-11 labeled muscarinic antagonists with better properties (Mulholland et al., 1988; Frey et al., 1990). Although not an easy task, there is, however, sufficient background in the development of site-directed radiotracers to guide future development.

The Model for Development and Validation of a Biochemical Probe: 2-[^{18}F]-Fluoro-2-Deoxyglucose.

One of the first biochemical probes, 2-[^{18}F]fluoro-2-deoxy-D-glucose (FDG), set a high standard for the development of other probes to be used *in vivo* (Kennedy, 1991). Using [^{14}C]deoxyglucose in studies spanning three decades, Dr. Louis Sokoloff developed a method of measuring glucose utilization in differing behavioral states, seizures, during development, during sensory stimulation, and following administration of drugs (Reivich et al., 1971). This brilliant integration of disciplines started with methods for the measurement of cerebral blood flow developed by Seymour Kety, including quantitative autoradiographic procedures using metabolically inert diffusible tracers for measuring local metabolic rates (Landau et al., 1955). This laid the groundwork for the crucial development: the determination of glucose utilization as a measure of energy metabolism and functional activity. Since glucose metabolism is rapid, such studies necessitate very short measurement times. Although [^{14}C]glucose can be used, metabolism leads to [^{14}C]CO_2 which makes quantification difficult. Based upon mechanistic studies, the metabolically trapped 2-deoxyglucose was determined to exhibit better properties. This substrate is phosphorylated like glucose by hexokinase but the product is not converted to fructose-6-phosphate, the next step in the glycolytic pathway. 2-Deoxyglucose-6-phosphate accumulates in brain to relatively high concentrations because it is a poor substrate for most enzymes present in brain and because glucose-6-phosphatase activity is very low in brain. Thus, [^{14}C]2-deoxyglucose could be used as a tracer for glucose with the autoradiographic technique that had been devised for the local cerebral blood flow method. Drawing on his experience with enzyme kinetics, Sokoloff developed a quantitative model to measure local glucose utilization. As a result new relationships were revealed and the full constellation and degree of participation of structures simultaneously activated or inhibited as a result of a given behavior or response to stimulation were demonstrated for the first time. Shortly thereafter, the method was adapted for use in human subjects and has become the most important technique in measuring brain function by external detection using Positron Emission Tomography (PET) (Reivich et al., 1979). It is rare to be able to characterize a physiological process after administration of a radiolabeled compound *in vivo*. Nevertheless, this measurement has been extended to humans in such neurological disorders as epilepsy, Huntington's disease, and Parkinsonism using FDG and PET. The method has also been used to localize and grade gliomas without brain biopsy. No single radiotracer has contributed more significantly to the growth in research on applications of non-invasive imaging using radiopharmaceuticals. However, metabolically trapped substrates or drugs are rare with the primary examples being FDG and the suicide inhibitors described by Joanna Fowler in Chapter 10. The preponderance of radiotracers which have been developed interact reversibly with target macromolecules and are free to diffuse in and out of tissue.

Problems with Early Attempts with Radioligands In Vivo

The difficulties in design of site-directed radiotracers can be well demonstrated by studies on radioiodinated estradiol receptor radioligands. The first radioiodinated estrogens were those that were easiest to radioiodinate using electrophilic iodination. Many of the syntheses were carried out with low specific activity radioiodine. In addition to being at low specific activity, the radioiodinated compounds readily deiodinated *in vivo* (for a review of these early studies, see Katzenellenbogen et al., 1982). In 1977, Komai et al. published pertinent data on four radioiodinated derivatives using a no-carrier-added technique. These included 2-iodoestradiol, iodohexestrol, and two iodinated derivatives that were being used in *in vitro* radioimmunoassays. None of the four showed specific binding either *in vitro* or *in vivo*. Iodination on the phenyl ring and the substitutions in the 7 position of estradiol reduce the affinity of the products below that which can provide specific binding. In the case of iodohexestrol, the apparent affinity was significantly reduced by high affinity interactions with other proteins than the estradiol receptor. Given the similarity of iodohexestrol to thyroxine, this hormone was used to block the binding of iodohexestrol to serum proteins. Under these conditions, iodohexestrol showed specific *in vitro* and *in vivo* receptor binding. The uterus to blood ratio was increased from 2 in the absence of additional drugs to 10.4 when thyroxine was injected intraperitoneally prior to the injection of radioligand. Katzenellenbogen et al., (1975) had previously shown that [^{3}H]iodohexestrol bound saturably to prealbumin and that specific binding *in vitro* could only be observed using sucrose gradient ultracentrifugation. These early studies led to the development of successfully iodinated estrogens based on several observations: i) high affinity ligands at high specific activity are necessary; and ii) when the molecule contains a phenol and is iodinated ortho- to the hydroxy group, the molecule deiodinates easily and specific binding to the thyroxine binding protein is often observed. As a result, researchers turned to estrogen derivatives that did not contain the iodophenol-structure. The most successful estrogens are the 17α–iodovinylestrogens (Jagoda et al., 1984) and 16α–fluoroestrogens (Pomper et al., 1990).

Validation of Site-Directed Radioligands - General Principles

The first question that should be answered concerning the synthesis of a site-directed radiotracer is what features are necessary for the external visualization of a receptor-ligand complex? Can any receptor (or enzyme) be imaged by any radioligand, or are there specific strictures which much be satisfied before information on the system can be obtained via external imaging? If the concentration of a radioligand in blood is greater than that in a target tissue, e.g., the myocardium, an image of the receptors in the myocardium will be obscured by the radioactivity in the blood pool. Therefore, imaging a receptor requires a specific signal above that of background radioactivity, either that in the blood or non-specifically bound. For a receptor-specific signal to exist, the radioligand must have a high affinity for its receptor. The exact requirements for that affinity are minimally determined by the concentration of receptor present in the tissue. The concentration of receptor present in a target tissue is a given, which leaves the affinity of the radioligand as the parameter which we can manipulate. The first step in site-specific radiotracer design is to determine what affinity is needed to provide the necessary target-to-blood ratio.

Great strides have been made in choosing receptor binding radiotracers for *in vivo* studies by using only the affinity constant, determined *in vitro*, to predict the likelihood that a radiotracer will provide a receptor-specific signal large enough to detect externally. The target-to-blood ratio can be easily calculated from the equilibrium expression developed by Scatchard (1949):

$$\text{Bound/Free} = \text{Target to Nontarget} = R_oK_A - BK_A$$

where R_o is the total receptor concentration, K_A is the affinity constant and B is the concentration of radioligand bound to receptor. When the concentration of radioligand bound is small compared to the concentration of receptor, the second term becomes insignificant and the maximum target-to-blood ratio is predicted by the product of the receptor concentration and the affinity constant, R_oK_A. The advantage of using very low concentrations of radioligand are two-fold. Firstly, as receptor-occupancy increases, the B/F ratio decreases which reduces the likelihood that a receptor-specific signal can be imaged. Secondly, low receptor occupancy is less likely to cause a pharmacological effect. For a radiotracer to provide an indication of receptor concentration or information on occupancy by agonist or a competing drug, the system should not be perturbed by the radiotracer. Since there is no pharmacological effect, the studies required for a radiopharmaceutical IND are only subacute toxicology and radiation dosimetry.

There have been many suggestions on how to systematically evaluate a putative receptor binding radiotracer. One such approach (Eckelman, 1992a) outlines sixteen categories in developing and testing a receptor binding radiotracer:

1. The choice of a receptor system.
2. Use of a mathematical model to choose potential receptor-specific radiopharmaceuticals.
3. The determination of the K_A for the parent compound.
4. Use of Tritium-labeled parent compound to determine distribution *in vivo*.
5. The preparation of the non-radioactive derivative.
6. The determination of *in vitro* stability of the non-radioactive derivative.
7. The determination of the K_A for the non-radioactive derivative.
8. The evaluation of various physical parameters of the non-radioactive derivative (Structure-Distribution-Relationship, SDR).
9. *In vivo* displacement of the Tritiated compound with the nonradioactive derivative.
10. Preparation of the radioactive derivative (radiotracer).
11. Chromatographic separation.
12. The determination of the K_A for the radiotracer.
13. Correlation of distribution of the radioatracer in animals with receptor location.
14. Use of preinjection, coinjection or postinjection to decrease effective specific activity, i.e., block specific binding, of the radiotracer.
15. Use of active and inactive stereisomers of the radiotracer.
16. Confirmation of animal distribution of the radiotracer in humans.

These sixteen steps were developed to guide investigators interested in developing a diagnostic agent for a particular receptor system. Many of these steps will have been accomplished by prior investigations in the natural course of drug discovery.

The choice of receptor to be imaged is dictated by project needs, e.g., a drug to reduce blood pressure or attenuate the symptoms of Alzheimer's disease. If the potential site-directed radiotracer is planned for studies in humans, we need to know the concentration of the drug binding site in the target tissue of humans. Such knowledge in humans is often difficult to obtain since such data is usually obtained from *in vitro* assays of autopsy samples from patients with specific diseases who have been extensively treated. Either the treatment or the disease may change receptor concentrations. For example, the concentration of the dopamine D_2 receptor in the basal ganglia of patients diagnosed with schizophrenia may be increased as a result of the disease or as the result of therapy with D_2 receptor antagonists such as haloperidol. There have been a number of reviews on the change in receptor concentration as a function of disease as obtained from autopsy samples (Gibson, 1982; Wagner, 1982). It is outside the scope of this chapter to review those data. Another source of information on receptor concentrations in target tissues is to use data obtained from animal studies. During the development and validation of a new site-directed radiotracer, we will need to know the concentration of receptor in the non-human species which will be used as models. However, these data can not be extended to humans without confirming studies since receptor concentrations may vary significantly from species to species.

Once knowledge of receptor concentration is known, the use of a simple mathematical model, such as the Scatchard equation presented above (Eckelman, 1979), can define the minimum affinity needed for a potential receptor-specific radiopharmaceutical. It should be noted that the product K_AR_o provides the maximum target-to-blood ratio. Significant reductions in that ratio may result from extensive binding of the radioligand to plasma proteins, rapid metabolism and poor access to the receptor (poor diffusion across the blood-brain-barrier or plasma membranes). Attempts to account for some of these factors has met with modest success (Katzenellenbogen, 1982).

An additional complexity is that K_AR_o only predicts the maximum *receptor specific* target-to-blood ratio, but the radiotracer may exhibit significant non-specific binding to sites other than the desired receptor. For example, Eckelman et al. (1980) reported good heart-to-blood ratios in rat heart using an [^{125}I]iodotyramine derivatives of the β_1–adrenoceptor antagonist practolol, but the binding was not the result of an interaction with the β-adrenoceptor (Jiang et al., 1978; Eckelman et al., 1980). The avid non-receptor binding was species specific and did not occur in rabbits. Finally, the affinity constant of the radioligand for the concentration of β-adrenoceptor in rat heart was not high enough to provide a receptor- specific signal. The question then is: what value of $K_A R_o$ is needed? We do not have a simple answer since a systematic study using radioligands has not been made. However, Atkins and Goodenough (1982) in a study with Technetium-99m using phantoms, showed that a target to non-target (blood and surrounding tissue) ratio of at least 10 was needed for planar imaging of a 1 cm in diameter target sphere. For Single-Photon-Emission-Computed-Tomography (SPECT) and PET imaging, a ratio of 4 is needed (Goodenough and Atkins, 1988). The affinity constants of several adrenergic receptor antagonists and their predicted maximum bound-to-free ratios are presented in Table 1. Using [^{3}H]alprenolol, we could not demonstrate a receptor specific heart-to-blood ratio in rat (Francis et al., 1982). By contrast, [^{3}H]carazolol localized in rat heart with a heart-to-blood ratio of 5, of which 66% was a receptor-specific signal (Francis et al., 1982). Both [^{123}I]Iodocyano-pindolol (ICYP) and [^{11}C]CGP 12177 have been used to study the β-adrenergic receptor *in vivo* in heart (Sisson et al., 1991; Delforge et al., 1991). Although it may be possible to obtain a receptor-specific planar images with ratios less than 10 (the heart in either dog or man is much larger than a 1 cm diameter sphere), these results suggest that the higher the ratio, the more likely one will obtain the image of interest. We emphasize, however, that high affinity is a necessary attribute which may not be sufficient to provide a site-directed radiotracer.

COMPOUND	K_A (M^{-1})	K_A*R_o
Alrenolol[a]	1.3×10^8	0.65
Carazolol[a]	1.6×10^9	8
ICYP[b]	1.0×10^{11}	500
CGP 12177[c]	3.3×10^9	17

a) Affinity constant from Francis et al. (1982).
b) Affinity constant from Bjornerheim et al. (1991).
c) Affinity constant from Hertal et al. (1983).

TABLE 1. Maximum bound-to-free ratios of four radioligands for the β–adrenergic receptor in heart (R_o = 5 nM; from Francis et al., 1982).

The affinity constants of a large number of compounds are available in the Structure-Activity-Relationship files for various projects within a pharmaceutical company. A search of such files will usually turn up many compounds with equilibrium dissociation constants ($K_d = 1/K_A$) in the desired range (subnanomolar). Since unlabeled compounds are screened, there is no method to adequately determine the free ligand concentration; all such affinities are less accurate than those determined by saturation studies with a radioligand. We indicate the affinity determined using the unlabeled ligand as $K_{A(app)}$. Although the radioligand may give a better estimate of K_A, $K_{A(app)}$ may be more informative since mechanisms which greatly reduce free ligand concentrations in *in vitro* assays, e.g., high lipophilicity, will also influence the *in vivo* characteristics of the ligand. Nonetheless, the physical properties of these ligands or lack of a convenient synthetic method to incorporate a positron or gamma emitting nuclide will eliminate many compounds from consideration.

In most cases an antagonist has been chosen as the potential receptor binding radiotracer because the affinity of the antagonist is more likely to be in the range to give an appropriate target -to-nontarget ratio than the affinity of an agonist. Notable exceptions have appeared in radiopharmaceutical research, such as [^{11}C]carfentanil, a mu-opiate agonist with very high affinity (Frost et al., 1990), and 16α–[^{18}F]fluoroestradiol (Kiesewetter et al., 1984) which has been used to image estrogen dependent breast tumors (Mintun et al., 1988).

It is essential that the affinity constants be determined in a physiologically relevant milieu. For example, the opiate agonists (Frost and Wagner, 1984) exhibit higher affinity in low salt buffers and in the absence of GTP. Similary, MK-801, an antagonist of the glutamate N-methy-D-aspartate (NMDA) receptor/channel, exhibits high affinity in low salt buffers (Ransom and Stec, 1988). Using physiologically relevant buffer lowers the affinity by 10-fold ($K_i \approx 100$ nM). This affinity is sufficiently low that we would predict that MK-801 should not work well.

The most efficient method to determine if a particular antagonist is suitable for radiolabeling with a positron or gamma-emitting isotope is to study Tritium labeled compounds. The specific activity of tritiated antagonist (a maximum specific activity of 30 Ci/mmol for one Tritium per molecule) is high enough to provide a good estimate of the maximum B/F ratio in *in vivo* experiments. It should be noted that Carbon-14 labeled compounds, frequently generated for metabolism studies, have specific activities too low to test as site-directed radiotracers. These experiments are clearly more sophisticated than the use of a simple mathematical model since the actual distribution of the ligand includes all the complicating factors assumed to be negligible in the modeling. Many tritiated ligands are commercially available for specific receptors, enzymes or transport proteins, which can provide a potential model ligand system to test the feasibility of imaging with a site-directed radiotracer. For example, studies by Eckelman et al. (1984) using tritiated adrenoceptor antagonists showed that [^{3}H]carazolol exhibited properties sufficient for imaging β–adrenoceptors in heart and lung. The feasibility of imaging the sigma receptor could have been tested with [^{3}H]haloperidol, a dopamine D_2 receptor antagonist which exhibits high affinity for the sigma receptor (Contreras et al., 1987). The realization that [^{18}F]haloperidol provides images of the sigma receptor resulted after images of the specific radiotracer distribution were demonstrated (Logan et al., 1989).

Once a particular receptor ligand has been deemed satisfactory either by *in vitro* receptor studies or preferably by *in vivo* distribution studies using the Tritium-labeled compound, consideration must be given to the means of incorporating a positron or gamma emitting nuclide into the candidate site-directed radiotracer. The best method is to isotopically replace an atom in the structure with Carbon-11, or Fluorine-18 or Iodine-123 if these atoms are present in the structure. For example, a very selective sigma receptor ligand, L-687,384 (**1**) was radiolabeled with Carbon-11 in the methylene residue of the benzyl moiety (Wong et al., 1991). Similarly, [^{18}F]haloperidol (**2**) can be prepared with the radionuclidic fluorine substituted for the stable fluorine which is already present in the drug.

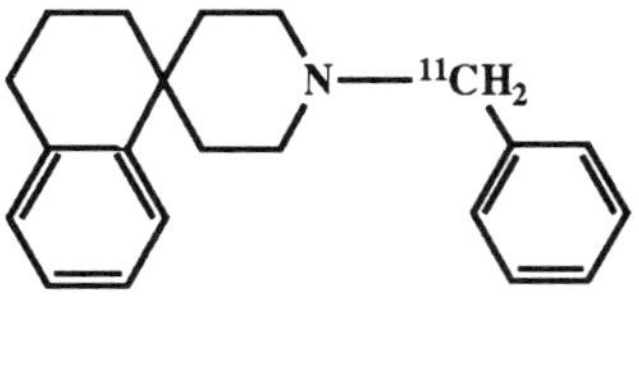

1 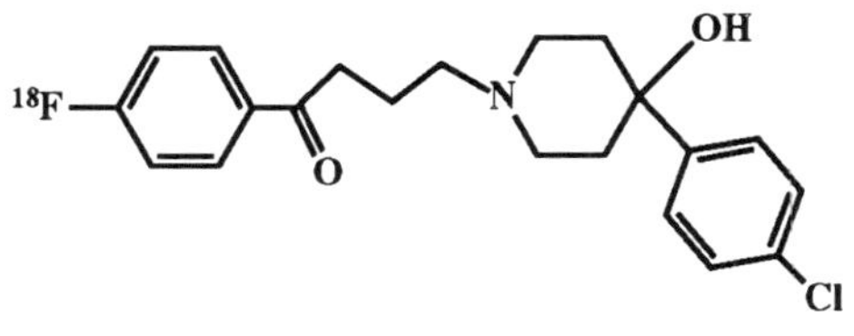 **2**

However, the structure of the candidate may not lend itself well to the rapid syntheses required for isotopic replacement and iodine-containing drugs are rare. Therefore, a close analogue of the candidate, with similar affinity and physicochemical properties, may be prepared. For example, N-[^{11}C]methylspiperone (**3**) is an easily synthesized derivative of spiperone, a dopamine D_2 receptor antagonist clinically available in Europe, which provides excellent images of the D_2 receptor in man (Wagner et al., 1983). 8-[^{11}C]Methoxy-MK-801 (**4**) is an isolipophilic analogue of MK-801 (Burns et al., 1989; Wong et al.,

1989), which is much easier to synthesize than [^{11}C]MK-801 (Kiesewetter et al., 1990). An iodinated derivative of MK-801, 3-[^{123}I]MK-801, exhibits properties that are better than the positron-emitting products; notably, the affinity of the iodinated derivative does not exhibit the extreme sensitivity to the ionic milieu as observed with MK-801 (Ransom et al., 1990). While the synthesis of the muscarinic receptor antagonist, 3-quinuclidinyl [^{11}C]benzilate (**5**) has been reported (Prenant et al., 1989), the incorporation of an iodine in the para-position of one of the benzilate phenyl rings (4IQNB, **6**) led to a ligand with *in vivo* properties that are better than QNB (Gibson et al, 1984; Gitler

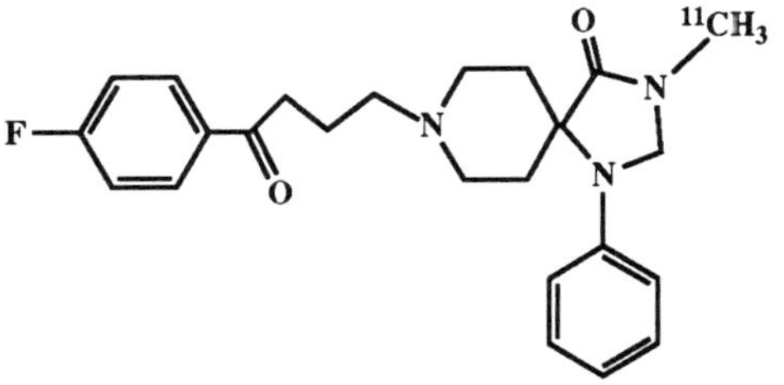

3

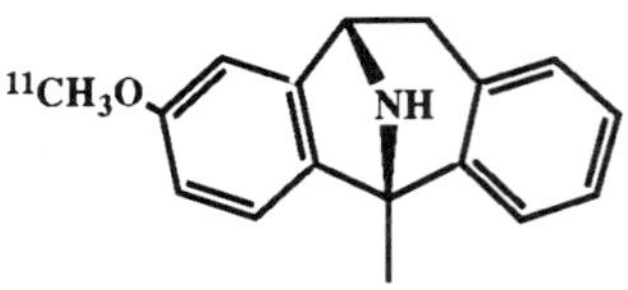

4

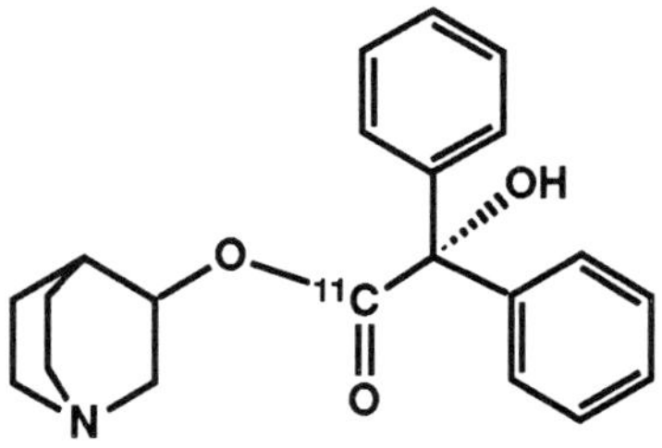

5

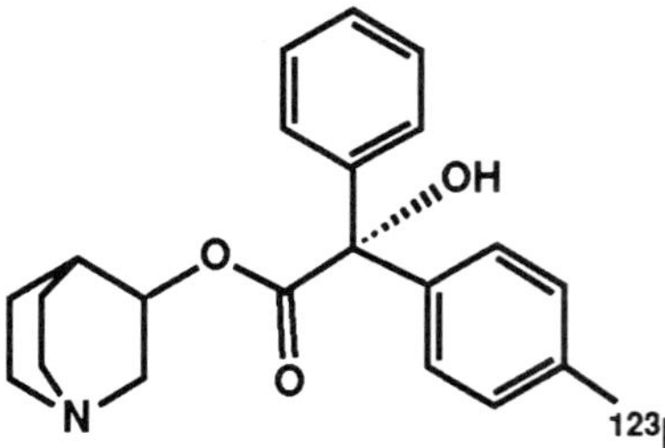

6

et al., 1992). However, analogues may not have the same affinity as the parent ligand. Therefore, the putative analogues should be prepared first in unlabeled form and the affinity for receptor or enzyme tested. This avoids the problems associated with a no-carrier-added synthesis until the derivative is shown to be a true tracer for the parent ligand. This also produces a reference compound for the radiotracer. The nonradioactive compound can be characterized using the classical analytical methods of elemental analysis, and IR and NMR spectroscopy. With high specific activity radiotracers chromatography is usually the only analytical tool available, and the characterized non-radioactive compound provides the necessary chromatrographic reference.

If stability of the putative radiotracer is of concern, particularly with halogenated compounds, the stability of the non-radioactive analogue may be determined prior to radiosynthesis. The difference in R_f's of the parent compound and the halogenated derivative are usually sufficiently large that thin-layer chromatography is adequate to test the stability. One caution when using the nonradioactive compound for stability testing is that the large concentrations used will mostly likely result in a second order reactions whereas the high specific activity ligand will most likely lead to pseudo-first order reactions. Additionally, such tests are not necessary when using halogenated derivatives which are known to be quite stable. For example, iodine substitution in a phenyl ring is, in most instances, sufficiently stable that one can proceed with confidence to the determination of affinity for the halogenated derivative. The stability should be tested before determining the affinity of the halogenated derivative for the receptor or enzyme target. If the dehalogenated product exhibits a high affinity for the target, perhaps higher than the halogenated analogue, the apparent affinity would be quite misleading.

If the halogenated derivative is known to be or shown to be stable, the preparation of the nonradioactive form of the putative receptor binding radiotracer leads to the determination of the $K_{A(app)}$ for the nonradioactive derivatives. The effect of positional isomers is most effectively determined using nonradioactive halogen derivatives. For example, we determined the affinities of the 2', 3', and 4'-halogen derivatives of QNB using tissue preparations of the muscarinic acetylcholine receptor from the corpus striatum and myocardium of rat. These results showed that the 4'-derivative exhibited the highest affinity and the 2'-derivative exhibited the lowest affinity for the m-AChR (Rzeszotarski et al., 1982; Gibson et al., 1983).

The choice of radionuclide is determined by several factors. The positron emitting radionuclides, Carbon-11, Fluorine-18 and Nitrogen-13, in general cause less perturbation when incorporated into a biochemical probe. As a result, many biochemical probes have been labeled with positron emitting radiotracers (Stöcklin, 1992; Kilbourne, 1990). The short half-life of these positron-emitting isotopes require a dedicated cyclotron. Additionally, if the half-life of the process to be measured is long compared to the half-life of the isotope, positron-emitting isotopes may not prove useful. Therefore, single-photon isotopes are frequently the nuclides of choice.

The radionuclides of choice for planar or SPECT imaging are Technetium-99m and Iodine-123; higher energy gamma-emitters with longer half-lives are Indium-111, Iodine-131 and Bromine-77.

Technetium-99m is by far the best radionuclide from an imaging point of view because of the ideal nuclear properties for the standard gamma camera and the ready availability afforded by the Molybdenum-99/Technetium-99m generator system. However, the incorporation of Technetium-99m into a receptor antagonist is a formidable synthetic task which, in most cases, will perturb the interaction of the receptor ligand with the receptor. Technetium chelates are typically the size of a phenanthrene moiety with molecular weights of 200 to 300. The addition of a [^{99m}Tc]chelate is thus a considerable modification to a small drug with molecular weight of 400 to 600. Another important consideration is the preparation of a neutral chelate to better match the lipophilicity of most receptor binding ligands and to cross cell membranes in the case of cerebral receptors and peripheral intracellular receptors. Neutral Technetium-99m chelates have been prepared by Yokoyama et al. (1976) with the KTS chelate, Lever et al. (1985) with the diaminodithio ligands, Troutner et al. (1980) with the amineoxime ligands and Nunn et al. (1987) with the seven coordinate tris dioximes. Technetium-99m chelates have been bound to various biochemicals, particularly peptide ligands which are not as adversely affected by the chelates and maintain high receptor affinity (Vera et al., 1984; see Flanagan, chapter 11) Additional successes have been the incorporation of a diaminodithio ligand into a ribonucleoside (Chen and Janda, 1992) and a progestin receptor antagonist RU 486 (DiZio et al., 1992). While these studies are promising, radiohalogens are more likely prospects for initial studies. The subject of radiohalogentation has received several good reviews which we recommend (Kabalka and Varma, 1989; Eckelman, 1992b).

The *in vitro* and *in vivo* stability of the radiohalogenated product should be determined prior to *in vivo* distribution studies. For example, a gamma-emitting estradiol receptor radiotracer can be prepared by the direct iodination of 17α-ethynylestradiol. The bond energy of iodine and an acetylenic carbon is about the same as that of a primary alkane, so the stability of the compound was tested *in vitro* by incubation with rat plasma. The iodine was rapidly lost (Mazaitis et al., 1980). By contrast, the iodine-olefinic bond is considerably stronger, so a 17α-[^{123}I]iodo-vinyl derivative of estradiol is stabile in plasma *in vitro* and *in vivo* upon injection in rats (Natkatsuka et al., 1984). All iodine-alkane bonds are not as labile. For example 19-[^{131}I]iodocholesterol (Counsell et al., 1973) and 16α-[^{125}I]iodoestradiol (Hochberg, 1979) are stable radiotracers because substitution and elimination reactions are sterically inhibited by the steroid structure.

Just as structure-activity relationships are the foundation of classic drug design, structure-distribution studies are the backbone of radiopharmaceutical development. An extensive review of this topic has been presented previously (Eckelman, 1992a). For ligands which must cross membranes, perhaps the most critical physical property for a high affinity site-directed radiotracer is the lipophilicity. Too frequently, the substituents that lead to high affinity are also likely to increase the lipophilicity. Ligands which are hydrophilic do not cross membranes well, while those with lipophilicities which are too high will bind so tightly to serum proteins that the free fraction available to cross the membrane is severely reduced. Very high lipophilicities also usually tend to provide compounds with excessive non-receptor binding in target tissues. Generally, compounds with partition coefficients (log P) < 0 or > 4 are poor candidates as *in vivo* imaging agents.

An interesting example of structure-distribution relationships is provided by a study in which unlabeled analogues of the muscarinic receptor antagonist, QNB, were used to block the binding of [^{3}H]QNB (Eckelman et al., 1984). When the apparent *in vivo* affinities of these analogues for m-AChR in the brain was plotted vs. the *in vitro* affinities in a logit plot, the correlation was reasonable. However, several compounds did not correlate well. One such compound was 3-quinuclidinyl 8-xanthene carboxylate (QNX), The affinity of QNX for the M_1-receptor is the same as that of QNB (Gibson et al., 1983), but QNX apparently does not cross the blood-brain barrier because its lipophilicity is too low (log $P < 0$). This study exhibits characteristics similar to that reported by Katzenellenbogen et al. (1981) on a series of estrogens in which the binding of ligand is a function of affinity *and* lipophilicity.

Animal distribution studies using the radiolabeled ligand are the most telling experiments. Besides the requirement that the radioactivity be present in the target organ with a target-to-nontarget ratio of at least 10, receptor binding radiotracers must also show receptor specific binding. The most important studies to conduct are *in vivo* blockade studies similar to those conducted with [^{3}H]QNB mentioned above. To demonstrate that a radioligand interacts specifically with a specific site *in vitro* requires blockade by the pharmacologically relevant ligands. The same is necessary for validation of *in vivo* binding. Also, if the radioligand exhibits stereochemistry, correct stereoselectivity should be demonstrated. Thus, [^{123}I]Iododexetimide shows high affinity and good specific binding to muscarinic receptors *in vivo* while the relatively inactive stereoisomer, [^{123}I]Iodolevetimide, provides an indication of the non-specific binding (Wilson et al., 1989; Muller-Gartner et al., 1992).

Once a newly synthesized radioligand has been subjected to the validation required to demonstrate specificity and, when necessary, sensitivity to binding site concentrations, much of the data needed to write a radiopharmaceutical IND are in hand so studies can be readily extended to humans. There is a clear advantage if the radioligand is an isotopically substituted drug candidate which has already received extensive toxicological studies. However, if the site-directed radiotracer is an analogue, only subacute toxicology in two species is needed for such an IND, and if the studies will significantly advance subsequent processing of an NDA, such studies are not costly.

CONCLUSION

SPECT and PET are rapidly expanding their sphere of influence from diagnosis, which has been the primary focus to date, to cerebral mapping and drug testing. The power of this noninvasive technique lies in its unique ability to "map" biochemical changes with high sensitivity. Many radiopharmaceuticals are currently available, some of which have been desribed in this and other chapters. However, the penchant for pharmaceutical companies to search for new targets for drugs, e.g., the angiotensin II receptor antagonists for treatment of hypertension, continually create areas in which new radiotracers may need to be developed. The sigma receptor provides an interesting example of a new site which may provide novel neuroleptic drugs, but for which there is no convenient physiological measure to determine receptor saturation, e.g., the blood pressure doesn't change. Thus drug-receptor saturation can only be rigorously demonstrated using imaging with an agent such as [^{11}C]L-687,384. Another potential use of [^{11}C]L-687,384 is to determine receptor concentration in various patient populations. If the *in vivo* localization of the radiotracer is sensitive to receptor concentration, such studies may indicate changes in receptor concentrations associated with pathologies other than schizophrenia. The uses of these radiotracers is therefore limited only by one's creativity.

REFERENCES

Aronstam RS (1982): Receptor Binding Studies: General considerations. In: *Receptor Binding Radiotracers Vol 1*, Eckelman WC, ed. Boca Raton: CRC Press, Inc.

Atkins FB and Goodenough DJ (1982): Simulated uptake ratio requirements for spherical lesions imaged with a conventional scintillation camera. In: *Receptor Binding Radiotracers Vol 2*, Eckelman WC, ed. Boca Raton: CRC Press, Inc.

Bjornerheim R, Golf S and Hansson V (1991): Specific non-beta-adrenergic binding sites for [^{125}I]Iodocyanopindolol in myocardial membrane preparations: a comparative study between human, rat, and porcine hearts. *Cardiovasc Res* 25: 764-773.

Burns HD, Eng W-s, Dannals RF, Wong DF, Guilarte TR, Wilson AA, Ravert HT, Gibson RE, Britcher SF, Frost JJ, Wagner HN Jr and Solomon HF (1989): Design, synthesis and preliminary evaluation of (+)-[C11]-8-Methoxy-MK-801: A potential radiotracer for NMDA receptor imaging via PET. *J Nucl Med* 30: 930.

Contreras PC, Quirion R, Gehlert DR, Contreras ML and O'Donohue TL (1987): Autoradiographic distribution of non-dopaminergic binding sites labeled by [^{3}H]haloperidol in rat brain. *Neurosci Lett* 75: 133-140.

Counsell RE, Ranade VV, Kulkarni PG and Afiatpour PJ (1973): Potential organ or tumor-imaging agents. 12. Esters of 19-radioiodinated cholesterol. *J Nucl Med* 14: 777-780.

Delforge J, Syrota A, Lançon J-P, Nakajima K, Loc'h C, Janie M, Vallois J-M, Cayla J and Crouzel C (1991): Cardiac beta-adrenergic receptor density measured in vivo using PET, CGP 12177 and a new graphical method. *J Nucl Med* 32: 739-748.

DiZio JP, Anderson CJ, Davison A, Erhardt GJ, Carlson KE, Welch MJ and Katzenellenbogen JA (1992): Technetium- and rhenium-labeled progestins: synthesis, receptor binding and *in vivo* distribution of an 11β-subtituted progestin labeled with technetium-99 and rhenium-186. *J Nucl Med* 33: 558-569.

Farrow JT and O'Brien RD (1973): Binding of Atropine and Muscarone to rat brain fractions and its relation to the acetylcholine receptor. *Mol Pharmacol* 9: 33-40.

Eckelman WC, Reba RC, Gibson RE, Rzeszotarski WJ, Vieras F, Mazaitis JK and Francis B (1979): Receptor binding radiotracers: a class of potential radiopharmaceuticals. *J Nucl Med* 20: 350-357.

Eckelman WC, Gibson RE, Vieras F, Rzeszotarski WJ, Francis B and Reba RC (1980): *In vivo* receptor binding of iodinated beta adrenoceptor blockers. *J Nucl Med* 21: 436-442.

Eckelman WC (1982): Receptor-Specific radiopharmaceuticals. In: *Emission Computed Tomography*, Ell PJ and Holman BL, eds. Oxford: Oxford University Press.

Eckelman WC, Grissom M, Conklin J, Rzeszotarski WJ, Gibson RE, Francis B, Jagoda E, Eng R and Reba RC (1984): *In vivo* competition studies with analogues of quinuclidinyl benzilate. *J Pharm Sci* 73: 529-533.

Eckelman WC (1992a): The testing of putative receptor binding radiotracers *in vivo*. In: *Radiopharmaceuticals and brain pathology with PET and SPECT*, Diksic M and Reba RC eds. Boca Raton: CRC Press.

Eckelman WC (1992b): The development of single-photon emitting, receptor-binding radiotracers. In: *The Chemistry and Pharmacology of Radiopharmaceuticals*, Nunn A, ed. New York: Marcel Dekker Inc.

Francis B, Eckelman WC, Grissom MP, Gibson RE and Reba RC (1982): The use of tritium labeled compounds to develop gamma-emitting receptor-binding radiotracers. *Int J Nucl Med Biol* 9: 173-179.

Frey KA, Koeppe RA, Mulholland GK and Kuhl DE (1990): Quantitation of regional cerebral muscarinic receptors in human brain with the use of [C-11]tropanyl benzilate and positron emission tomography. J *Nucl Med* 31: 885.

Frey KA, Koeppe RA, Mulholland GK, Jewett D, Hichwa R, Ehrenkaufer RLE, Carey JE, Wieland DM, Kuhl DE and Agranoff BW (1992): In vivo muscarinic cholinergic receptor imaging in human brain with [^{11}C]scopolamine and positron emission tomography. *J Cereb Blood Flow Metab* 12: 147-154.

Frost JJ and Wagner HN Jr (1984): Kinetics of binding to opiate receptors in vivo predicted from in vitro parameters. *Brain Res.* 305: 1-11.

Frost JJ, Mayberg HS, Sadzot B, Dannals RF, Lever JR, Ravert HT, Wilson AA, Wagner HN Jr and Links JM (1990): Comparison of [C-11]diprenorphine and [C-11]carfentanil binding to opiate receptors in humans by positron emission tomography. *J Cereb Blood Flow Metab* 10: 484-492.

Gibson RE (1982): Quantitative changes in receptor concentration as a function of disease. In: *Receptor Binding Radiotracers Vol. II*, Eckelman WC, ed. Boca Raton: CRC Press, Inc.

Gibson RE, Rzeszotarski WJ, Eckelman WC, Jagoda EM, Weckstein DJ and Reba R (1983): Difference in the affinities of muscarinic acetylcholine receptor antagonists for brain and heart receptors. *Biochem Pharmacol* 32: 1851-1856.

Gibson RE, Weckstein DJ, Jagoda EM, Rzeszotarski WJ, Reba RC and Eckelman WC (1984): The characteristics of I-125 QNB and H-3 QNB *in vivo* and *in vitro*. *J Nucl Med* 25: 214-222.

Gitler MS, Zeeberg BR and Reba RC (1992): *In vivo* muscarinic receptor binding: Pharmacokinetic sensitivity of [I-125]IQNB localization within different regions of the rat brain. *J Nucl Med* 33: 883.

Goodenough DJ and Atkins FB (1988): Theoretical limitation of tumor imaging. In: *Radiolabeled Monoclonal Antibodies for Imaging and Therapy*, Srivastiva SC, ed. New York: Plenum Publishing.

Hertal C, Muller P, Portenier M and Stachelin M (1983): Determination of the desensitization of β-adrenergic receptors by [^{3}H]CGP 12177.

Biochem J 216: 669-674.

Hochberg RB (1979): Iodine-125-labelled estradiol: a gamma-emitting analogue of estradiol that binds to the estrogen receptor. *Science* 205: 1138-1140.

Jensen EV and Jacobson HI (1962): Basic guides to the mechanism of estrogen action. *Recent Prog Horm Res* 18: 387-408.

Jiang V, Gibson RE, Rzeszotarski WJ, Eckelman WC and Reba RC (1978): Radio-iodinated derivatives of beta adrenoceptor blockers for myocardial imaging. *J Nucl Med* 19: 918-924.

Kabalka GW and Varma RS (1989): The synthesis of radiolabeled compounds via organometallic intermediates. *Tetrahedron* 21: 6601-6621.

Katzenellenbogen JA, Hsiung HM, Carlson KE, McGuire WL, Kraay RJ and Katzenellenbogen BS (1975): Characterization of the binding and estrogenic activity of iodinated hexestrol derivatives in vitro and in vivo. *Biochem* 14: 1742-1750.

Katzenellenbogen JA (1981): The development of gamma-emitting hormone analogs as imaging agents for receptor-positive tumors. In: *The Prostatic Cell: Stucture and Function*, Murphy GP, Sandberg AA and Karr JP, eds. New York: Alan R. Liss, Inc.

Katzenellenbogen JA, Heiman DF, Carlson KE and Lloyd JE (1981): *In vitro* and *in vivo* steroid receptor assays in the design of estrogen radiopharmaceuticals. In: *Receptor Binding Radiotracers Vol I*, Eckelman WC, ed. Boca Raton: CRC Press.

Kennedy C (1991): At three score and ten. *J Cereb Blood Flow Metab* 11: 885-889.

Kiesewetter DO, Kilbourn MR, Landvatter SW, Heiman DF, Katzenellenbogen JA and Welch MJ (1984): Preparation of four fluorine-18-labeled estrogens and their selective uptakes in target tissues of immature rats. *J Nucl Med* 25: 1212-1221.

Kiesewetter DO, Finn RP, Rice KC and Monn JA (1990): Synthesis of [^{11}C]labeled (±)-5-methyl-10,11-dihydro-5H-dibenzo(8,d)cyclo-heptene-5,10-imine (± [^{11}C]MK-801). *Int J Radiat Appl Instr (Part A)* 41: 139-142.

Kilbourne M (1990): Radiotracers for PET studies of neurotransmitter binding sites: design considerations in *in vivo* imaging of neurotransmitter functions. In: *Brain, Heart and Tumors,* Kuhl D, ed. Washington, DC: American College of Nuclear Physicians.

Komai T, Eckelman WC, Johnsonbaugh RE, Mazaitis A, Kubota H and Reba RC (1977): Derivatives for the external localization of estrogen dependent malignancy. *J Nucl Med* 18: 360-366.

Landau WH, Freygang WH, Rowland LP, Sokoloff L and Kety SS (1955): The local circulation in the living brain: values in the unanesthetized and anesthetized cat. *Trans Am Neurol Assoc* 80: 125-129.

Lever SZ, Burns HD, Kervitsky TM, Goldfarb HW, Woo DV, Wong DF, Epps LA, Kramer AV and Wagner HN Jr (1985): Design, preparation and biodistribution of a technetium-99m triaminedithiol complex to assess regional cerebral blood flow. *J Nucl Med* 26: 1287-1294.

Logan J, Dewey SL, Shiue C-Y, Fowler JS, Wolf AP, Christman DR, Bendriem B and Volkow N (1989): Kinetic analysis of [^{18}F]haloperidol binding in baboon and human brain. *J Nucl Med* 30: 898.

Mazatis JK, Gibson RE, Komai T, Eckelman WC, Francis B and Reba RC (1980): Radioiodinated estrogen derivatives. *J Nucl Med* 21: 142-146.

Mintun MA, Welch MJ, Siegel BA, Mathias CJ, Brodack JW, McGuire AH and Katzenellenbogen JA (1988): Breast Cancer: PET imaging of estrogen receptors. *Radiol* 169: 45-48.

Mulholland GK, Jewett DM, Otto CA, Kilbourn MR, Sherman PS and Kuhl DE (1988): Synthesis and regional brain distribution of [C-11]N-mmethyl-4-piperidyl benzilate ([C-11]NMPB) in the rat. *J Nucl Med* 29: 373-379.

Muller-Gartner HW, Wilson AA, Dannals RF, Wagner HN Jr and Frost JJ (1992): Imaging muscarinic cholinergic receptors in human brain *in vivo* with SPECT, [^{123}I] 4-iododexetimide and [^{123}I]levetimide. *J Cereb Blood Flow Metab* 12: 562-570.

Nakatsuka I, Ferreira NL, Eckelman WC, Francis BE, Rzeszotarski WJ, Gibson RE, Jagoda EM and Reba RC (1984): Synthesis and evaluation of (17α,20E)-21-[^{125}I]Iodo-19-norpregna-1,3,5(10),20-tetraene-3,17-diol and (17α,20E)-21[^{125}-I]Iodo-11ß-methoxy-19-norpregna-1,3,5(10),20-tetraene-3,17-diol (17α-(iodovinyl)estradiol derivatives) as high specific activity potential radiopharmaceuticals. *J Med Chem* 27: 1287-1291.

Nunn AD, Feld TA and Treher EN (1987): Boronic acid adducts of technetium-99m dioxime complexes. US Patent No. 4705849, Nov. 10.

Parascandola J (1986): The development of receptor theory in discoveries in pharmacology. In: *Pharmacological Methods, Receptors & Chemotherapy*, Parnham MJ and Bruinvels J, eds. Elsevier Science Publishers B.V.

Paton WDM and Rang HP (1966): The uptake of atropine and related drugs by intestinal smooth muscle of the guinea-pig ileum in relation to acetylcholine receptors. *Proc Roy Soc Ser B Biol Sci* 163: 2-44.

Pomper MG, VanBrocklin H, Thieme AM, Thomas RD, Kiesewetter

DO, Carlson KE, Mathias CJ, Welch MJ and Katzenellenbogen JA (1990): 11β-Methoxy-, 11β-ethyl- and 17α-ethynyl-subtituted 16α-fluoroestradiols: Receptor-based imaging agents with enhanced uptake efficiency and selectivity. *J Med Chem* 33: 3143-3155.

Prenant C, Barre L and Crouzel C (1989): Syntehsis of [^{11}C]-3-quinuclidinyl benzilate (QNB). *J lablled Compd Radiopharm* 27: 1257-1265.

Ransom RW and Stec NL (1988): Cooperative modulation of [H-3]MK-801 binding to the N-methyl-D-aspartate receptor-ion channel complex by l-glutamate, glycine and polyamines. *J Neurochem* 51: 830-836.

Ransom RW, Eng W-s, Burns HD, Gibson RE and Solomon HF (1990): (+)-3-[^{123}I]Iodo-MK-801: Synthesis and characterization of binding to the N-methyl-D-aspartate receptor complex. *Life Sci* 46: 1103-1110.

Reivich M, Sano N and Sokoloff L (1971): Development of an autoradiographic method for the determination of regional glucose consumption. In: *Brain and Blood Flow*, Ross-Russell RW, ed. London: Pitman.

Reivich M, Kuhl D, Wolf A, Greenberg J, Phelps M, Ido T, Cassella V, Fowler J, Hoffman E, Alavi A, Som P and Sokoloff L (1979): The [^{18}F]fluoro-deoxyglucose method for the measurement of local cerebral glucose utilization in man. *Circ Res* 44: 127-137.

Rzeszotarski WJ, Gibson RE, Eckelman WC, Simms DA, Jagoda EM, Ferreira NL and Reba RC (1982): Analogs of 3-Quinuclidinyl Benzilate. *J Med Chem* 25: 1103-1106.

Scatchard G (1949): The attractions of proteins for small molecules and ions. *Ann NY Acad Sci* 51: 660-672 .

Sisson JC, Wieland DM, Koeppe RA, Normolle D, Frey KA, Bolgos G, Johnson J Van Dort ME and Gildersleeve DL (1991): Scintigraphic portrayal of beta receptors in the heart. *J Nucl Med* 32: 1399-1407.

Stöcklin G (1992): Tracers for metabolic imaging of brain and heart. Radiochemistry and radiopharmacology. *Eur J Nucl Med* 19: 527-551.

Troutner DE, Simon J, Ketring AR, Volkert WA and Holmes RA (1980): Complexing 99mTc with cyclam. *J Nucl Med* 21: 443-448.

Vera DR, Krohn KA, Stadalnik RC and Sheibe PO (1984): [^{99m}Tc]Galactosyl-neoglycoalbumin: *in vitro* characterization of receptor-mediated binding. *J Nucl Med* 25: 779-787.

Wagner HN Jr. (1982): Introduction: The role of receptors in disease. In: *Receptor Binding Radiotracers Vol II*, Eckelman WC, ed. Boca Raton: CRC Press, Inc.

Wagner HN Jr, Burns HD, Dannals RF, Wong DF, Langstrom B,

Duelfer T, Frost JJ, Ravert HT, Links JM, Rosenbloom SB, Lukas SE, Kramer AV and Kuhar MJ (1983): Imaging dopamine receptors in human brain by positron tomography. *Science* 221: 1264-1266.

Wilson AA, Dannals RF, Ravert HT, Frost JJ and Wagner HN Jr (1989): Synthesis and biological evaluation of [^{125}I]- and [^{123}I]-4-Iododexetimide, a potent muscarinic cholinergic receptor antagonist. *J Med Chem* 32 : 1057-1062.

Wong DF, Burns HD, Solomon HF, Dannals RF, Villemagne V, Goldman S, Young T, Eng W, Ravert HT, Wilson AA, Gibson RE, Iversen L, Guilarte T and Wagne HN Jr (1989): Imaging of NMDA Receptor Sites with 8-[C-11]Methoxy-MK-801 *in vivo* by PET. *J Nucl Med* 30: 741.

Wong DF, Gibson R, Burns HD, Dannals RF, Shaya E, London ED, Brenner N, Wilson AA, Ravert HT, Chen C, Minkin E and Wagner HN Jr (1991): *In vivo* imaging of signa receptors in primate brain. *J Nucl Med* 32: 1071.

Yamamura HI and Snyder SH (1974): Muscarinic cholinergic binding in rat brain. Proc Nat Acad Sci 71: 1725-1729.

Yokoyama A, Terauchi Y, Horiuchi K, Tanaka H, Odori T, Morita R, Mori T and Torijuka K (1976): Technetium-99m-kethoxal-bis(thiosemicarbazone), an uncharged complex with a tetravalent ^{99m}Tc state and its excretion into bile. *J Nucl Med* 17: 816-819.

APPLICATIONS OF AUTORADIOGRAPHY TO DRUG DISCOVERY

Raymond E. Gibson and Holly T. Beauchamp
Radiopharmacology
Merck Research Laboratories

Susan Iversen
Neuroscience Research Center
Merck Research Laboratories

Barry Everitt
Department of Anatomy
Cambridge University

James McCulloch and Christopher Wallace
Wellcome Surgical Institute
University of Glasgow

INTRODUCTION

Compared to the Nuclear Medicine methodologies described in other chapters of this book, autoradiography is an invasive technique; but, many of the features which define the usefulness of Nuclear Medicine in drug discovery are applicable to autoradiographic techniques. Additionally, autoradiographic studies, both *in vitro* and *in vivo*, have been used to characterize the properties of radiolabeled drugs used in nuclear medicine (Huang and Phelps, 1986; Kung et al., 1988a; Kung et al., 1988b; Gibson et al., 1992). Thus, results from autoradiographic

Nuclear Imaging in Drug Discovery, Development, and Approval
H.D. Burns, R. Gibson, R. Dannals, Peter Siegl, Editors

studies can be used to adequately define the characteristics of a radiotracer system before embarking upon the more costly, and technically challenging non-invasive imaging study. For example, the imaging of glucose metabolism in man using 2-[^{18}F]fluoro-2-deoxy-D-glucose has relied heavily on *ex vivo* autoradiographic studies conducted in animals (Kennedy et al., 1975; McCulloch, 1982; Huang and Phelps, 1986). One of the major uses of autoradiography in the pharmaceutical industry is to determine drug disposition (usually Carbon-14 labeled) by sectioning rats (whole body studies) at various times after injection. These studies are required for drug approval in Japan, but are not currently required by the Food and Drug Administration in the United States. Although gross distribution of radioactivity and drug metabolism are important applications of autoradiography, our discussion will focus on the uses of autoradiography to provide mechanistic or morphologic information that may be otherwise difficult to obtain.

We have divided this discussion into application of autoradiography using sections of isolated organs or tissues vs. large sledge or whole body autoradiography (Som and Oster, Chapter 7). This division is based in part on differences in resolution. Inherently, there is no reason why whole body studies should exhibit lower resolution than that obtained with sections cut on a small cryomicrotome; but in practice, the thickness of sections from whole body studies is greater. This increased thickness most greatly affects the quality of the histology used for identifying structures of interest. However, resolution of the autoradiographs will also be poorer in the thicker sections used in whole body studies when the radioisotope used penetrates tissue well (e.g., Iodine-125, Technetium-99m, and positron-emitting isotopes; Lear, 1986). Finally, we do not endeavor to provide a manual of the techniques used in autoradiography since these techniques have been adequately described elsewhere (Kuhar, 1986).

We have applied autoradiography to studies on two receptor systems. The studies provide not only basic knowledge concerning the presence of receptor subtypes and distribution, but also suggest that radiopharmaceuticals should be developed to study these receptors in humans.

NMDA RECEPTOR

The N-methyl-D-aspartate (NMDA) receptor is a subtype of the glutamate receptor. During ischemia, this excitatory amino acid receptor may cause cell death by allowing increased Ca^{+2} influx (Faden et al., 1989; Rothman and Olney, 1986; Meldrum, 1985). Non-competitive antagonists, such as MK-801, have been shown to be neuroprotective in several animal models, apparently by binding to a site in the activated ion channel (Foster et al., 1987; Ozyurt et al., 1988). Thus, the kinetics of binding of [^{3}H]MK-801 are increased during activation by glutamate. For a drug such as MK-801 to be an effective intervention to reduce the sequela of stroke, the optimum time to administer the drug is during the period of ion-channel activation which may be problematic since many stroke patients do not seek medical assistance for many hours after the stroke. One means of determining the time course of activation of the NMDA ion channel is by labeling the activated sites *in vivo* with an appropriately labeled analogue of MK-801. Although several positron-emitting analogues of MK-801 have been prepared (Wong et al.,1989; Blin. et al., 1991), the properties of these ligands were not apparently suitable for *in vivo* receptor/ion channel imaging.

We have prepared a radioiodinated analogue of MK-801, (±)-3-[^{123}I or ^{125}I]Iodo-5-methyl-10,11-dihydro-5H-dibenzo[a,d]cycloheptene-5,10-imine and the (+)-stereoisomer exhibits high affinity for the NMDA-receptor channel *in vitro* (Burns et al., 1989; Ransom et al., 1990). 3-Iodo-MK-801 exhibits binding properties preferable to those of MK-801 (or the positron-emitting analogues), principally, lower sensitivity to the ionic milieu of the binding site (Ransom et al., 1990). We were thus able to demonstrate receptor specific binding *in vivo* in the brains of rats (Gibson et al., 1992). However, in normal animals, the amount of receptor-specific binding represents a small fraction (40%) of the total radioactivity found in the brain (Fig. 1). Non-specific binding is very high and we had to wait four hours for enough of the non-specifically bound radioligand to clear which then permitted visualization of the receptor-specific component. However, this represents the interaction of the radioligand with a receptor which is not excessively activated by, for example, ischemic insult. When 3-[^{125}I]Iodo-MK-801 was injected into rats in which the middle cerebral artery was ligated, the radioligand binding was significantly increased with that increase reflecting the receptor-specific interaction (Fig. 2). We were thus able to determine the time course of activation of the NMDA receptor/ion channel complex after ischemia was induced (Fig. 3). Although the highest binding is evident at the earliest time point examined (15 min. post-surgery), enhanced binding of the radioligand is observed at four hours (the last time examined in this study). If the kinetics of activation in man are similar, these results suggest that a reasonable window exists during which drug intervention would benefit stroke patients.

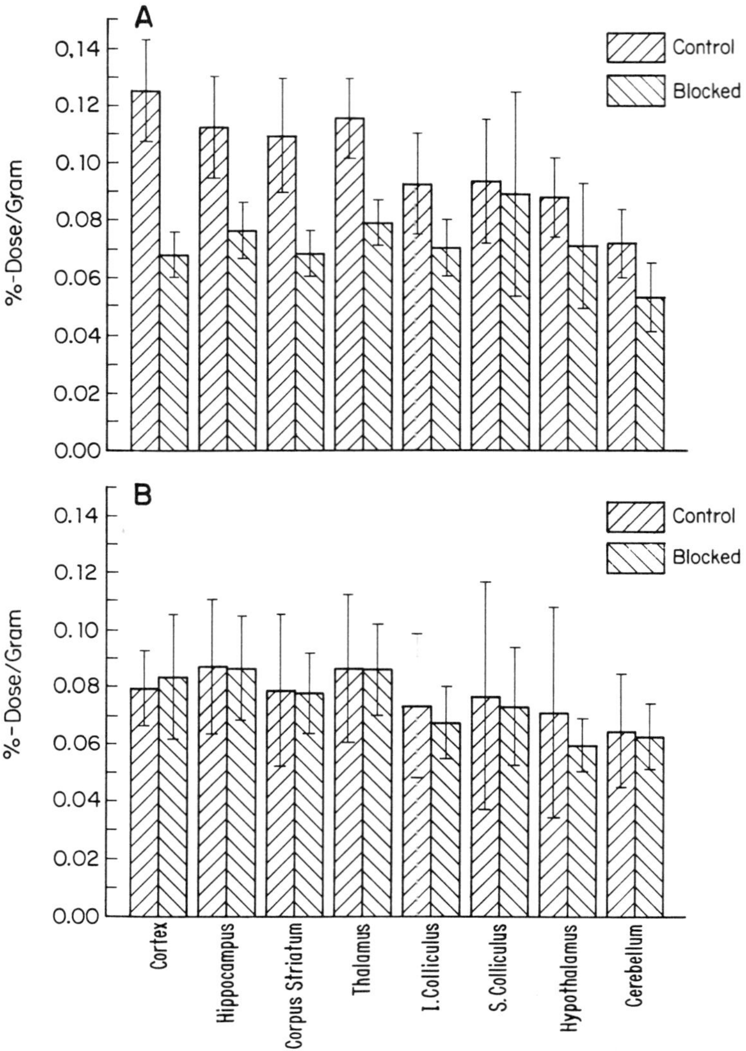

FIGURE 1. Specificity of *in vivo* distribution of (+)-3-Iodo-MK-801 (A) and (-)-3-Iodo-MK-801 (B) in CNS. Error bars represent the 95% confidence interval. To determine the amount of radioligand specifically bound to receptor, 4 mg/Kg of MK-801 was coinjected with the radioligand and the animals killed at 4 hrs. Data are expressed as %-dose/gram of tissue (wet weight). Reproduced with permission from Gibson et al. (1992).

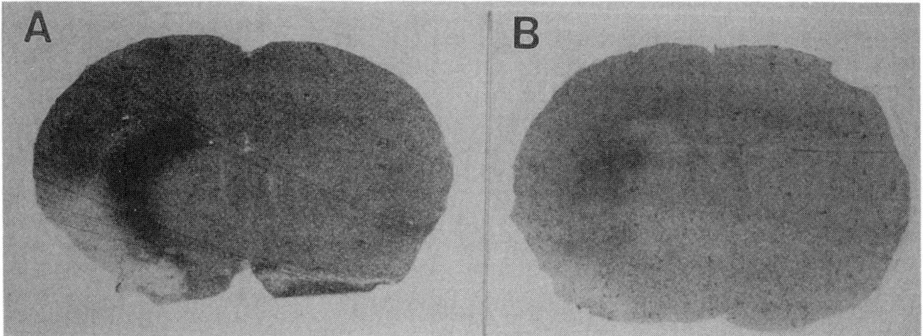

FIGURE 2. Specific localization of (+)-3[^{125}I]Iodo-MK-801 in rat brain in which the middle cerebral artery was ligated. (A) Radioligand injected only; (B) coinjection with 1 mg/kg MK-801. Radioligand was injected 15 min after surgery, the brain removed after 60 min, snap frozen and sectioned for autoradiography.

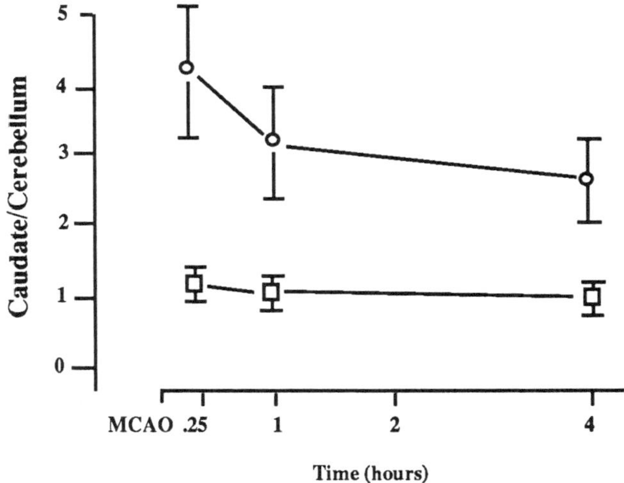

FIGURE 3. Time course of NMDA receptor/channel activation in rat brain, as measured by (+)-3-[^{125}I]Iodo-MK-801, following middle cerebral artery ligation. Ischemic caudate (○); contralateral caudate (□).

ANGIOTENSIN II RECEPTOR

Control of hypertension by inhibiting the renin-angiotensin system has been extremely successful with the advent of angiotensin-converting enzyme (ACE) inhibitors, e.g., captopril and enalapril. Recent efforts on this system have focused on the angiotensin II (AngII) receptor since ACE inhibitors have infrequent, but bothersome, side-effects (Ferner et al., 1987; Slatter et al., 1988). In principle, blockade of the AngII receptor with a specific antagonist should provide an anti-hypertensive agent as effective as ACE inhibitors. Additionally, the AngII receptor has been shown to exhibit subtypes which have been classified AT_1 and AT_2 on the basis of differences in the affinity of non-peptide antagonists (Chiu et al., 1989; Chang and Lotti, 1990). Since ACE inhibitors should be functionally non-selective, the presence of receptor subtypes provides an opportunity to design receptor antagonists which may exhibit fewer side-effects. This hypothesis has been enthusiastically embraced by the pharmaceutical industry and has already yielded many AT_1 and AT_2 selective ligands.

For example, DuP 753 (losartan) exhibits high affinity and selectivity for the AT_1 receptor while PD-121981 (≡WL-19 in Chang and Lotti, 1990) exhibits high affinity and selectivity for the AT_2 receptor (Chang and Lotti, 1990). To date, all of the pharmacological responses of interest are influenced by the AT_1 receptor, and no convincing evidence has been presented for a function of the AT_2 receptor. Nonetheless, the existence of the AT_2 receptor is seen as both a potential target for future drugs and a potential problem. Selective AT_1-blockade will increase circulating AngII in the plasma without blocking the AT_2 receptor. In order to determine the potential for AT_2 selective drugs or flag potential problems, we determined the localization of AT_1 and AT_2 receptors in several target tissues autoradiographically. We mapped the distribution of receptors using a non-selective radioiodinated peptide antagonist, $[^{125}I]$[Sar,Ile]angiotensin II ($[^{125}I]$[Sar,Ile]AngII), and determined the receptor subtype by competition with the selective receptor ligands, DuP 753 and PD-121981.

Kidney

One significant target for AngII receptor antagonists is the kidney in which AngII exerts a variety of actions, including modulation of renal blood flow (Navar and Langford, 1974), glomerular filtration rate (Navar and Rosivall, 1984), sodium and water resorption (Barraclough et al., 1967; Levens et al., 1980), and inhibition of renin release (Davis and Freeman, 1976). The distribution of AngII receptors in rat and monkey kidney has been reported in autoradiographic studies using $[^{125}I][Sar^1,Ile^8]$AngII (Mendelsohn et al., 1987): high receptor concentrations are present over the vasa recta bundles and the highest concentration of receptors in the renal cortex are associated with glomerular mesangial cells. In our studies (Gibson et al., 1991), we found that the AT_1 represents >90% of the total receptor population in rat kidney and the presence of the AT_2 is not readily apparent. By contrast, we were able to identify significant AT_2 receptor populations in Rhesus monkey kidney, particularly in the juxtaglomerular (JG) apparatus (i.e., JG cells and the afferent and efferent arterioles) associated with the glomerular tuft (Gibson et al., 1991). We have also observed radioligand binding associated with the JG apparatus in dog renal cortex which is blocked with the AT_2 selective ligand, PD-121981. By blocking all of the AT_1 receptor in rat kidney with DuP 753 at 10^{-5} M and exposing the radiolabeled sections to film four times longer than in studies of the AT_1 receptor, we were able to visualize the AT_2 receptor in rat kidney cortex which also appears to be associated with the JG apparatus. Thus, the presence of the AT_2 receptor associated with the JG apparatus appears to be a general observation, but the relative concentration of these receptors varies considerably among the species examined.

The association of the AT_2 receptors with the JG apparatus prompted us to determined the effect of the AT_2 receptor antagonist on renin release by measuring increases in plasma renin activity (PRA). PRA was increased equally well by DuP 753 and the ACE inhibitor enalapril. When maximally effective doses of DuP 753 and enalapril were added together, there was no additional increase in PRA. Lastly, PD-121981 did not have an affect on PRA. Thus, despite the presence of the AT_2 receptor associated with the JG apparatus, the AT_2 receptor antagonist exerted no measurable influence. This does not exclude a function for this subtype of receptor in the kidney, but we have not yet identified one.

Brain

The second site of interest in which we determined the distribution of subtypes is in the rat brain. AngII receptors are known to be in high concentration associated with periventricular structures (e.g., subfornical organ and OVLT) in which AngII is known to exert central nervous system (CNS) control over blood pressure and electrolyte balance (Fitzsimmons, 1980; Long et al., 1983). We found the AT_1 and AT_2 receptors to be associated with specific structures (Fig. 4). We were thus able to determine the receptor subtype present in 30 different structures

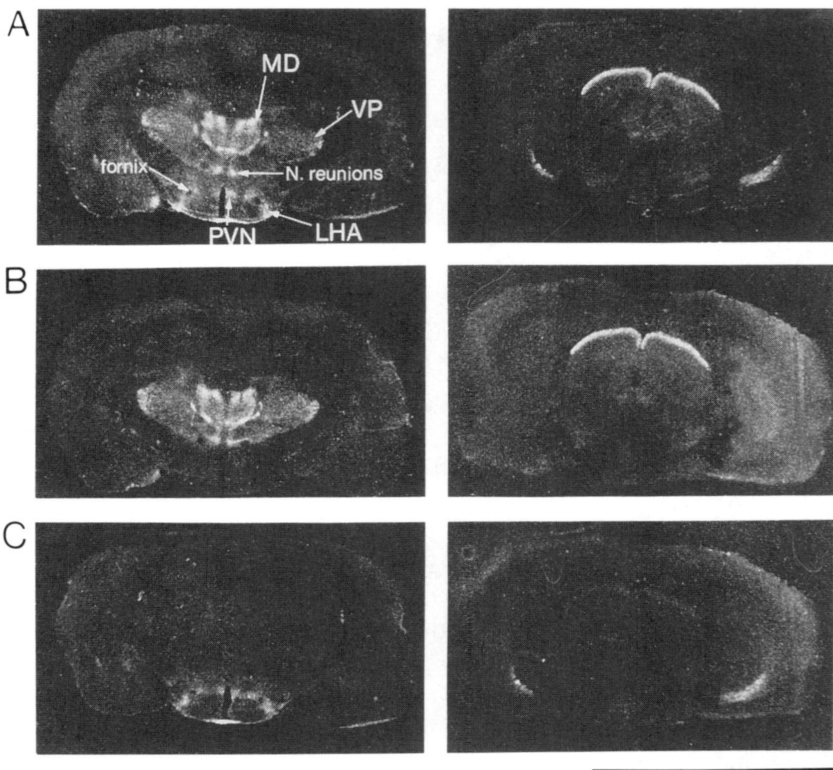

FIGURE 4. Distribution of [^{125}I][Sar^1,Ile^3]AngII binding subtypes in coronal sections of rat brain. Left panels: section at level of mid-thalamus. Right panels: more posterior sections at level of superior colliculus. A. Total [^{125}I][Sar^1,Ile^3]AngII binding. B. Sections treated with the AT_1 specific antagonist DuP 753. C. Sections treated with the AT_2 specific antagonist PD-121981. Abbreviations: MD - mediodorsal thalamic nucleus; VP - ventral posterior thalamic nucleus; LHA - lateral hypothalamic area; PVN - paraventricular hypothalamic nucleus. Calibration bar = 5 mm.

in the rat brain (Table 1). This study does not preclude the presence of mixed populations of AT_1 and AT_2 receptors in these regions, but one subtype dominates. These results are in general agreement with recently reported results (Millan et al., 1991).

Functional studies suggest that a range of AngII rich sites in brain (e.g., medial preoptic area and median preoptic nucleus) and structures outside the blood-brain-barrier accessible to circulating AngII (e.g.,

Structure	Relative Conc.	Subtype
Anterior olfactory nucleus	+++	1
N. lateral olfactory tract	+++	1
Pyriform cortex	++++	1
Endopyriform nucleus	++	1
Corticomedial amygdala	+++	1
Lateral septum	+	2
Ansa lenticularis/posterior palladium	++	2
OVLT	+++	1
Subfornical organ	++++	1
Median preoptic nucleus	+++	1
Suprachiasmatic nucleus	++++	1
Median eminence/arcuate n.	++++	1
Lateral hypothalamus/perifornical	+++	1
Paraventricular hypothalamic nucleus	+++	1
Periventricular hypothalamic nucleus	+	1
Ventrolateral/ventromedial thalamus	++	2
Medialis dorsalis nucleus thalamus	++	2
Ventroposterolateral thalamus	++	2
Gustatory nucleus	++	2
Lateral posterior thalamic nucleus	+	2
Centromed/centrolat thalamic nucleus	+++	2
Posterior thalamus	++	2
Nucleus reuniens	+++	2
Superior colliculus	++	2
Ventrolateral/dorsolateral geniculate n.	+	2
Medial geniculate nucleus	+++	2
Dentate gyrus	++	1
Sub-thalamic nucleus	+++	2
Presubiculum/subiculm	+++	1
Locus coeruleus	++	2

TABLE 1. [^{125}I][Sar^{1},Ile^{8}]AngII Binding Site Subtypes in Rat Brain. Concentrations correspond to ++++: >2 fmole/mg tissue; +++: >1 and <2 fmole/mg tissue; ++: >0.5 but <1 fmole/mg tissue; and +: <0.5 fmole/mg tissue. AT_1 inhibition by DuP 753; AT_2 inhibition by PD-121981.

subfornical organ and OVLT) are involved in drinking behavior and water balance. The binding at these sites was blocked by DuP 753 and is classified as AT_1. In the paraventricular and suprachiasmatic nuclei (SCN), where AngII coexists with vasopressin (VP) and where AngII infusion releases VP resulting in pressor responses, the receptors are also of the AT_1 type. AT_1 receptors appear, therefore, to be especially present in the hypothalamus, both in areas concerned with neuroendocrine control (preoptic, periventricular, and paraventricular nuclei) and in structures involved with circadian rhythms (e.g., SCN). However, AT_1 receptors are not restricted to these areas and were also demonstrated to be present in a number of olfactory structures (anterior olfactory nucleus, pyriform cortex, and corticomedial amygdala). The subiculum, a major efferent region of the hippocampal formation, was also shown to contain a moderate density of AT_1 receptors, but no AngII binding was seen in the cornu ammonis.

A striking observation is the widespread, often dense, distribution of AT_2 receptors in the thalamus, particularly in sensory nuclei (ventroposterolateral and ventroposteriomedial thalamic nuclei, gustatory, and lateral and medial geniculate nuclei). However, binding was not restricted to such nuclei. The mediodorsal nucleus is especially rich in AT_2 sites, throughout its three sub-divisions; the overlying habenula nuclei were, however, devoid of AngII receptors. The data indicate that the AT_2 receptor may be important in modulating thalamic projections to sensory and pre-frontal cortex. Several of the midline "non-specific" nuclei were also rich in AT_2 receptors including nucleus centromedialus and nucleus reunions (ventral part) consistent with a role of AT_2 receptors in modulating cortical activity.

In the brain stem three structures showed intense AT_2 binding; the sub-thalamic nucleus, a motor outflow structure which is part of the striatopallidal feedback circuitry (Gimenez-Amaya and Graybiel,1990; Graybiel 1990); the superficial layers of the superior colliculus, which, like the LGN and SCN receives input directly from the retina; and the locus coeruleus, which is the source of the major noradrenergic innervation of the cerebral cortex and is strongly implicated in arousal processes, particularly in a stress-related context (Seldon et al., 1990).

These findings agree with several recent autoradiographic studies on receptor subtypes in rat brain (Gehlert et al., 1990; Rowe et al., 1990; Wamsley et al., 1990). Our findings suggest that AngII receptors in the brain have functions which transcend those concerned with water and electrolyte balance. But it is impossible at this stage to propose a unifying hypothesis and indeed angiotensin probably serves a number of different sensory, motor, and integrative functions in the CNS.

CONCLUSION

The results from these autoradiographic studies have defined several directions for pursuit using Nuclear Medicine methodologies. Using *in vitro* autoradiography, we have shown that both AT_1 and AT_2 are present in distinct structures in the kidney and brain. The autoradiographic studies in kidney led to physiologic studies designed to specifically test for a pharmacological function of the AT_2 receptor in these tissues; we have not been able to identify a function as yet, but we also have not identified any difficulty that may arise from selective blockade of only the AT_1 receptor. The presence of the AT_2 receptor in the CNS suggests a role for that receptor other than that of fluid and electrolyte balance. To demonstrate what function this receptor subtype may have, the best model to test for significant beneficial CNS effects is in humans. A radiotracer of the AT_2 receptor could be used to determine changes in the receptor concentration as a function of disease or to provide a measure of receptor occupancy by a drug candidate. If the influence of the AT_2 receptor is subtle, such studies can better define the dose needed to saturate the AT_2 receptor for phase II IND studies.

The NMDA receptor antagonist, 3-[^{123}I]Iodo-MK-801, or an analogue with even better properties, can be used to demonstrate which clinical situations would best benefit from treatment with antagonists to this subtype of the glutamate receptor. If, in man, the sequelae from stroke or head injury involve significant and prolonged activation of the NMDA receptor, as demonstrated by increased radiotracer localization, therapy with NMDA receptor antagonists should prove efficacious. The autoradiographic studies reported for the NMDA receptor provide a good example of the type of studies needed to validate a receptor binding radiotracer before embarking upon *in vivo* imaging studies.

REFERENCES

Barraclough MA, Jones NF and Marsden CD (1967): Effect of angiotensin on renal function in the rat. *Am J Physiol* 212: 1153-1157.

Blin J, Denis A, Yamaguchi T, Crouzel C, MacKenzie ET and Baron JC (1991): PET studies of [^{18}F]methyl-MK-801, a potential NMDA receptor complex radioligand. *Neurosci Letters* 121: 183-186.

Burns HD, Eng W-S, Gibson RE, Ransom RW, Thorpe H, Fioravanti C, Magill CA and Solomon HF (1989): Synthesis of [I-123]- and [I-125]-(+)-3-Iodo-MK-801: potential radiotracer for SPECT and autoradiographic imaging of the NMDA receptor-ion channel

complex. *J Nuc Med* 30: 923.
Chang RSL and Lotti VJ (1990): Two distinct angiotensin II receptor binding sites in rat adrenal revealed by new selective nonpeptide ligands. *Molec Pharmacol* 29: 347-351.
Chiu AT, Herblin WF, McCall DE, Ardecky RJ, Carini DJ, Duncia JV Pease LJ, Wong PC, Wexler RR, Johnson AJ and Timmermans PBMWM (1989): Identification of angiotensin II receptor subtypes. *Biochem Biophys Res Commun* 163: 284-291.
Davis JO and Freeman RH (1976): Mechanisms regulating renin release. *Physiol Rev* 56: 1-56.
Faden AI, Demediuk P, Panter SS and Vink R (1989): The role of excitatory amino acids and NMDA receptors in traumatic brain injury. *Science* 244: 798-800.
Ferner RE, Simpson JM and Rawlins MD (1987): Effects of intradermal bradykinin after inhibition of angiotensin converting enzyme. *Br Med J* 294: 1119-1120.
Fitzsimmons JT (1980): Angiotensin stimulation of the central nervous system. *Rev Physiol Biochem Pharmacol* 87: 117-167.
Foster AC, Gill R, Kemp JA and Woodruff GN (1987): Systemic administration of MK-801 prevents N-methyl-d-aspartate-induced neuronal degeneration in rat. *Neurosci Let* 76: 307-311.
Gehlert DR, Gackenheimer SL, Reel JK, Lin H and Steinberg MI (1990): Non-peptide angiotensin II receptor antagonists discriminate subtypes of ^{125}I-angiotensin II binding sites in the rat brain. *Eur J Pharmacol* 187: 123-126.
Gibson RE, Thorpe HT, Cartwright ME, Frank JD, Schorn TW, Bunting PB and Siegl PKS (1991): Angiotensin II receptor subtypes in renal cortex of rats and rhesus monkeys. *Am J Physiol* 261: F512-F518.
Gibson RE, Burns HD, Thorpe HT, Eng W-s, Ransom R and Solomon H (1992): Autoradiographic Imaging of the *In Vivo* Binding of (+)-3-[^{125}]Iodo-MK-801 to the NMDA-Receptor/Channel Complex in Rat Brain. *Nucl Med Biol* 19: 319-326.
Gimenez-Amaya JM and Graybiel AM (1990): Compartmental origins of the striatopallidal projection in the primate. *Neurosci* 34: 111-126.
Graybiel AM (1990): Neurotransmitters and neuromodulators in the basal ganglia. *TINS* 13: 244-254.
Huang S-c and Phelps ME (1986): Principles of tracer kinetic modeling in positron emission tomography and autoradiography. In: *Positron Emission Tomography and Autoradiography*, Phepls MW, Mazziota JC and Schelbert HR, eds. New York: Raven Press.
Kennedy C, Des Roseris M, Reivich M, Sharp F, Jehle JW and Sokoloff L (1975): Mapping of functional neural pathways by

autoradiographic survey of local metabolic rate with [^{14}C]deoxyglucose. *Science* 187: 850-853.

Kuhar MJ (1986): Quantitative Receptor Autoradiography: An Overview. In : *Quantitative Receptor Autoradiography*, Boast CA, Snowhill EW and Alter CA, eds. New York: Alan R Liss, Inc.

Kung HF, Guo Y-Z, Billings JJ, Xu X, Mach RH, Blau M and Ackerman R (1988a): Preparation and biodistribution of [^{125}I]IBZM: A potential CNS D-2 dopamine receptor imaging agent. *Nucl Med Biol* 15: 195-201.

Kung HF, Billings JJ, Guo Y-Z and Mach RH (1988b): Comparison of *in vivo* D-2 dopamine receptor binding of IBZM and NMSP in rat brain. *Nucl Med Biol* 15: 203-208.

Lear JL (1986): Principles of single and multiple radionuclide autoradiography. In: *Positron Emission Tomography and Autoradiography*, Phelps MW, Mazziota JC and Schelbert HR, eds. New York: Raven Press.

Levens LG, Peach MJ and Carey RM (1980): Role of the intrarenal renin-angiotensin system in the control of renal function. *Circ Res* 48: 157-167.

Long RE, Unger T, Rascher N and Garten D (1983): Brain angiotensin. In: *Handbook of Psychopharmacology, Vol 16*, Iversen L, Iversen S and Snyder S, eds. New York: Plenum Publishing Corp.

McCulloch J (1982): Mapping functional alterations in the CNS with [^{14}C]deoxyglucose. In: *Handbook of Psychopharmacology, Vol 15*, Iversen L, Iversen S and Snyder S, eds. New York: Plenum Publishing Corp.

Meldrum B (1985): Excitatory amino acids and anoxic-ischemic brain damage. *Trends Neurosci* 8: 47-48.

Mendelsohn FAO, Millan M, Quirion R, Aquilera RG, Chou S-T and Catt KJ (1987): Localization of angiotensin II receptors in rat and monkey kidney by *in vitro* autoradiography. *Kidney Int* 31: S40-S44.

Millan MA, Jacobwitz DM, Catt KJ and Agvilera G (1991): Distribution of angiotensin II receptors in the brain of nonhuman primates. *Peptides*, 11: 342-253.

Navar LG and Langford HG (1974): Effects of angiotensin on the renal circulation. In: *Angiotensin*, Page IH, Allman D and Bumpus FM, eds. New York: Springer-Verlag.

Navar LG and Rosivall L (1984): Contribution of the renin-angiotensin system to the control of intrarenal hemodynamics. *Kidney Int* 25: 857-868.

Ozyurt E, Graham DI, Woodruff GN and McCulloch J (1988):

Protective effect of the glutamate antagonist, MK-801 in focal cerebral ischemia in cat. *J Cereb Blood Flow Metab* 8: 138-143.

Ransom RW, Eng W, Burns HD, Gibson RE and Solomon HF (1990): (+)-3-[^{125}I]Iodo-MK-801: Synthesis and characterization of binding to the N-methyl-D-aspartate receptor complex. *Life Sci* 46: 1103-1110.

Rothman SM and Olney JW (1986): Glutamate and the pathophysiology of hypoxic ischemic brain damage. *Ann Neurol* 19: 105-111.

Rowe BP, Grove KL, Saylor DL and Speth RC (1990): Angiotensin II receptor subtypes in the rat brain. *Eur J Pharmacol* 186: 339-342.

Seldon NRW, Robbins TW and Everitt BJ (1990): Enhanced behavioral conditioning to context and imparied behavioral and neuroendocrine responses to conditional stimuli following ceruleocortical noradrenergic lesions: support for an attentional hypothesis of central noradrenergic function. *J Neurosci* 10: 531-539.

Slatter EE, Merrill DD, Guess HA, Roylance PJ, Cooper WD, Inman WHW and Ewan PW (1988): Clinical profile of angioedema associated with angiotensin-converting enzyme inhibition. *J Am Med Assoc* 260: 967-970.

Wamsley JK, Herblin WF, Alburges ME and Hunt M (1990): Evidence for the presence of angiotensin II-type 1 receptors in brain. *Brain Res Bull* 25: 397-400.

Wong DF, Burns HD, Solomon HF, Dannals RF, Villemagne V, Goldman S, Young T, Eng W, Ravert HT, Wilson AA, Gibson RE, Iversen L, Guilarte T and Wagner HN Jr (1989): Imaging of NMDA receptor sites with 8-[C-11]Methoxy-MK-801 *in vivo* by PET. *J Nucl Med* 30: 741.

Quantitative Whole Body Autoradiographic Microimaging for Pharmaceutical Research

Prantika Som
Medical Department
Brookhaven National Laboratory

Zvi H. Oster
Department of Radiology
State University of New York at Stony Brook

INTRODUCTION

Autoradiographic (ARG) microimaging is probably the highest spatial resolution method for imaging the distribution of radiolabeled compounds. This method can be used in sections of whole small animals such as mice, rats and even small monkeys, or in parts of animals, using the new large-block cryomicrotomes. Radiolabeled compounds and their metabolites have been studied in normal animals as well as in animal models of human disease. By sacrificing animals at various time intervals after the injection of the radiocompounds, the dimension of time can be introduced, thus generating data showing the temporo-spatial distribution and biokinetic pathways of compounds. This technique can be used to study the biodistribution of new compounds, the toxicity of substances, and the effects of pharmacological interventions of various active compounds (Som and Sacker, 1992; Ullberg, 1954; Berlin and Ullberg, 1963a; Berlin and Ullberg, 1963b; Cohen and Hood, 1969;

Nuclear Imaging in Drug Discovery, Development, and Approval
H.D. Burns, R. Gibson, R. Dannals, Peter Siegl, Editors

Busch, 1977, Cross, 1977a; Cross, 1977b; Som et al., 1983; d'Argy et al., 1984; Yonekura et al., 1985).

When comparing ARG microimaging with the conventional tissue-sampling method for evaluating the distribution of gamma or positron emitters, the following advantages of ARG microimaging should be noted:

1) ARG microimaging provides a high resolution pictorial map of the radiotracer localization in the whole body. Coronal, axial, and sagittal sections can be obtained, depicting the distribution in three dimensions.
2) ARG microimages can be digitized along with standards so the data can be quantitated. Three-dimensional computer reconstructions of organs can be performed with commercially available systems.
3) Given the high resolution of the method, which is the less than 1 mm, ARG microimaging can detect and localize areas of tracer accumulation which are well below the size of tissue samples that can be dissected and weighed accurately.
4) The long exposure times that can be used in ARG results in increased sensitivity as compared to scintillation counting. The photographic emulsion of the film exposed to tissue sections containing radiotracers integrates the incident radiation emitted over long periods of time (days to weeks) well beyond what is practical in tissue-counting methods or in scintigraphy.
5) The rather complex process of assaying the distribution of ß-emitting compounds in tissues is well known. The samples have to be completely dissolved and, preferably, a colorless and clear solution should be obtained for counting, a very tedious and time-consuming task. On the other hand, ARG micro-imaging of ß-emitting compounds is very similar to that used for gamma emitters and, therefore, is much simpler and less complex than scintillation counting.
6) Tissue distribution studies can rarely encompass all organs or all parts of the body while whole body ARG microimaging will depict the global, as well as the discrete, minute sites of tracer accumulation which can be missed by non-pictorial sampling methods.

METHODOLOGY

The procedure of ARG microimaging involves the *in vivo* administration of radiolabeled compounds to normal animals or to animals with models of human disease. Pharmacological interventions can be performed to investigate the effect of receptor blockers or of other compounds on changing the normal distribution of the radiocompound. Animals such as mice and rats and small monkeys up to 4 lbs. in weight can be studied after they are sacrificed at predetermined time intervals. The methods of sacrifice should be carefully selected since it can impact on the final distribution of the radiotracer. The animals are quickly frozen in hexane/solid CO_2 or in liquid nitrogen and embedded in carboxymethylcellulose. Sections of the whole body or body parts are made with a large-blade cryomicrotome, mounted on scotch-tape, and exposed on X-ray film along with step-wedge standards. After the film is developed, visual inspection and quantitation by digital videodensitometry can be performed. Compounds can be labeled with negatron (β^-), positron (β^+), or gamma (γ) emitters. Isotopes with longer half-lives are more convenient to work with since smaller amounts of tracer can be used and compensated by longer exposures. Examples of radioisotopes in common use are Iodine-125, Carbon-14, Tritium, Iron-59. However, because of their long half-life, the problem of disposal is a hindrance. Short-lived isotopes such as Fluorine-18 and Carbon-11 can also be used for ARG microimaging. The procedure should be modified to include rapid drying of the sections and fast exposure on X-ray film. The high-energy and very low-energy radionuclides may pose problems that can be solved by using special X-ray films.

Quantitation of ARG first introduced by Berlin and Ullberg (1963a) was based initially on a "radioactive scale" which was matched with areas in the ARG. This method was later improved by using manually operated analog densitometers (Appelgren, 1967; Cross et al., 1974). Semi-quantitative analysis of ARGs has been performed by many investigators (Sokoloff et al., 1977; Irons and Gross, 1981; Masukoa and Pfeiffer, 1977) and later a system was developed to quantitate ARGs using a computer-coupled scanning densitometer (Gooche et al., 1980). The latest methods involve capturing the image from the film by a digital scanner or by a high-resolution TV camera, digitizing the image, i.e., assigning numerical values to the spectrum of the gray scale in the ARG image, and comparing it to an array of standards of various concentrations exposed on film under the same conditions (Yonekura et al., 1983). By quantitating data from animals sacrificed at various time-intervals, one can determine the temporo-spatial distribution of the radiotracer. Systems currently available commercially are simple to operate and enable image enhancement, normalization, pseudo-coloring, and separation of isotopes, as well as three-dimensional reconstructions.

ARG microimaging can be performed with a single isotope or with dual or multiple tracers (Som and Sacker, 1992). Separation of the tracers can be accomplished by taking advantage of the different energies of the emissions and of the different half-lives. Thus, low-energy emissions can be blocked off by interposing thin, attenuating sheets between the tissue sections and the film. When using long- and short-lived isotopes simultaneously, the initial brief exposure of the section on film will represent the photons originating from the isotope with the short half-life. The subsequent exposure of the tissue section on film (after a minimum decay of 10 half-lives of the short-lived isotope) will represent the distribution of the long-lived tracer. Thus, studies of blood flow and metabolism or receptor binding can be performed simultaneously in the same animal. Compounds carrying the radioactive label at different sites on the molecule can be injected at the same time, and the differences in metabolic pathways can be studied.

EXAMPLES OF ARG MICROIMAGING APPLICATIONS

The monograph by Ullberg et al. (1977) contains a plethora of the basic information on whole body ARG methodology with particular emphasis on toxicological applications. Some other examples of ARG application are listed below.

The Effects of Pharmacological Intervention on Somatic and Cerebral Iodoamphetamine (IMP) Distribution

Radiolabeled iodoamphetamine (IMP) has been shown to be a radiotracer which provides information on regional cerebral blood flow (Winchell et al., 1980; Kuhl et al., 1982). Iodine-131 or Iodine-125 labeled IMP whole-body distribution, and the effect of some pharmacological maneuvers were studied in several species (Som et al., 1985). High [^{131}I]IMP uptake is seen in the chorio-retinal and ciliary body of the eye in the marmoset, as well as in the cortical gray matter and the adrenal cortex (Fig. 1). In the lungs, IMP uptake was patchy. Pretreatment with potassium iodide did not significantly alter the distribution of IMP in any of the organs examined. In mice, [^{125}I]IMP concentrated heavily in the brain, olfactory bulbs, lungs, kidneys, and adrenals at 5 min post-inection. In delayed ARG (not shown), there is more accumulation in the chorio-retina and liver with decreased activity in the olfactory bulbs and cerebellum. The liver and kidney activity of [^{125}I]IMP increases in time. A band of high uptake at the junction of the renal cortex and the medulla, which probably represents the juxtamedullary region, was noted. This finding would have been missed by conventional tissue distribution or scintigraphy because of its small dimensions. The distribution of the radiotracer in rat brain, which is

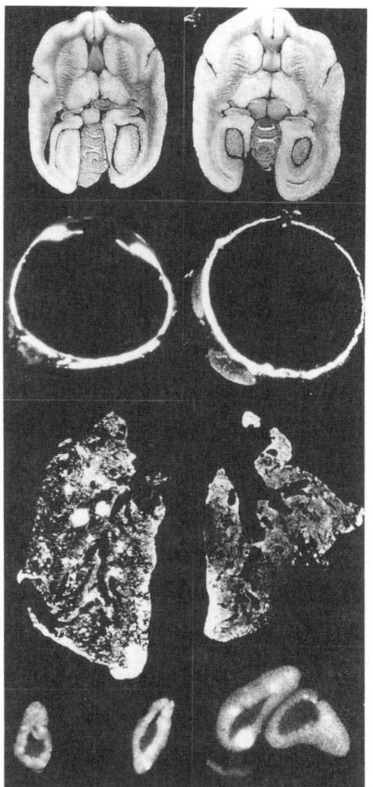

FIGURE 1. ARG of brain, (top frame) eye (2nd frame), lungs (3rd frame), and adrenals (bottom frame) of marmoset following injection of [^{131}I]IMP after KI (left panels) and in controls (right panels). Notice the uptake of IMP in cortex, basal ganglia, chorio-retinal layers, lungs and adrenals. In the lungs, uptake is not uniform which may indicate areas of preferential IMP binding.

related to cerebral blood flow, is presented in the tranverse ARG sections of the rat head 60 min post-injection of [^{125}I]IMP (Fig. 2). The exqusite detail which can be obtained with ARG microimaging is demonstrated by the brain anatomy in Fig. 1 and 2. For example, the dimensions of the rat head are approximately 2 x 3 cm (Fig. 2), so structures that are less than 1 mm are identified easily. Such high resolution pictorial representations should challenge the designers of scintigraphic imaging devices to improve imaging systems. It seems that the poorer resolution of the scintigraphic images is not caused by "fuzzy" distribution of radiotracers, but is inherent in the imaging device. Comparison of

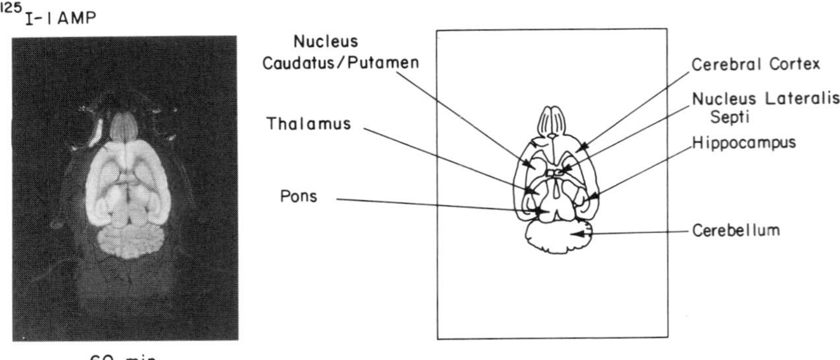

FIGURE 2. Axial section of rat brain ARG after [^{125}I]IMP (≡ ^{125}I-IAMP). Note the exquisite resolution showing structures less than 1 mm in size (e.g., fibers of the olfactroy bulbs).

quantitative data obtained by ARG quantitation correlated well with the data from tissue dissection and counting (routine tissue distribution).

Dual Isotope Studies using Two Glucose Analogues and the Effects of Chlorpromazine

We have shown that whole body dual tracer ARG microimaging can be used to study the effect of chlorpromazine on glucose metabolism using two glucose analogues in the same animal simultaneously, before and after chlorpromazine (Som et al., 1983). Carbon-14 labeled 2-deoxy-D-glucose ([^{14}C]2DG) was given first as a baseline followed by chlorpromazine and [^{18}F]2-fluoro-2deoxy-D-glucose ([^{18}F]FDG). Thus, [^{14}C]2DG depicts normal glucose metabolism and the [^{18}F]FDG shows the effects of chlorpromazine on glucose metabolism. Chlorpromazine treatement was shown to decrease glucose metabolism in muscle and brown-fat compared to the saline control (Fig. 3). By using a digital videodensitometric method, a 2 x 2 cm field of X-ray film was digitized as a 128 x 128 array, each pixel representing 156 x 156 μm. This provides a quantitative comparison of the activity from the two radiotracers (Fig. 4) in fairly small structures (≈ 1mm).

Triple Isotope Studies of Myocardial Perfusion and Energy Substrate Utilization in Hypertensive Heart Disease

Changes in perfusion and fatty acid utilization were studied in salt-sensitive hypertensive rats on either a high salt diet (hypertensive) vs. a

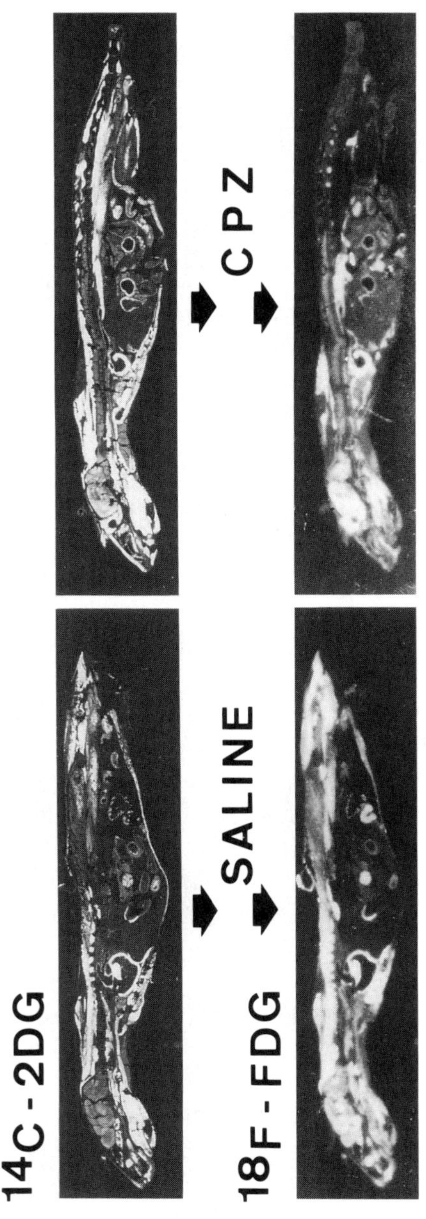

FIGURE 3. The effect of cholrpromazine (CPZ) on glucose uptake using the dual isotope technique. [^{14}C]2-DG depicts the baseline state and [^{18}F]FDG depicts the effect of CPZ as compared to control (saline).

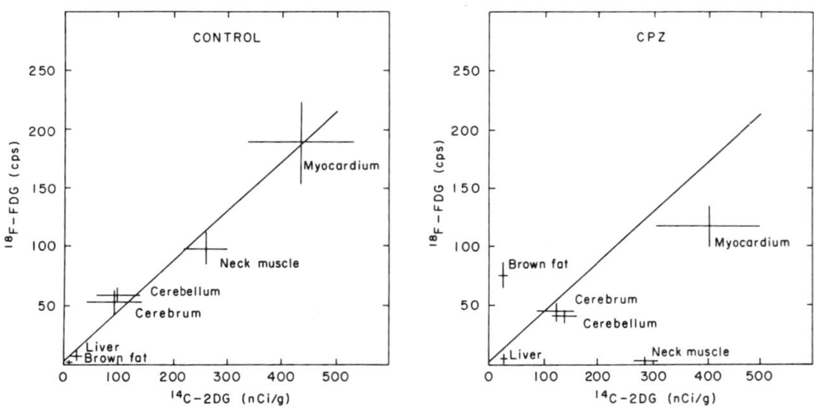

FIGURE 4. Results of quantitation of uptake from Fig. 3 depicting the [^{18}F]FDG uptake vs. [^{14}C]2-DG uptake in controls (left panel) as compared to CPZ treated animals (right panel). Small structures such as basal ganglia could be identified and quantitated.

low salt diet (normotensive). We used the early distribution of Thallium-201 (5-min post-injection) as a marker of perfusion, [^{14}C]2DG as a marker of glucose metabolism, and [^{14}C]β-methylheptadecanoic acid, a synthetic fatty acid analogue, as a marker of fatty acid metabolism (Yonekura et al., 1985) . In normotensive rats, the distribution of the three compounds was homogeneous throughout the entire thickness of the left ventricular wall. The ARG microimages revealed that, in the hypertensive heart, Thallium-201 distribution remains uniform whereas the fatty acid analogue accumulation decreased in the endocardium and in the free wall of the left ventricle. In these areas of decreased fatty acid uptake, the triple-isotope study showed a definite increase in glucose utilization. This shift from preferential myocardial use of fatty acid to glucose is known to be an indicator of hypoxia. This shift from aerobic to anaerobic substrate utilization was not detectable by changes in the distribution of Thallium-201, the perfusion maker, which remained unchanged.

The Effects of Therapeutic Interventtion: Myocardial Metabolism and Blood Flow Studies in Cardiomyopathy

An animal model of congestive cardiomyopathy (B53.58 strain of the Syrian golden hamster) was studied by ARG microimaging using Thallium-201, [^{14}C]2DG, and [^{131}I]dimethyliodophenylpentadecanoic acid (DMIPP) as the marker of fatty acid metabolism (Kubota et al.,

1988). The animals were followed from early age, when cardiomyopathy is not evident, through the mid-stage of disease, and to the full-fledged end-state of the disease (Fig. 5). Note the distribution of the three tracers in normals (upper panel), in early cardiomyophaty (middle panel) , and the advanced stage (lower panel). The first indicator of cardiomyopathy in the early stage was the decrease of fatty acid uptake. In the late stage, uptake of all tracers is decreased. Abnormalities in glucose uptake were very small initially, while fatty acid uptake abnormalities were much more pronounced. ARG microimaging revealed that Thallium-201 uptake, representing perfusion, became

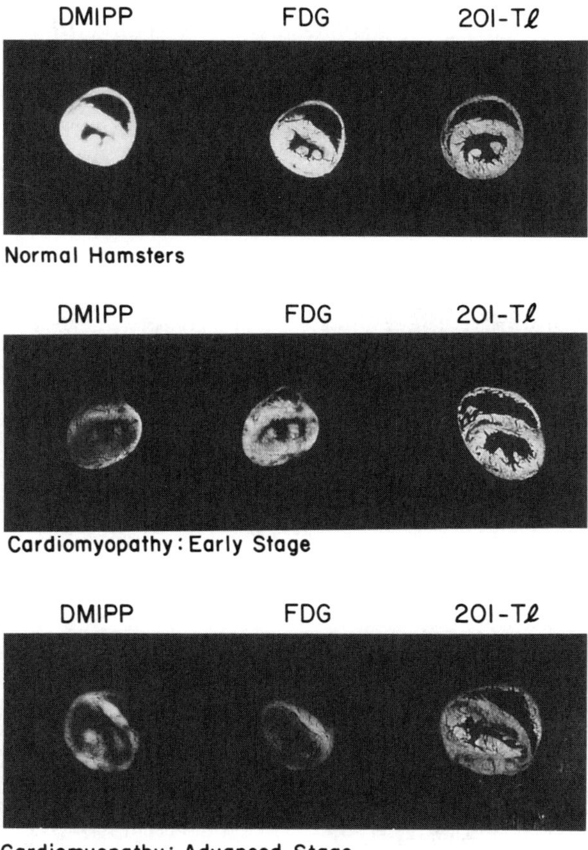

FIGURE 5. By using the triple-isotope method, the coronory flow (Thallium -201), glucose uptake ([^{18}F]FDG), and fatty acid uptake ([^{131}I]DMIPP) were studies at the three stages of cardiomyopathy in the golden Syrian hamster.

abnormal only in the end stage cardiomyopathic hamsters. In parallel, genetically identical hamsters were given verapamil, a calcium channel-blocking agent starting at 3 weeks of age. The verapamil-treated hamsters did not show perfusion or metabolic abnormalities (results with the fatty acid analogue presented in Fig. 6). These findings were confirmed by histologic studies which indicated that, in the verapamil-treated animals, the typical histological features of cardiomyopathy did not develop. Thus, the autoradiographic microimaging study suggests the possibility of developing a test for humans with cardiomyopathy using radiolabeled fatty acids and gamma scintigraphy which may also be useful in monitoring disease progression or the effects of treatment.

ARG Studies of Addictive Substances: Brain and Somatic Effects

Psychoactive substances lend themselves to be studied by timed, quantitative ARG imaging. Thus, the time-course distribution of [^{3}H]phencyclidine (PCP, "angel dust") was studied in mice (Fand et al.,

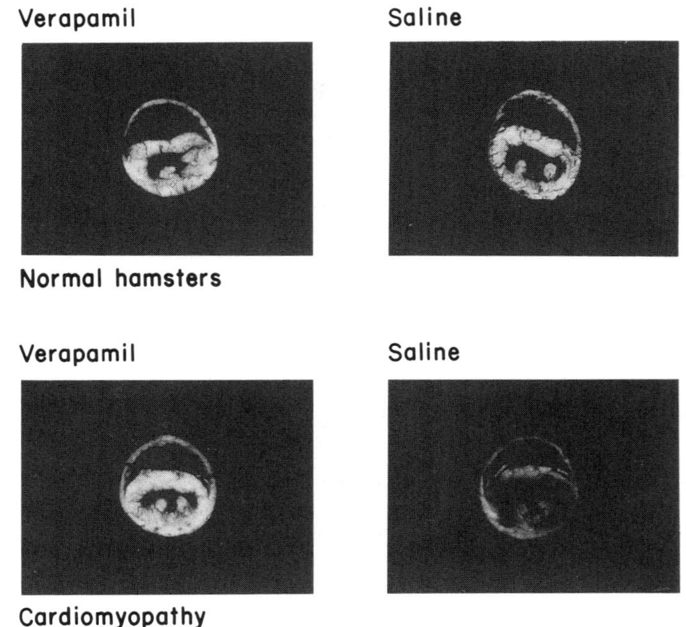

FIGURE 6. The effect of verapamil treatment in preventing the development of cardiomyopathy, as indicated by [^{131}I]DMIPP uptake, is shown in the left panels as compared to non-treated hamsters (right panels).

1988). Immediately after injection, [^{3}H]PCP is found in the pituitary, thyroid, lung, and stomach. At 2-min post-injection, there is uptake in the adrenals and, in later views, accumulation of PCP in the gastrointestinal tract.

The time course of intravenous cocaine has been studied extensively (Som et al., 1989). However, most previous studies focused mainly on the brain. Our ARG microimaging studies using [^{14}C]cocaine revealed the kinetics of cocaine in the whole body (Som et al., 1990). Within 2-3 min intensive [^{14}C]cocaine uptake in the brain was seen (Fig. 7A and B).

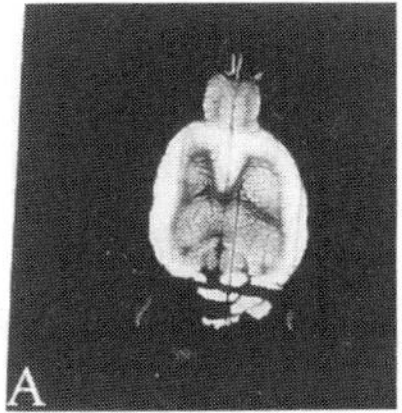

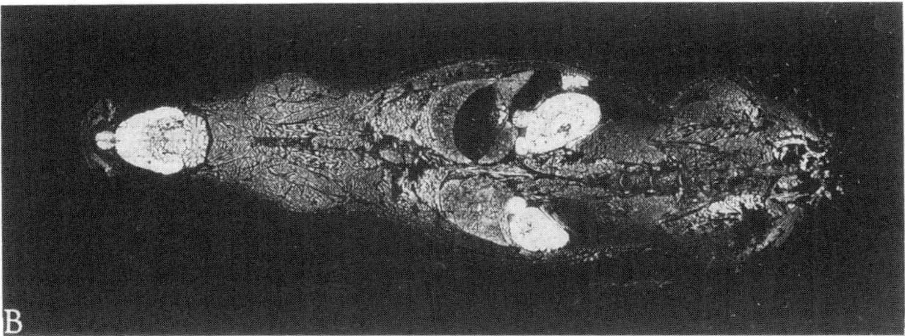

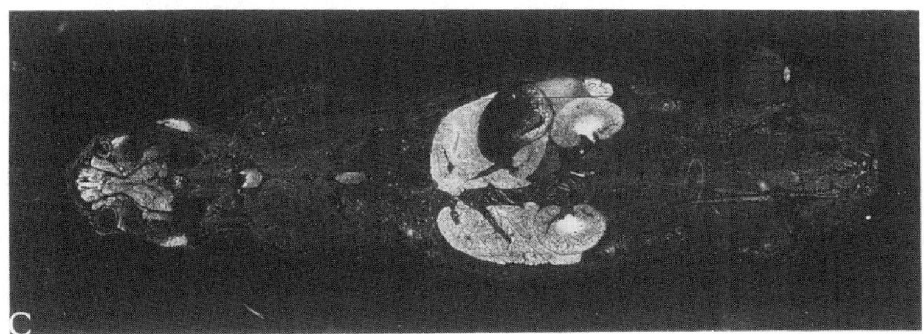

FIGURE 7. Early distribution of [^{14}C]cocaine in rat showing intense uptake in cortex, basal ganglia, and cerebellum (A,B) and in the adrenals and renal parenchyma (B). In ARG obtained 15 min after radiotracer injection, the activity in brain and adrenals decreased, while activity in the liver and renal pelvis increased (C).

These findings concur with the clinical symptoms after cocaine, i.e., the initial feeling of "high", the euphoric state, elevation in pulse rate and blood pressure, and perspiration. Over the next 10-15 min, there is gradual washout of cocaine from the "primary target organs" with progressive accumulation in the kidneys, liver, and gastrointestinal tract probably representing the excretory pathways (Fig. 7C). At the early time, there is also intense accumulation of cocaine in the heart and adrenals (Fig. 8A). Pretreatment with desipramine, a norepinephrine/serotonin reuptake blocker, decreased [^{14}C]cocaine uptake in the spinal cord, cerebellum, and to a greater extent in the heart and adrenals (Fig 8B). Pretreatment with GBR 12909, a dopamine reuptake blocker, caused a decrease of [^{14}C]cocaine uptake in the brain (striatum, cortex), heart and adrenals (Fig. 8C) (Som et al., 1990; Som et al., 1992).

Studies of the Skeletal System and Bone Tumor with Radioactive Tin Compounds

The use of ARG microimaging has also been applied to bone-seeking agents. Tin-117m diethylenetriamine pentaacetic acid ([^{117m}Sn] (4+) DTPA) was shown to bind to normal bone and to a transplanted osteosarcoma (Oster et al., 1984). This isotope has very good imaging properties (158.6 keV gamma rays) and the 13.6-day half-life makes it a promising candidate for therapy of bone malignancies and as an imaging agent for very delayed follow-up of the tracer. We have also shown that the uptake of this compound in the bone marrow is lower than Phosphorus-32, thus reducing the bone marrow radiation dose.

Targeting of Radiolabeled Monoclonal Anti-Tumor Antibody: Visualization by ARG

Radiolabeled anti-colon/ovarian tumor antibody was studied in athymic nude rats bearing human colon cancer xenografts (Zamora et al., 1988). Whole body ARG complemented the infomation obtained by tissue distribution. Intense accumulation of radioiodinated antibody was seen at the site of the implanted tumor. Radioactivity also accumulated in the region of the thyroid. However, using an immunohistochemical analysis for the antibody, no acuumulation could be demonstrated in the thyroid thus indicating that the activity seen on the ARG represented free iodine (the animal was not pretreated with Lugol's solution).

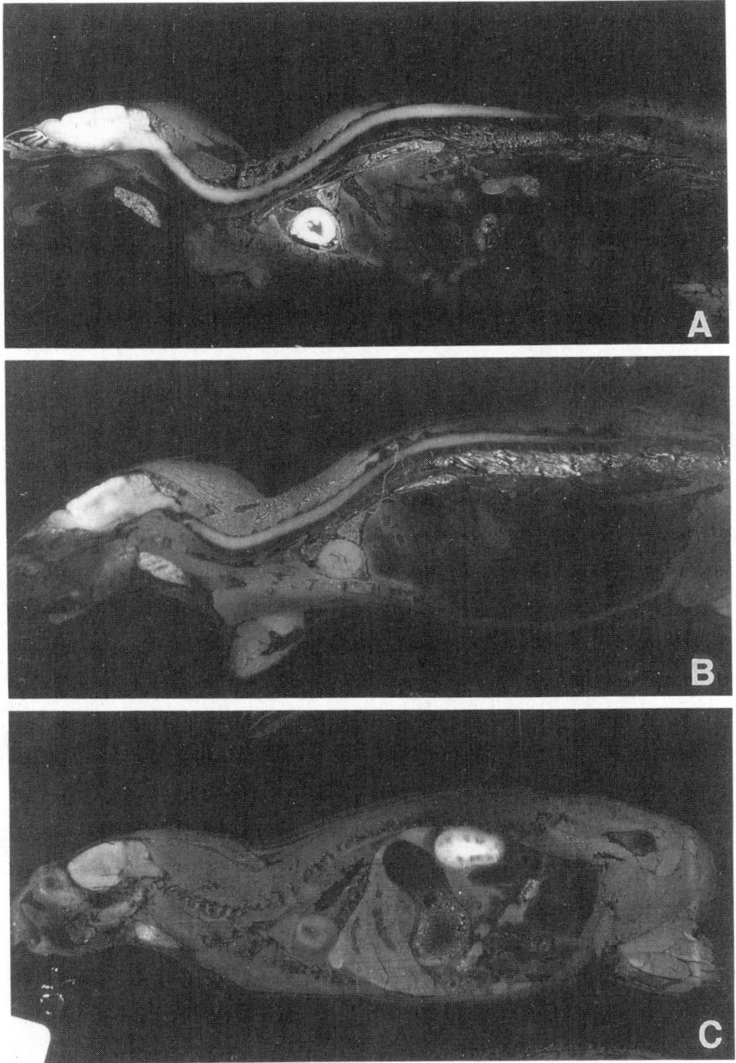

FIGURE 8. Sagittal sections of rat ARG after [^{14}C]cocaine. Note the intense uptake in brain, spinal cord and heart (A). Pretreatment with desipramine (B) resulted in small changes in [^{14}C]cocaine uptake in the cerebellum and spinal cord, but significantly reduced activity in heart and adrenals. Pretreatment with GBR 12909 (C) reduced uptake of [^{14}C]cocaine in brain and heart with most of the activity accumulating in the renal collecting system.

Technetium-99m Labeled Human Polyclonal Gamma Globulin Studies of Abscesses and Tumors

Intracerebral glioma and turpentine abscess in small rodents were clearly delineated 2 hrs after the injection of Technetium-99m labeled IgG. This information complemented the data obtained by tissue distribution, blood clearance, and scintigraphy (Oster et al., 1992).

Studies of Placental Transfer of Compounds

Studies by others have shown the value of ARG in studying the placental transfer of nutrients, drugs and toxic compounds (Ullberg et al., 1961; Ullberg et al., 1967; Ullberg and Dencker, 1977; van der Kleijn et al., 1977; Dencker, 1976). We performed preliminary studies on the placental transfer of [^{14}C]cocaine in early pregnant animals (unpublished observation), but could demonstrate only slight transfer to the embryo which may be greater later in pregnancy, as suggested from newborns with symptoms of cocaine dependency.

The Effect of Peroral Prussian Blue on Thallium-201 Distribution

It was shown that, immediately after intravenous injection of Thallium-201 in animals (and probably also in humans), the radiotracer is excreted into the gastrointestinal (GI) tract and reabsorbed into the blood stream via the enterohepatic circulation. Prussian Blue traps Thallium-201 in the GI tract preventing reabsorption. This increases excretion thus reducing the radiation burden, particularly to the GI tract (Oster et al., 1982). The phenomenon is vividly depicted in tissue sections with the blue-colored Prussian Blue in the bowel and the intense Thallium-201 accumulation seen on the ARG in the same areas.

Chemical Analysis of Samples

The whole body ARG technique can be used to visualize the distribution of a compound and, subsequently, to identify the species of radioactive compound (parent vs. metabolite) in the same tissue by punching out small pieces from the whole body section and subjecting them to chromatographic analysis (d'Argy , 1977). Utilizing a more sophisticated technique such as the three-step whole body ARG, total radioactivity in an organ of volatile and non-volatile metabolites as well as non-extractable metabolites can be determined (Bergman and Tjalve, 1977). In these studies, tissue sections are exposed at different temperatures to separate different types of metabolites (volatile and non-volatile compounds). Subsequently, sections are punched out and subjected to differential extractions procedures.

CONCLUSION

The examples listed above demonstrate that whole body microimaging, which is a pictorial mapping of the temporo-spatial distribution pattern of compounds, is a powerful tool for evaluation of new radiolabeled drugs. It allows the simultaneous measurements of different physiological parameters, for studies of the effects of therapeutic interventions, and for the characterization of receptor binding sites. Subsequent chemical analysis can be used to identify the radiospecies in various organs. Since these studies are often difficult or impossible with imaging studies in SPECT and PET, whole body autoradiography provides very valuable data needed to interpret in *vivo* imaging studies.

ACKNOWLEDGEMENTS

This research was supported in part by the U.S. Department of Energy under Contract DE-ACO2-76CH00016.

REFERENCES

Appelgren LE (1967): Sites of steroid hormone for median. Autoradiographical studies using labelled precursors. *Acta Physiol Scand*, 301(Suppl): 1-108.

Bergman K and Tjalve H (1977): Three-step autoradiography of organic and plastic monomers to register total radioactivity, non-volatile metabolites, and non-extractable metabolites. *Acta Pharmacol Toxicol* 41(Suppl. 1): 22-23.

Berlin M and Ullberg (1963a): The fate of Cd^{109} in the mouse. An autoradiographic study after a single intravenous injection of $Cd^{109}Cl_2$. *Arch Environ Hlth* 7: 686-693.

Berlin M and Ullberg (1963b): Accumulation and retention of mercury in the mouse. *Arch Environ Hlth* 6: 589-616.

Busch U (1977): Wholebody autoradiography (WBARG): use for pilot studies of pharmacokinetics in rats. *Acta Pharmacol Toxicol* 41: 28-29.

Cohen EM and Hood M (1969): Application of low temperature autoradiography to studies of the uptake and metabolism of volatile anaesthetics in mouse. *Anesthesiol* 30: 306-314.

Cross SAM (1977a): The localization of metiamide and cimetidine using autoradiographical techniques. *Proc Eur Soc Toxicol* 18: 288-290.

Cross SAM (1977b): Localization of histamine H_2-receptor antagonist in the gastric mucosa. *Histochem J* 9: 619-644.

Cross SAM, Groves AD and Hesselbo T (1974): A quantitative method for measuring radioactivity in tissues sectioned for whole-body autoradiography. *Int J Appl Radiat Isot* 25: 381-386.

d'Argy R (1977): Tissue pieces from wholebody sections used for chromatographic separation of metabolites. *Acta Pharmacol Toxicol* 41 (Suppl. 1): 16-17.

d'Argy R, Ullberg S, Stalacke C-G and Langstorm B (1984): Whole-body autoradiography using ^{11}C with double-tracer applications. *Int J Appl Radiat Isot* 35: 129-134.

Denker L (1976): Tissue localization of some teratogens at early and late gestation related to fetal effects. *Acta Pharmac* 39: 1-131.

Fand I, McNally WP, Koul O, Yonekur Y, Som P, Brill AB and Deutch DG (1988): Whole-body autoradiographic localization of {^3H}Phenylcycline and its metabolites in mice. *Toxicol Appl Pharmacol* 93: 421-432.

Gooche C, Rasband W and Sokoloff L (1980): Computerized densitometry and color coding of (^{14}C) deoxyglucose autoradiography. *Ann Neurol* 7: 359-370.

Irons RD and Gross EA (1981): Standardization and calibration of wholebody autoradiography for routine semiquantitatives analysis of the distribution of ^{14}C-labeled compounds in animal tissues. *Toxicol Appl Pharmacol* 59: 250-256.

Kubota K, Som P, Oster ZH, Brill AB, Goodman MM, Knapp FF Jr, Atkins HL and Sole MJ (1988): Detection of cardiomyopathy in an animal mode using quantitative autoradiography. *J Nucl Med* 29: 1697-1703.

Kuhl DE, Barrio JR, Huang S-C, Selin C, Acerkmann RF, Lear JL, Wu JL, Lin TH and Phelps PE (1982): Quantifying local cerebral blood flow by N-isopropyl-p-[^{123}I]iodoamphetamine (IMP) tomography. *J Nucl Med* 23: 196-203.

Masukoa DT and Pfeiffer E (1977): An *in-vivo* brain autoradiographic technique and a digital microdensitometer for the quantitation of autoradiograms . *Acta Pharmacal Toxicol* (Suppl. 1) 41: 54-55.

Oster ZH, Som P, Brill AB, Sacker DF and Atkins HL (1982): Enhancement of radiopharmaceutical excretion by chemical interventions In: *Proceeding of the 3rd World Congress of Nuclear Medicine and Biology, Vol 4*, Raynaud C, ed. Paris: Pergamon Press.

Oster ZH, Som P, Rhodes BA, Wong CTC, Cabahug C, Sacker, DF,

Wang G-J and Meinken GE (1992): A one-step Kit for ^{99m}Tc-labeling of non-specific immunoglobulin (HIG): Imaging of inflammation and tumor *J Nucl Med* 33: 1028.

Oster ZH, Som P, Srivastava SC, Fairchild RG, Meinken GE, Tillman DY, Sacker DF, Richards P, Atkins HL, Brill AB, Knapp FF Jr and Butler TA (1985): The development and *in vivo* behavior of tin-containing radio-pharmaceuticals. II. Autographic and scintigraphic studies in normal animals and in animal models of bone bone disease. *Int J Nucl Med Biol* 12: 175-184.

Sokoloff L, Revich M, Kennedy C, Des Rosier MH, Patlak CS, Pettigrew KD, Sakurado O and Shinohara M (1977): The (^{14}C) deoxyglucose method for measurement of local cerebral glucose utilization: Theory, procedure and normal values in the conscious and anaesthetized albino rats. *J Neurochem* 28: 897-916.

Som P, Oster ZH, Yamamoto K, Meinken GE, Srivastava SC, Yonekura Y, Ebner SA, Atkins HL, Brill AB, Fawwaz RA, Alderson PO, Coffey J, Carlton E and Hubner KF (1985): Some factors affecting the cerebral and extracerebral accumulation of N-isopropyl-p-iodo-amphetamine (IAMP). *Int J Nucl Med Biol* 12: 185-196.

Som P, Oster ZH, Volkow ND, Sacker DF and Weber DA (1989): Studies on wholebody distribution and kinetics of cocaine. *J Nucl Med* 30: 831.

Som P, Oster ZH, Volkow ND, Sacker DF and Weber DA (1990): Organ variations in cocaine receptors. *J Nucl Med* 31: 886.

Som P, Oster ZH, Wang G-J, Volkow ND and Sacker DF (1992): Wholebody autoradiographic microimaging studies of the spatial and temporal distribution of cocaine. *Synapse* , in press.

Som P and Sacker DF (1992): Whole-Body autoradiographic microimaging: Applications in radiopharmaceutical and drug research. In: *Using Autoradiography and Correlative Imaging in vitro and in vivo.* Stumpf W and Solomon H, eds. San Diego: Academic Press, in press.

Som P, Yonekura Y, Oster ZH, Meyer MA, Pelletteri ML, MacGregor RR, Russell JA, Wolf AP, Fand I, McNally WP and Brill AB (1983): Quantitative autoradiography with radiopharmaceuticles. Part II: Applications in radiopharmaceutical research. *J Nucl Med* 24: 238-244.

Ullberg S, Kristofferson H, Flodh H, and Hanngren Å (1967): Placental passage and fetal accumulation of labelled vitamin B_{12} in the mouse. *Arch Int Pharmacodyn Ther* 167: 431-449.

Ullberg S, Sorbo B, and Clemedson C-J (1961): Distribution of radioactive iron in pregnant mice studied in whole-body autoradiography. *Acta Radiol* 55: 145-155.

Ullberg S and Dencker L (1977): Localization in fetal tissues. *Acta Pharmacol Toxical* 41 (Suppl. 1) 170-171.

Ullberg S (1954): Studies on the distribution and fate of ^{35}S-labelled benzyl penicillin in the body. *Acta Radiol*, Suppl. 118: 1-10.

Ullberg S (1977): Conference on Autoradiography in Pharmacology and Toxicology. In: *Acta Pharmacol Toxicol 41, suppl. 1*. Ullberg S, Applegren LE, Dencker L, Hammarstrom L, Hansson E and Tjalve H, eds. Copenhagen: Munksgaard Publ.

van der Kleijn K, Rijntjes NVM, Vree TB, Schobben F, Westenberg H, Jonkman J, Knop HJ and van Geyn H (1977): Similarities and differences in the kinetics of distribution and of placental transfer of anti-epileptic drugs of extremely varying physico-chemical properties. *Acta Pharmacol Toxicol* 41 (Suppl. 1): 168-169.

Winchell HS, Baldwin RM and Lin TH (1980): Development of I-123 labeled amines for brain studies: localization of I-123 iodophenylalkyl amines in rat brain. *J Nucl Med* 21: 940-946.

Yonekura Y, Brill AB, Som P, Yamamoto K, Srivastava SC, Iwi J, Elmaleh DR, Livni E, Strauss HW, Goodman MM and Knapp FF Jr (1985): Quantitative autoradiographic measurement of regional myocardial substrate utilization in hypertensive rats. *Science* 227: 1494-1446.

Yonekura Y. Brill AB, Som P, Bennett GW and Fand I (1983): Quantitative autoradiography with radiopharmaceuticals. Part I: Digital film analysis system by videodensitometry. *J Nucl Med* 24: 231-237.

Zamora PO, Pant KD, Shah VO, Sass KS, Newell KD, Shao HS, Oster ZH and Som P (1988): Anti-Colon/ovarian tumor antigen: Localization of colon cancer Xenograft in athymic rats. *Int J Nucl Med Biol* 15: 261-270.

CEREBRAL METABOLIC RATES OF 2-[^{18}F]FLUORO-2-DEOXY-D-GLUCOSE IN THE PRESENCE OF OFLOXACIN A GABA$_A$ RECEPTOR ANTAGONIST

Edwaldo E. Camargo, Zsolt Szabo, Robert F. Dannals and Henry N. Wagner, Jr.
Department of Radiology and Radiological Science
The Johns Hopkins Medical Institutions

INTRODUCTION

The involvement of the γ-aminobutyric acid (GABA) receptors in the mediation of inhibitory activity in the central nervous system has been well documented (Nistri and Constanti, 1979; Tallman and Gallager, 1985). Agents that block GABA receptors are potentially excitatory, in that they decrease inhibitory activity in the central nervous system. The reported hyperexcitability caused by some quinolones may be explained on the basis of their action as GABA receptor-blocking agents (Tsuji et al., 1988).

Ofloxacin is a new synthetic antibacterial quinolone which has been reported to cause insomnia and excitation in fewer than 1% of patients in clinical trials (Monk and Campoli-Richards, 1987). By using a [^{36}Cl]chloride influx assay in rat brain synaptoneurosomes, we have recently demonstrated that ofloxacin is a GABA$_A$ receptor antagonist, but it is weaker than the GABA$_A$ receptor antagonist bicuculline (Guilarte et al., 1989).

We hypothesized that changes in cerebral glucose metabolism might occur as a consequence of the hyperexcitation caused by GABA receptor

Nuclear Imaging in Drug Discovery, Development, and Approval
H.D. Burns, R. Gibson, R. Dannals, Peter Siegl, Editors

blocking agents or antagonists in the central nervous system. Depending on their magnitude, these metabolic changes might be detectable by Positron Emission Tomography (PET) imaging by using 2-[^{18}F]fluoro-2-deoxy-D-glucose ([^{18}F]FDG).

The purpose of this investigation was to evaluate the possible effects of ofloxacin on the global and regional cerebral metabolic rates of glucose in normal subjects by using [^{18}F]FDG and PET imaging.

MATERIALS AND METHODS

Twelve healthy normal male volunteers (mean age, 26.7 years) were studied in a double-blind, placebo-controlled protocol. Exclusion criteria included allergy or serious adverse reactions to quinolones; significant renal, hepatic, cardiovascular, hematologic, gastrointestinal, neurologic, psychiatric, respiratory, or metabolic disease; hospitalization within 1 month prior to the start of the study; alcoholism or drug abuse; and use of any medication within 7 days prior to entrance into the study.

Three weeks prior to the study, all volunteers signed a consent form and had a complete medical history, physical examination, a standard 12-lead electrocardiogram, a chest radiograph, and a series of laboratory tests, including hematology profile, blood chemistry, urinalysis, and urine toxicology. The entire protocol had a duration of 11 days (Fig. 1). Ofloxacin (400 mg every 12 hrs) or placebo was given by a crossover design. A placebo washout period started on the evening of day 3 and continued through day 7. Plasma samples for monitoring compliance with ofloxacin administration were obtained on days 3 and 10.

Study Protocol

Procedure	Prestudy	Day										
		0	1	2	3	4	5	6	7	8	9	10
History	X											
Physical Exam	X											X
Electrocardiogram	X											
Lab Tests	X	X			X							X
PET Imaging		X			X				X*			X
Ofloxacin or Placebo			X	X	X					X	X	X
Placebo Washout					X	X	X	X	X			
Plasma Samples					X							X

*If questionable on day 3

FIGURE 1. Study Protocol.

PET imaging was done on days 0, 3, and 10 in all volunteers and on day 7 in six of them. At the end of the protocol, 12 baseline (day 0), 18 placebo (days 3 or 10 and day 7), and 12 ofloxacin (days 3 or 10) studies were performed, for a total of 42 PET imaging studies.

All subjects were confined overnight prior to the PET imaging study. The next morning they had a light breakfast that was low in glucose, 2 to 4 hr prior to intravenous injection of [^{18}F]FDG. In a quiet room with dim light, two intravenous lines were inserted. One intravenous line, on the dorsum of a hand, was maintained with saline and heparin and was covered with a warming pad at a comfortable temperature (44°C) for arterialization of venous blood (Huang et al., 1980; Duara et al., 1987; Metz et al., 1989). The other in the opposite forearm, was maintained with a saline drip for tracer injection. Earplugs were then placed and the subjects blindfolded 5 min prior to tracer injection and so maintained for the next 35 to 40 min. In each study, the volunteers were injected with 4 to 5 mCi (148 to 185 MBq) of [^{18}F]FDG. The radiopharmaceutical [^{18}F]FDG was synthesized from [^{18}F]fluoride, which was produced in a biomedical cyclotron (MC-16F; Scanditronix, Uppsala, Sweden) by the (p,n) reaction on 98% enriched [^{18}O]H_2O by the method of Hamacher et al. (1986). The radiochemical purity of the final product, which was determined in every study using thin-layer chromatography or high-performance liquid chromatography on an amino acid column eluted with aqueous acetonitrile, was greater than 98%. All preparations were determined to be sterile and apyrogenic. Arterialized venous blood sampling was begun immediately after tracer injection and was continued for 90 min. Prior to imaging, a previously made thermoplastic face mask was placed on the subject, and the subject was positioned on the scanner table. The inferior orbitomeatal line was marked on the mask with orthogonal laser beams. This line was used as a reference for precise repositioning in repeat studies. These conditions, including the time of injection, were reproduced as closely as possible for repeat studies in the same individual. Imaging was started 45 to 60 min after radiopharmaceutical injection by using a NeuroECAT PET scanner with an in-plane resolution of 8 mm. The metabolic rate of glucose (in milligrams per minute per 100 g) was obtained by using the equation proposed by Huang et al. (1989). Twenty-four regions of interest were selected at various levels of the brain, and their metabolic rates were compared from one study to another in the same individual and among individuals.

The differences between group means were tested by one-way analysis of variance and repeat measures analysis of variance (Cochran, 1941). Homogeneity of variances was tested by the Bartlett-Box F test (Bartlett, 1937; Box, 1953). Individual group means were compared by the multiple range test of Tukey (1949).

RESULTS

Figures 2 to 4 illustrate how the absolute metabolic rate of glucose varied with the region of the brain chosen in the baseline, placebo, and ofloxacin studies, respectively. They also show intersubject variations in each region of the brain. The magnitudes of the intersubject and regional fluctuations of glucose metabolism are similar among Fig. 1 to 3.

The mean regional metabolic rate of glucose for the three studies is shown in Fig. 5. There was a tendency for higher values in the baseline study than in the placebo study, but this did not occur in every subject. Therefore, the effect of nonfamiliarity with the study (Metz et al., 1989) on the metabolic rate of glucose could not be confirmed in this population.

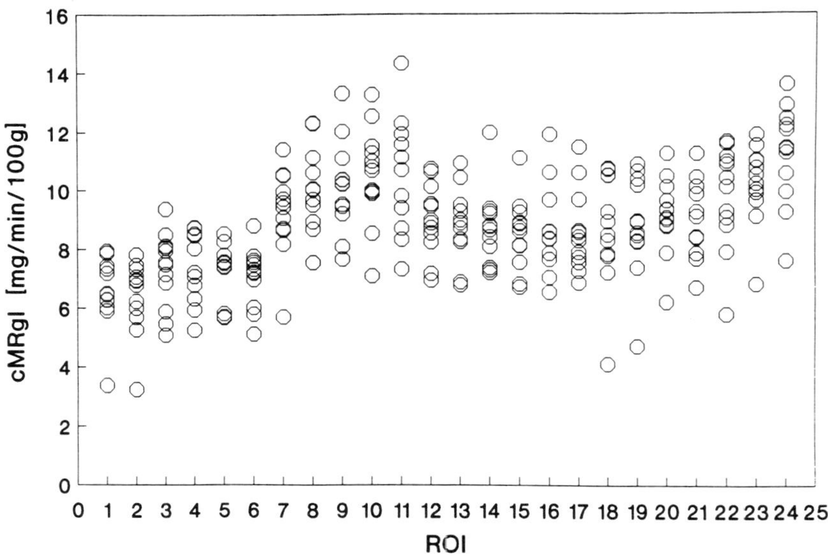

FIGURE 2. Cerebral metabolic rate of glucose (cMRgl) in each of 24 regions of each volunteer in the baseline study. ROI, Region of interest.

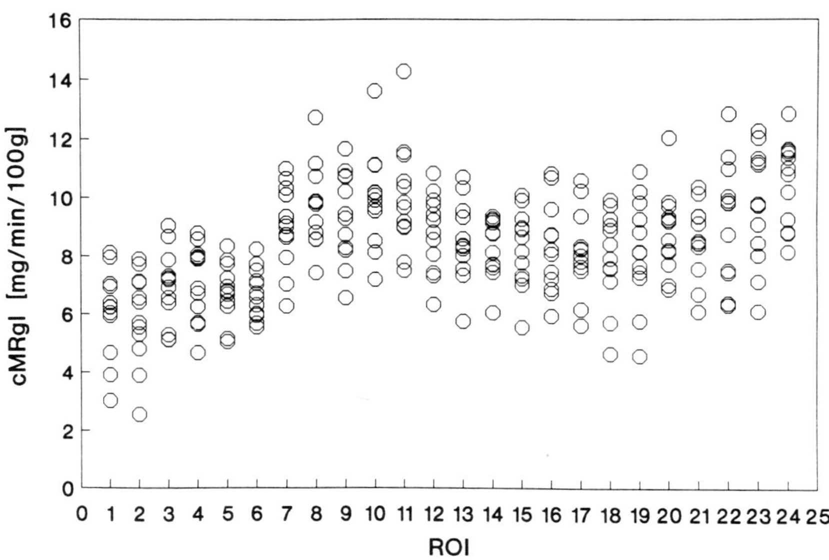

FIGURE 3. Cerebral metabolic rate of glucose (cMRgl) in each of 24 regions of each volunteer in the placebo study. ROI, Region of interest.

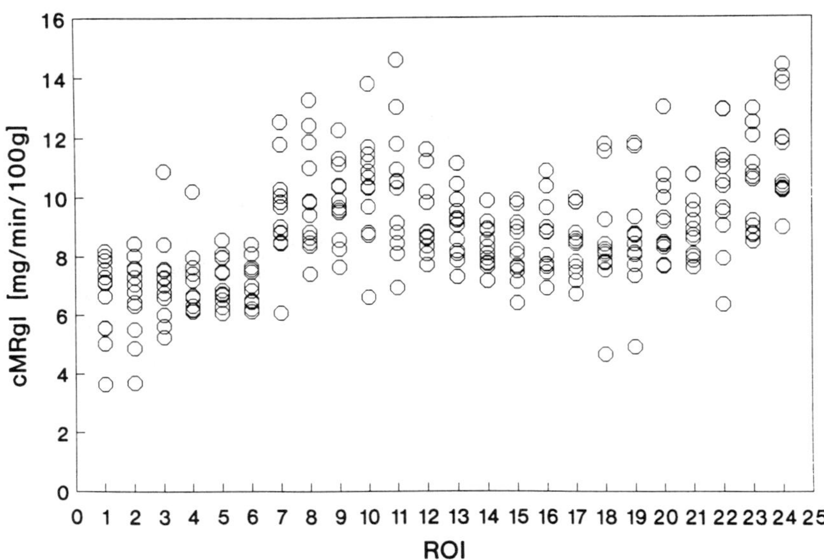

FIGURE 4. Cerebral metabolic rate of glucose (cMRgl) in each of 24 regions of each volunteer in the ofloxacin study. ROI, Region of interest.

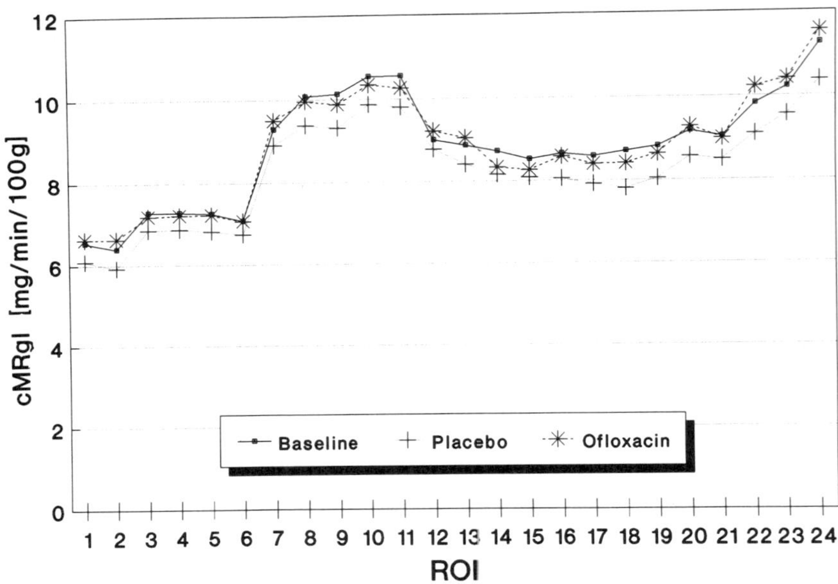

FIGURE 5. Mean regional cerebral metabolic rate of glucose (cMRgl) for the baseline, placebo, and ofloxacin studies. ROI, Region of interest

The percent differences between the baseline and placebo studies and between the ofloxacin and placebo studies in each individual are given in Table 1. The mean global differences between the baseline and placebo studies and between the ofloxacin and placebo studies were 5.1 and 6.6%, respectively, by using relative values or 11.6 and 10%, respectively, by using absolute values. The global cerebral metabolic rates of glucose for all individuals in the baseline, placebo, and ofloxacin studies are also given in Table 1. There was no statistically significant difference from one study to another, at the 0.05 level. Table 2 shows the mean, standard deviation, analysis of variance F, and probability estimates of regional metabolic rates of glucose in these three studies. There was no significant difference in the variances of these three groups at the 0.05 level. Repeat measures analysis of variance also showed no significant differences for the various regions of the brain.

DISCUSSION

Quinolones are synthetic antibiotics with different potencies that

Subject	Global Metabolic Rate (mg/min/100 g) Baseline (B)	Placebo (P)	Ofloxacin (O)	100[(B-P)/B]	100[(O-P)/P]
1	6.42	6.97	7.05	-8.6	1.1
2	8.65	6.03	8.83	30.3	46.4
3	9.28	8.62	7.87	7.1	-8.7
4	8.84	9.40	9.20	-6.3	-2.1
5	8.06	9.92	10.70	-23.1	7.9
6	9.69	9.17	9.13	5.4	0.4
7	8.88	8.50	8.97	4.3	5.5
8	9.89	9.99	9.30	-1.0	-6.9
9	7.07	7.42	7.74	5.0	4.3
10	10.60	7.18	7.74	32.3	7.8
11	9.34	8.65	10.92	7.4	26.2
12	9.10	8.31	8.12	8.7	2.3
Mean	8.82	8.24	8.79	5.1	6.6
± SD	1.17	1.17	1.18		
cv[a] (%)	13	14	15		

a) Coefficient of variation.

TABLE 1. Global metabolic rates and percent differences between baseline and placebo studies and between ofloxacin and placebo studies.

cause inhibition of DNA gyrase. All members of the family, which includes nalidixic acid, cinoxacin, norfloxacin, ciprofloxacin, enoxacin, ofloxacin, pefloxacin, and others, are rapidly bactericidal at lower rather than higher concentrations. The newer members of the family are more active and have a broader spectrum. Several of the fluorinated derivatives have been synthesized over the last decade. One of them, ofloxacin, which was first developed by Sato et al (1982), is as effective as, or better than norfloxacin, depending on the microorganism studied. In infected mice, ofloxacin is two to seven times more effective than norfloxacin (Sato et al., 1982). The newer member, ciprofloxacin, is about four times more active than other quinolone derivatives, including ofloxacin (Van Caekenberghe and Pattyn, 1984).

Region[a]	Regional metabolic rate (mg/min/100 g) Baseline[b] (n=12)	Placebo[b] (n=18)	Ofloxacin[b] (n=12)	Analysis of Variance	P
1. Left cerebellum(I)	6.54±1.22	6.09±1.69	6.63±1.40	0.59	0.56
2. Right cerebellum	6.39±1.25	5.92±1.62	6.63±1.39	0.91	0.41
3. Left temporal	7.28±1.28	6.85±1.14	7.18±1.47	0.48	0.62
4. Right temporal (II)	7.28±1.20	6.87±1.22	7.21±1.17	0.51	0.61
5. Left cerebellum(II)	7.25±0.98	6.82±0.93	7.22±0.81	1.05	0.36
6. Right cerebellum(II)	7.07±0.99	6.74±0.79	7.06±0.76	0.76	0.48
7. Cingulate	9.28±1.45	8.90±1.29	9.49±1.67	0.63	0.54
8. Left caudate	10.07±1.40	9.37±1.56	9.95±1.82	0.84	0.44
9. Right caudate	10.12±1.56	9.30±1.59	9.88±1.33	1.18	0.32
10. Left putamen	10.54±1.66	9.87±1.62	10.35±1.80	0.63	0.54
11. Right putamen	10.55±1.95	9.80±1.78	10.26±2.17	0.56	0.57
12. Left temporal(II)	9.00±1.21	8.77±1.28	9.21±1.23	0.44	0.64
13. Right temporal(II)	8.86±1.23	8.41±1.25	9.04±1.12	1.07	0.35
14. Left parietal	8.71±1.29	8.16±0.96	8.33±0.77	1.04	0.36
15. Right parietal	8.52±1.20	8.08±1.36	8.26±1.07	0.45	0.64
16. Left visual (I)	8.65±1.49	8.05±1.46	8.58±1.22	0.84	0.44
17. Right visual (I)	8.58±1.38	7.91±1.40	8.39±1.08	1.06	0.36
18. Left thalamus	8.70±1.92	7.80±1.52	8.41±1.86	1.05	0.36
19. Right thalamus	8.81±1.70	8.04±1.62	8.63±1.84	0.85	0.43
20. Left visual (II)	9.17±1.29	8.56±1.48	9.28±1.53	1.13	0.33
21. Right visual (II)	9.03±1.30	8.48±1.42	8.99±1.07	0.87	0.43
22. Left frontal	9.82±1.74	9.09±1.84	10.23±1.91	1.49	0.24
23. Right frontal	10.21±1.34	9.55±1.89	10.41±1.56	1.11	0.34
24. Visual (high)	11.24±1.68	10.37±1.37	11.54±1.75	2.25	0.12

a) I and II indicate levels of the same structure.
b) Values are means ± 1 standard deviation.

TABLE 2. Regional metabolic rates of glucose in the baseline, placebo and ofloxacin studies.

Most quinolones cause neurotoxic effects such as headache, dizziness, nausea, vomiting, and sometimes, seizures. With ofloxacin the incidence of central nervous system symptoms such as headache, dizziness, and sleep disturbances is 0.89%; hallucinations and psychotic reactions are very rare (Monk and Campoli-Richards, 1987).

The mechanisms of the central nervous system effects of quinolones is unknown. Tsuji et al. (1988) hypothesized that it might be similar to the convulsant action of penicillins and cephalosporins, which have been related to decreased GABA release from nerve terminals or inhibition of

GABA binding to receptor sites (Tsuji et al., 1988). GABA$_A$ receptors are competitively blocked by bicuculline and are coupled to an anion-selective ionophore, with chloride being the most important charge carrier *in vivo*. Barbiturates and benzodiazepines potentiate GABA-ergic transmission, whereas convulsants such as picrotoxin depress that transmission by acting at a site associated with the chloride ionophore (Tallman and Gallager, 1985). To test their hypothesis, Tsuji et al. (1988) investigated the inhibitory activities of various quinolones, including ofloxacin, on the specific binding of Tritium-labeled GABA to synaptic membranes from rat brains. They all caused inhibition in a concentration-dependent manner, but their potencies differed widely (Tsuji et al., 1988). In our laboratories, we further investigated the physiologic effect of ofloxacin at the GABA recognition site, using the GABA-mediated Chloride-36 influx assay in rat brain synapto-neurosomes. We demonstrated that ofloxacin is a GABA$_A$ receptor antagonist, but it is about two orders of magnitude weaker than bicuculline (Guilarte et al., 1989).

This physiologic action of ofloxacin was insufficient to cause detectable changes in global and regional cerebral glucose metabolism, as measured by PET imaging with [^{18}F}FDG, when compared with the physiologic actions in placebo and baseline studies in normal volunteers. Ideally, a true GABA receptor radiopharmaceutical agent should have been used, but none is available. Also, so far it has not been possible to label GABA itself with a radionuclide suitable for imaging purposes, not to mention the fact that such a molecule probably would not cross the blood-brain-barrier. We demonstrated that ofloxacin does not increase or decrease the cerebral metabolic rate of glucose beyond the limits of the variability of our method. However, this finding cannot be extended to other quinolones. Since quinolone potencies differ widely (VanCaekenberghe and Pattyn, 1984; Tsuji et al., 1988), it is theoretically conceivable that more potent quinolones may cause changes in the cerebral metabolic rate of glucose and that such changes may be detectable by using [^{18}F]FDG and PET imaging. However, it is important to mention that possible metabolic rate changes in a small brain region below the limit of resolution of our scanner (8 mm in-plane) may go undetected.

Regional variations in the cerebral metabolic rate of glucose were found in this study, irrespective of the presence of ofloxacin. Similar variations have been described previously (Huang et al., 1989; Duara et al., 1984; Heiss et al., 1984; Duara et al, 1987; Tyler et al., 1988). These variations are probably related to biological and technical factors and deserve further exploration.

ACKNOWLEDGEMENTS

This investigation was supported by a research grant from the Robert Wood Johnson Pharmaceutical Research Institute, Raritan, N.J. We thank Jean Reyes for typing the manuscript.

REFERENCES

Bartlett MS (1937): Some examples of statistical methods of research in agriculture and applied biology. *JR Statist Soc* (Suppl) 4: 137-170.

Box GEP (1953): Non-normality and tests on variances. *Biometrika* 40: 318-335.

Cochran WG (1941): The distribution of the largest of a set of estimated variances as a fraction of their total. *Ann Eugen (London)* 11: 47-52.

Duara R, Grady C, Haxby J, Ingvar D, Sokoloff L, Margolin RA, Manning RG, Cutler NR and Rapoport RI (1984): Human brain glucose utilization and cognitive function in relation to age. *Ann Neurol* 16: 702-713.

Duara R, Gross-Glenn K, Barker WW, Chang JY, Apicella A, Lowenstein D and Boothe T (1987): Behavioral activation and the variability of cerebral glucose metabolic measurements. *J Cereb Blood Flow Metab* 7: 266-271.

Guilarte TR, Miceli RC and Camargo EE (1989): Personal communication.

Hamacher K, Coenen HH and Stocklin G (1986): Efficient stereospecific synthesis of no-carrier-added 2-[^{18}F]fluoro-2-deoxy-D-glucose using aminopolyether supported nucleophilic substitution. *J Nucl Med* 27: 235-238.

Heiss WD, Pawlik G, Herholz K, Wagner R, Goldner H and Weinhard K (1984): Regional kinetic constants and cerebral metabolic rate for glucose in normal human volunteers determined by dynamic positron emission tomography of [^{18}F]-2-fluoro-2-deoxyglucose. *J Cereb Blood Flow Metab* 4: 212-223.

Huang SC, Phelps ME, Hoffman EJ, Sideris K, Selin CJ and Kuhl DE (1989): Non-invasive determination of local cerebral metabolic rate of glucose in man. *Am J Physiol* 238: E89-E69.

Metz JT, Odeh H and Cooper MD (1989): "First session" effects in PET studies. *J Nucl Med* 30: 899.

Monk JP and Campoli-Richards DM (1987): Ofloxacin. A review of its antibacterial activity, pharmacokinetic properties and therapeutic use. *Drugs* 33: 346-391.

Nistri A and Constanti A (1979): Pharmacological characterization of different types of GABA and glutamate receptors in vertabrates and invertebrates. *Prog Neurobiol* 13: 117-235.

Sato K, Matsuura Y, Inoue M, Une T, Osada Y, Ogawa H and Mitsuhashi S (1982): *In vitro* and *in vivo* activity of DL-8280, a new oxazine derivative. *Antimicrob Agents Chemother* 22: 548-553.

Tallman JF and Gallager DW (1985): The GABA-ergic system: a locus of benzodiazepine action. *Ann Rev Neurosci* 8: 21-44.

Tsuji A, Sato H, Kume Y, Tamai I, Okezaki E, Hagata O and Kato H (1988): Inhibitory effects of quinolone antibacterial agents of g-aminobutyric acid binding to receptor sites in rat brain membranes. *Antimicrob Agents Chemother* 32: 190-194.

Tukey JW (1949):. Comparing individual means in the analysis of variance. *Biometrics* 5: 99-114.

Tyler JL, Strother SC, Zatorre RJ, Alivisatos B, Worsley KJ, Diksic M and Yamamoto YL (1988): Stability of regional cerebral glucose metabolism in the normal brain measured by positron emission tomography. *J Nucl Med* 29: 631-642.

Van Caekenberghe DL and Pattyn SR (1984): *In vitro* activity of ciprofloxacin compared with those of other new fluorinated piperazinyl-substituted quinolone derivatives. *Antimicrob Agents Chemother* 25: 518-521.

POSITRON EMISSION TOMOGRAPHY, ENZYMES AND DRUG RESEARCH AND DEVELOPMENT

Joanna S. Fowler, *Nora D. Volkow and Alfred P. Wolf
Department of Chemistry and the *Medical Department
Brookhaven National Laboratory

INTRODUCTION

The use of Positron Emission Tomography (PET) to examine drug pharmacokinetics and pharmacodynamics and the relationship of these properties to the behavioral, therapeutic and toxic properties of drugs is emerging as a powerful new scientific tool (Fowler et al., 1990). PET provides a new perspective on drug research by virtue of its ability to directly assess different parameters such as absolute uptake, regional distribution, kinetics, pharmacological profile and metabolic effects of drugs in human subjects. For example, the labeled drug itself can be used to measure the absolute uptake, regional distribution and kinetics at its site of action in the body. Additionally, the labeled drug and whole body PET can be used to determine the target organs for the drug and its labeled metabolites, thus providing information on potential toxic effects as well as tissue half-lives. On the other hand, different labeled tracers can be used to assess the effects of a drug on particular physiological or functional processes. For example, with appropriate radiotracers, the effects of a drug on metabolism, neurotransmitter activity, blood flow, enzyme activity or other processes can be probed. Because of the short half-life of the positron emitters, these parameters can be assessed

Nuclear Imaging in Drug Discovery, Development, and Approval
H.D. Burns, R. Gibson, R. Dannals, Peter Siegl, Editors

directly in the human body both in normal controls and in patients and serial studies can be done where a subject serves as his own control. Moreover, multiple tracers can be used so that different parameters can be assessed in the same subject. An important point is that PET can be used to assess the behavior of a drug at its site of action directly in human subjects. This is relevant because drug pharmacokinetics and availability may vary across animal species. It also enables the assessment of drug behavior in diseases where there are no animal models. This information is not available with other methods and places PET in a unique position to streamline the process of drug development and evaluation.

Enzymes are common molecular targets in drug development (Sammes, 1990; Crout, 1989). Thus, the development of PET tracers which can be used to assess the effects of a drug on enzyme activity at the site of action of the drug in humans can be of potentially great importance in drug development. [^{18}F]2-fluoro-2-deoxy-D-glucose (FDG) is the most widely used PET tracer and is metabolically trapped by its reaction with the glycolytic enzyme hexokinase (Reivich et al., 1979). It has been used extensively in the study of therapeutic drugs and substances of abuse (Fowler et al., 1990). Although, a detailed discussion of FDG is beyond the scope of this chapter, it highlights the potential importance and widespread application of enzyme targeted tracers.

Except for FDG, there are only a small number of PET tracers which have been developed specifically to probe enzyme activity. Many of these have not been fully validated as tracers for a particular enzymatic reaction. Still fewer have actually been applied with a view to drug development. Nonetheless, there is great potential in the use of PET and appropriate radiotracers for the study of enzyme-targeted drugs.

In this chapter, we will focus on examples of PET tracers to probe enzyme activity. Most of the PET work in this area to date has focussed on monoamine oxidase. Thus research in this area will be a major focus of this chapter. However, PET tracers for several other enzymes are under development and these will also be summarized. Although, an attempt has been made to survey literature for potential PET tracers for application in the development of enzyme targeted drugs, it is certainly possible that some examples were overlooked. Therefore, if there are omissions which might be expected to be of future value, we would appreciate having them brought to our attention.

MONOAMINE OXIDASE, MONOAMINE OXIDASE INHIBITOR DRUGS AND TRACERS

Monoamine oxidase (MAO, E.C. 1.4.3.4) was discovered in 1928 (Hare, 1928). It is a flavin containing enzyme which is located on the outer mitochondrial membrane and is responsible for the oxidative deamination of endogenous neurotransmitter amines as well as amines from exogenous sources. MAO exists in two subtypes, MAO A and MAO B which are identified by their substrate selectivities and their sensitivity to different inhibitors (Fowler et al., 1981; Finberg and Youdim, 1983). MAO A oxidizes 5-hydroxytryptamine and is selectively inhibited by clorgyline (Johnston, 1968) whereas MAO B oxidizes benzylamine and is selectively inhibited by L-deprenyl (Knoll and Magyar, 1972). MAO formally oxidizes amines according to the following reaction scheme:

$$RCH_2NH_2 + O_2 + H_2O \xrightarrow{MAO} RCHO + H_2O_2 + NH_3$$

Interest in MAO as a therapeutic target was stimulated in the early 1950s when it was serendipitously discovered that iproniazid, a drug used to treat tuberculosis, caused mood elevation and behavioral activation in some patients. Subsequent biochemical studies revealed that iproniazid inhibited MAO (Zeller et al., 1955). This contributed to the hypothesis that monoamine regulation may play a role in mood and led to the use of MAO inhibitors in the treatment of depression (Pare, 1972). Currently the application of MAO inhibitors as antidepressants is mostly limited to the use of MAO A or mixed A and B inhibitors.

Following the use of MAO inhibitors as antidepressant drugs, the selective irreversible MAO B inhibitor, L-deprenyl was used in combination with L-DOPA therapy in Parkinson's disease to inhibit the degradation of dopamine. The combination of L-DOPA and L-deprenyl has been reported to have prolonged therapeutic efficacy in patients with Parkinson's disease when compared to the use of L-DOPA alone (Birkmayer and Riederer, 1984).

The success of L-deprenyl therapy and the rationale for its combined use with L-DOPA was generally attributed to its synergistic action with L-DOPA. However, in 1983, it was discovered that pretreatment with L-deprenyl could prevent the development of Parkinson's disease in animals treated with 1-methyl-4-phenyl-1,2,3,5-tetrahydropyridine (MPTP) (Heikkila et al., 1984). MAO B inhibition prevented neurotoxicity by inhibiting the conversion of MPTP to MPP+ (1-methyl-

4-phenylpyridinium), a neurotoxin. This observation lead to the hypothesis that the MAO B catalyzed activation of an environmental toxin may be associated with the development of Parkinson's disease and that treatment with MAO B inhibitors might have a neuroprotective effect. Subsequent clinical trials in early Parkinson's disease showed that L-deprenyl significantly retarded the progression of the disease as determined by assessing its effects on the time period between the early symptoms of the disease and the requirement for L-DOPA therapy (Tetrud and Langston, 1989; The Parkinson's Study Group, 1989; Langston, 1990).

The success of L-deprenyl has stimulated the development and examination of other MAO B inhibitors with different characteristics and has raised questions on its mechanisms of action (Cohen and Spina, 1989), on whether its effects are protective or symptomatic (Kofman, 1991) and on its duration of action (Friedhoff, 1990). PET is currently being used to address some of these issues. In the following section, progress in the development and evaluation of PET tracers for assessing MAO activity in the brain will be summarized.

PET Radiotracers for MAO

Three different approaches have been explored for selectively probing MAO: i) radiorespirometry (with Carbon-11 labeled aliphatic amines) (Gallagher et al., 1977); ii) the labeled substrate approach (with Carbon-11 labeled N,N-dimethylphenethylamine); and iii) the labeled suicide inactivator approach (with [^{11}C]L-deprenyl, [^{11}C]clorgyline and [^{11}C]pargyline). The discussion below will focus on the latter two approaches since radiorespirometry lacked the selectivity to probe MAO regionally.

[^{11}Cl]N,N-Dimethylphenethylamine (DMPEA, **1**) is a good substrate for MAO B (Inoue et al., 1985). Its oxidation by MAO B produces Carbon-11 labeled methylamine. Since methylamine exists predominantly in the protonated form at physiological pH, it remains trapped at its site of production. Although the other product of the reaction, phenylacetaldehyde undergoes further metabolism, it (and its

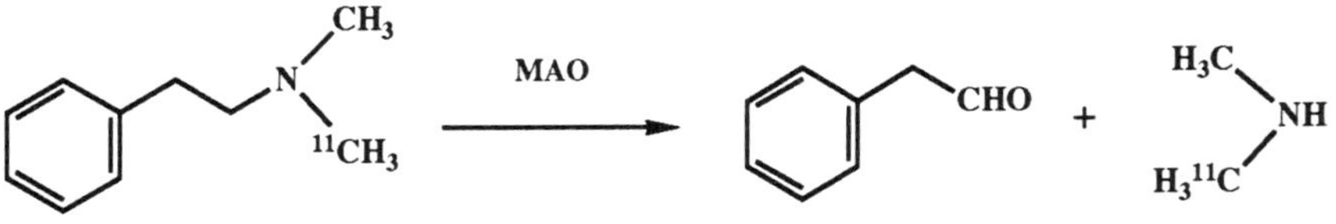

1

metabolites) is unlabeled and not detected. Mechanistic studies, including the demonstration of a deuterium isotope effect, and PET studies examining the effect of labeling on different positions of the DMPEA molecule (Halldin et al., 1989; Inoue et al., 1985; Hashimoto et al., 1986) set the stage for recent studies with PET which showed the trapping of radioactivity within MAO rich areas of the human brain (Shinotoh et al., 1987).

Clorgyline (**2**) (Johnston, 1968) and L-deprenyl (**3**) (Knoll, 1972) are irreversible inhibitors of MAO A and B which have been labeled with Carbon-11 for PET studies (MacGregor et al., 1988). Pargyline has also been labeled with Carbon-11 (Ishiwata et al., 1985). In unlabeled form, these compounds have been used as basic tools in the neurosciences (Finberg et al., 1983; Fowler et. al, 1981) and also as investigational drugs in therapeutic applications to diseases involving abnormalities in monoamine concentration or metabolism (Murphy et al., 1983). Clorgyline and L-deprenyl, along with the antihypertensive drug, pargyline, have a common structural feature, the N-propargyl group. The propargyl group is a latent reactive functional group which is unmasked within the enzyme-substrate complex during the catalytic step (Maycock et al., 1976). The catalytically activated substrate then forms a covalent bond to the flavin cofactor of the enzyme causing its irreversible inactivation. This process is commonly referred to as suicide inactivation (or mechanism based inactivation) because the enzyme catalyzes its own destruction (Abeles and Maycock, 1976).

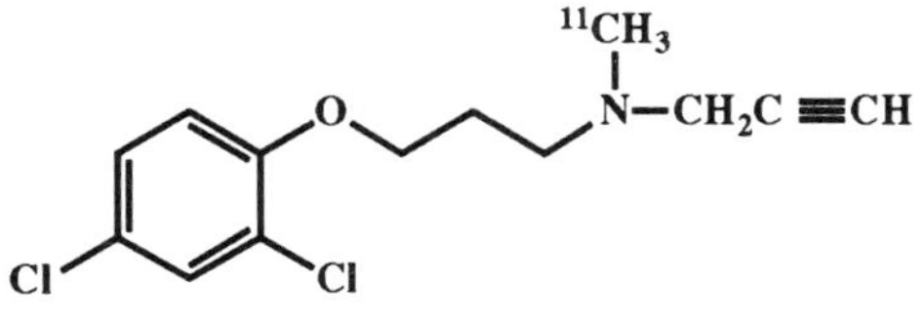

$^{11}CH_3$ O N$-CH_2C\equiv CH$ Cl Cl

2

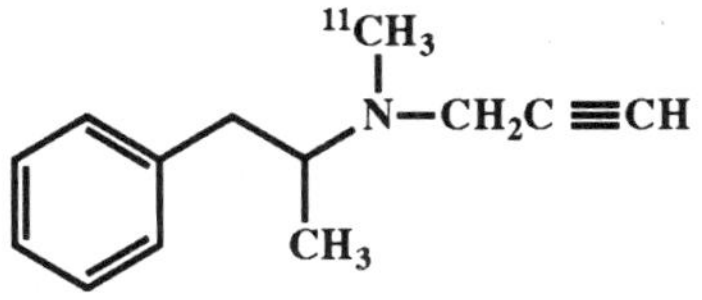

$^{11}CH_3$ N$-CH_2C\equiv CH$ CH_3

3

The rationale for using a suicide inhibitor to assess functional MAO activity *in vivo* with PET is based on the covalent attachment of the labeled suicide inactivator to the enzyme while it is in the enzyme-substrate complex. Because the covalent labeling of the enzyme requires that the labeled suicide inactivator be oxidized by the enzyme, the image of radioactivity distribution after the initial distribution phase represents functional enzyme activity. For this approach to be successful, the partition ratio must be small (i.e., the rate of enzyme inactivation must be greater than the rate of turnover) (Waley, 1985; Walsh, 1982).

The feasibility of using Carbon-11 labeled clorgyline and L-deprenyl to label MAO A and B *in vivo* and to map the regional concentration of MAO A and B in brain was initially demonstrated in mice (MacGregor et al., 1985) in experiments which demonstrated that Carbon-11 activity in different organs was covalently bound to protein, as one would expect if suicide inactivation and labeling of the enzyme had taken place. These studies formed the basis for baboon and human studies (Fowler et al., 1987) with the Carbon-11 labeled tracers. Although human PET studies with [^{11}C]clorgyline were consistent with MAO A activity, the major focus after these initial studies has been on the study of MAO B with [^{11}C]L-deprenyl. In validating the labeled suicide inactivator approach, studies in humans and baboons were used to examine the following parameters:

1) Pharmacology: PET studies in humans and in baboons have demonstrated that MAO inhibition reduces the uptake and retention of L-deprenyl (Fowler et al., 1987; Arnett et al., 1987).
2) Regional distribution: The regional distribution of Carbon-11 after the injection of [^{11}C]clorgyline or [^{11}C]L-deprenyl (Fowler et al, 1987) parallels the distribution of the two enzyme subtypes as determined by *in vitro* analysis of MAO activity in human autopsy material (Fowler et al.,1980) and immunohistochemical analysis of MAO in non-human primate brain (Westlund et al., 1985).
3) Stereoselectivity: Deprenyl contains an asymmetric carbon atom and the MAO inhibitory potency of the L-enantiomer of deprenyl has been shown to exceed that of the D-enantiomer approximately 25 times (Robinson 1985). A comparison study of [^{11}C]L and D-deprenyl in human brain showed stereoselective retention of the L-isomer (Fowler et al., 1987).
4) Deuterium isotope effects: The rate limiting step for oxidation by MAO is alpha-hydrogen abstraction (Belleau and Moran, 1963; Yu et al., 1981). Substitution of hydrogen by deuterium in a carbon-hydrogen bond is used as a mechanistic probe to determine whether the rate limiting step in a reaction involves the cleavage of this particular bond. A comparison of the uptake and retention of [^{11}C]L-deprenyl and [^{11}C]L-deprenyl-α,α-D_2 (**4**) in baboon brain, using a three compartment model to reveal a significant deuterium isotope effect, established that MAO catalyzed cleavage of the alpha carbon-hydrogen bond on the propargyl group is the rate limiting (or a major rate contributing) step in the retention of carbon-11 in brain (Fowler et al., 1988).

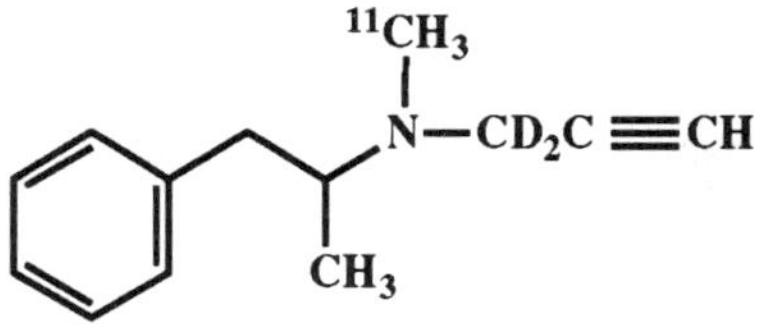

$^{11}CH_3$

$N-CD_2C\equiv CH$

CH_3

4

PET Study Design and Quantitation

One of the great advantages of PET and Carbon-11 labeled tracers is the ability to make repeated measurement on the same subject and to effectively use a subject as his own control. Accurate evaluation of the effects of a drug on a given subject requires a knowledge of the intra-individual reproducibility of repeated measures. Similarly, the intercomparison of values across individuals requires a knowledge of intersubject variability preferably within the same age, gender and disease group. This is especially important in the study of MAO B, where it is known that the concentration of the enzyme increases with age (Robinson et al., 1972).

Two approaches have been used to provide an index of enzyme activity from dynamic PET measurements of [^{11}C]L-deprenyl (tissue time-activity data and plasma input function). The first used the Patlak graphical analysis approach for tracers which are irreversibly trapped in tissue over the time course of the study (Patlak et al., 1983; Arnett et al., 1987). This approach provides the parameter K_i which is the blood-to-tissue influx constant. The second approach used a three compartment model which is a simplification of the branched pathway kinetic scheme for suicide inhibition (Fowler et al., 1988; Brust et al., 1991; Lammertsma et al., 1991). Solution of the differential equations corresponding to this model provides the parameter k_3E, where k_3 is the bimolecular rate constant (concentration^{-1} x min^{-1} of the enzyme substrate complex) and E is the concentration of the free enzyme. The use of the three compartment model rather than the graphical analysis for data from [^{11}C]L-deprenyl experiments in humans is preferred because the large value of the transport constant (K_1) influences the blood to brain influx constant calculated using the graphical analysis method (Lammertsma et al., 1991). Model assumptions and application have been described (Fowler et al., 1988).

APPLICATIONS OF PET IN DRUG RESEARCH AND DEVELOPMENT RELATIVE TO MAO B

Both the labeled substrate approach with [^{11}C]DMPEA or the labeled suicide inactivator approach with [^{11}C]L-deprenyl represent viable approaches to mapping MAO B in brain. The fact that L-deprenyl is an approved drug significantly facilitates its use as a tracer for human PET studies. The following applications can be envisioned for [^{11}C]L-deprenyl or other validated tracers for MAO B activity:

1) Determination of the regional distribution of MAO B activity in the normal and diseased brain.
2) Assessment of the activity of MAO B after exposure to acute or chronic drug administration.
3) Determination of the effective dose of another MAO B inhibitor.
4) Determination of the subtype selectivity of MAO inhibitor.
5) Measurement of the duration of MAO B inhibition after MAO inhibitor.
6) Determination of the synthesis rate for MAO B in the normal and diseased human brain after inhibition with an irreversible MAO inhibitor.

PET and [^{11}C]L-deprenyl have been used to measure the rate of recovery (synthesis) of MAO B in baboon brain after a single (1 mg/kg) dose of L-deprenyl (Arnett et al., 1987). The results of this study showed that MAO B synthesis in baboon brain is very slow with a half-life for recovery of about 4 weeks (Fig. 1). Another PET study in human subjects reported a similar half-time of MAO B synthesis after a single 5 mg dose of L-deprenyl (Brust et al., 1991). A third PET study of MAO B recovery in pig brain showed a half-time of 6.5 days (Orland, 1991).

Clinical trials of L-deprenyl showing that it retards the progression of Parkinson's disease have raised questions as to whether the washout time after the last dose of L-deprenyl and the time at which symptoms were assessed were adequate considering that L-deprenyl is an irreversible MAO B inhibitor (for a summary see Langston, 1990). In principle, a PET study of MAO B recovery is a quantitative assessment of how long the enzyme inhibiting effects of the drug persist in the brain after the last dose of the drug. Such a study protocol should include a baseline study on each subject to control for intersubject variability and two or more PET studies at the time of maximal blockade and at different times after the last dose of drug.

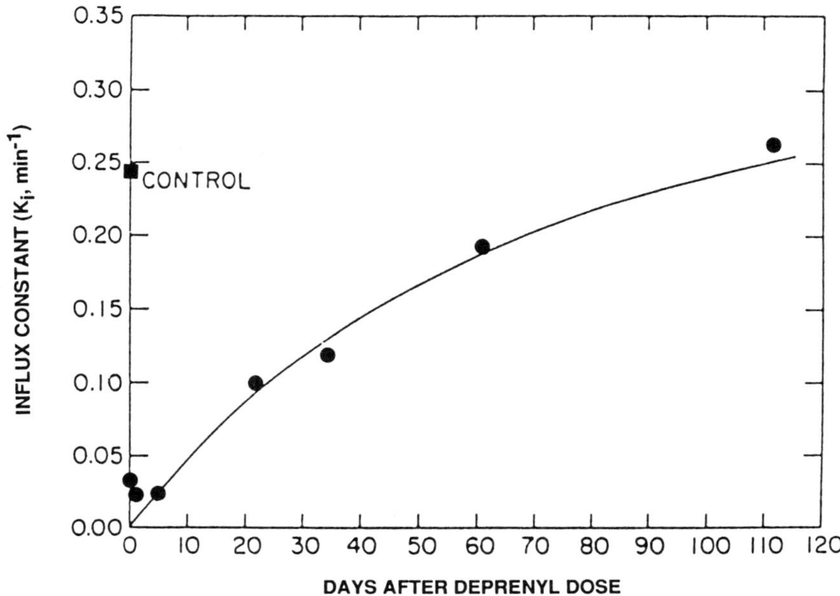

FIGURE 1. Effect of a single intravenous dose (1.0 mg/kg) of unlabeled L-deprenyl on the influx constants (K_i, in min^{-1}) for [^{11}C]L-deprenyl (adapted from Arnett et al., 1987).

RO 19 6327 (**5**) is a very selective, reversible MAO B inhibitor which is being developed to treat Parkinson's disease (Cesura et al., 1988). A recent publication describes the use of [^{11}C]L-deprenyl for determining the dose of RO 19 6327 required to inhibit brain MAO B in normal human subjects. The study protocol involved the administration of 6 different single doses of RO 19 6327 to six different subjects and performing a PET scan with [^{11}C]L-deprenyl 12 hours after the last dose of RO 19 6327. One unmedicated subject received a PET scan to serve as a baseline for the medicated subjects. Data was analyzed using a three compartment model to PET time-activity data and it was concluded that the drug dose of RO 19 6327 required to give 90% MAO B inhibition

O
$NHCH_2CH_2NH_2$
N
Cl

5

was 0.685 mg/kg (Bench et al., 1991).

ANGIOTENSIN CONVERTING ENZYME (ACE) DRUGS

The renin-angiotensin system is a target for antihypertensive drugs. Of particular interest are ACE inhibitors (Douglas, 1985). The importance of ACE stems from its involvement in the conversion of angiotensin I, a decapeptide with no biological activity, to angiotensin II, a vasoconstrictive octapeptide which releases aldosterone from the adrenal glands thereby conserving sodium and maintaining blood volume. ACE also catalyzes the cleavage of the naturally occurring vasodilator bradykinin. Although angiotensin I is slowly converted to angiotensin II in plasma, very rapid metabolism occurs *in vivo* due to the activity of tissue bound ACE on the luminal side of endothelial cells. For example, 20-40% of angiotensin I is converted to angiotensin II in the human lung in a single circulation. Extrapulmonary conversion of angiotensin I to angiotensin II also occurs.

Because of the potent pressor properties of angiotensin II, ACE inhibitors are an important class of drugs in the treatment of hypertension. Captopril (2D-methyl-3-mercaptopropanoyl-L-proline, K_i = 1.7 nM), the prototypical inhibitor of ACE, has been widely used to treat hypertension, and its use is expanding.

Recently, a Fluorine-18 labeled derivative of captopril (**6**) was synthesized and evaluated (Hwang et al., 1991). Its distribution in rats at 30 minutes showed high uptake in the lungs, kidneys and aorta, organs known to have high ACE concentrations. Uptake was saturable as determined by coinjection of captopril with [^{18}F]captopril and, in general, the expected binding capacity was obtained in lung, kidney, heart and aorta.

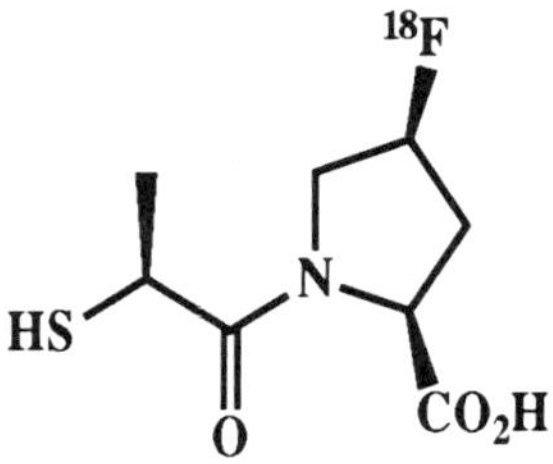

^{18}F

HS N O CO_2H

6

The suitability of [^{18}F]captopril to PET studies of ACE drugs was demonstrated in man where it was shown that displaceable activity was observed in lung and kidney (Fig. 2). One of the potential uses of

[18F]captopril is to study the kinetics of captopril and other ACE drugs *in vivo* in man where displacement studies will permit the determination of drug half-life.

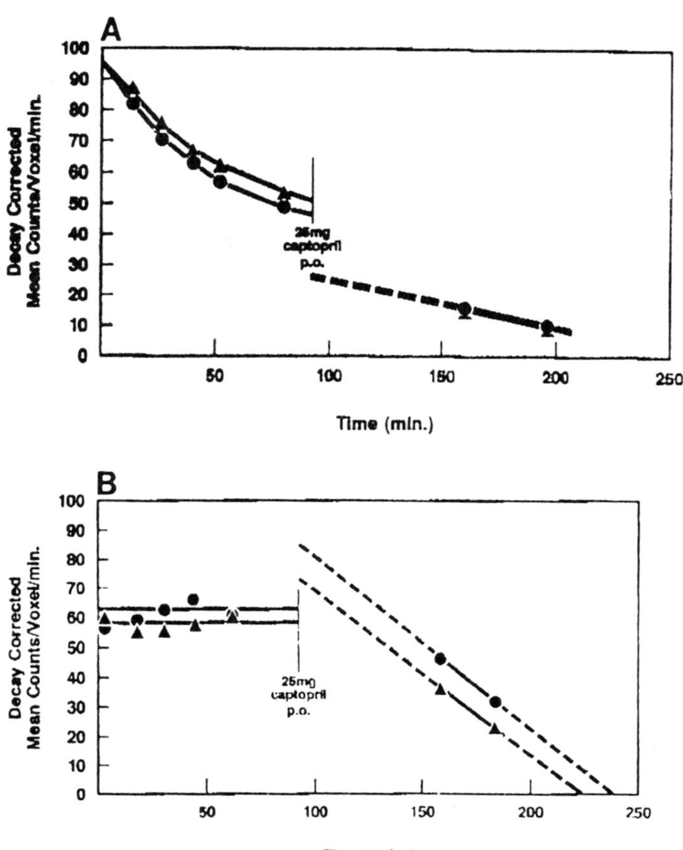

FIGURE 2. Decay-corrected time-activity curves generated from sequential PET images of the lung (A) and kidney (B) in a normal human subject following administration of [18F]captopril. ▲ designates the right organ and ● the left. Five images were taken prior to administration of unlabeled captopril and two were taken postadministration. Reprinted with permission of the Society of Nuclear Medicine from Hwang et al. (1991).

OTHER ENZYMES

In addition to the above examples, there are examples of other enzyme systems which are of potential relevance in drug research with PET and these are briefly described in this section. However, there is a caution that many of these have not been fully validated as markers for a particular enzyme and their application in drug research would require a careful assessment of their pharmacological and kinetic profiles to assure that their uptakes, regional distributions and kinetics reflect the specific process under investigation.

Ornithine Decarboxylase (ODC)

PET tracers designed for metabolic trapping have been prepared for examining ODC activity in tumors. Since ODC is expressed early in the deregulation of cellular activity and is one of the markers for rapid tissue growth, it has been a target for chemotherapeutic drugs. It has also been the target for tracers assessing increases in ODC activity. One of these is 5-[^{11}C]ornithine (**7**) (Ding et al., 1989). ODC catalyzed decarboxylation of 5-[^{11}C]ornithine would produce 1-[^{11}C]putrescine (**8**) which would be retained at the site of production. Note that labeling ornithine at the carboxyl carbon would result in a loss of label at the decarboxylation step. The tumor uptake of D,L-5-[^{14}C]ornithine and a comparative study of ODC activity in tumor with L[1-^{14}C] and L-[5-^{14}C]ornithine have recently been reported thus supporting this approach (Ishiwata et al., 1988; Elsinga et al., 1989).

$$\underset{\mathbf{7}}{NH_2{}^{11}CH_2CH_2CH_2\underset{\displaystyle CO_2H}{\underset{|}{C}}HNH_2} \xrightarrow{\text{ODC}} \underset{\mathbf{8}}{NH_2{}^{11}CH_2CH_2CH_2CH_2NH_2}$$

Glycosidase Enzymes

The synthesis and evaluation of 2-deoxy-2[^{18}F]fluoro-ß-mannosyl fluoride (**9**) as a mechanism based probe for glycosidase enzymes has recently been reported (Adam et al., 1992). This represents a PET labeled version of a class of covalent, mechanism based glycosidase inhibitors (Withers et al., 1987). This tracer was developed with a view toward imaging cerebral glycosidase activity with PET and ultimately to

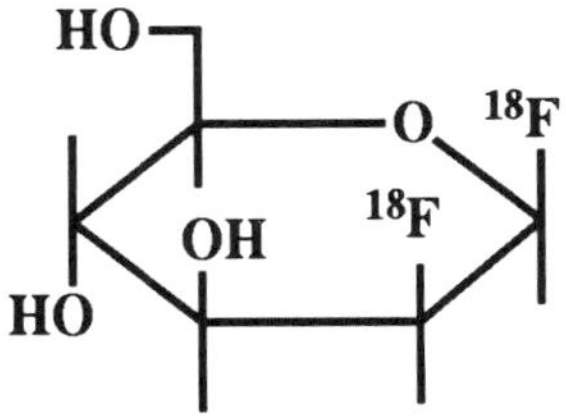

9

assess the activity of this enzyme in lysosomal storage diseases. This class of PET tracers may be of value in the future in the development of glucosidase inhibitors which are of interest in the treatment of obesity and diabetes because of their ability to control blood glucose levels.

Acetylcholinesterase

The development of strategies for increasing acetylcholine levels in brain is of current interest in the treatment of Alzheimier's disease (Sarter et al., 1988). One class of drugs being developed for this purpose is the acetylcholinesterase inhibitors like tacrine and physostigmine. The development of PET tracers for acetylcholinesterase is underway. One of these is the [^{11}C]N-methylamide (**10**) shown below (Mulholland et al., 1989). The unlabeled version of this drug has been shown to be a potent noncompetitive inhibitor of acetylcholinesterase (IC_{50} = 0.6 nM). A second potential tracer for acetylcholinesterase activity is [^{11}C]physostigmine (**11**) (Lours et al., 1992). Although both of these tracers were designed to assess acetylcholinesterase activity in Alzheimer's disease (and therefore to mark cholinergic terminals), one could envision the application of validated PET tracers of this class to the development and evaluation of drugs designed to enhance

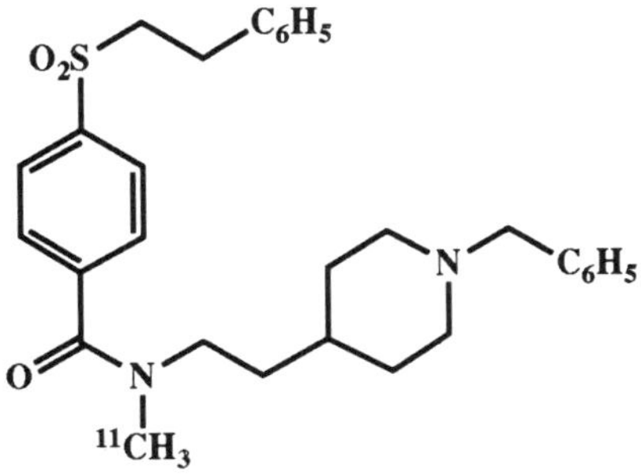

10

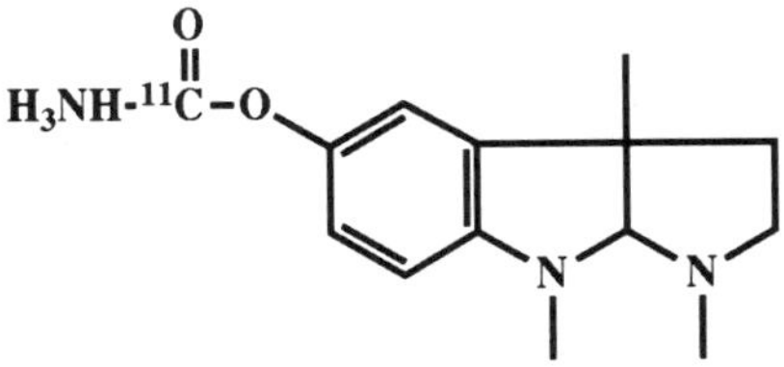

11

acetylcholine levels in brain. Another approach to assess the effectiveness of acetylcholinesterase inhibitor drugs in elevating acetylcholine levels would be to use a radioligand which binds reversibly to the cholinergic receptor and which is sensitive to changes in acetylcholine concentration.

L-Aromatic Amino Acid Decarboxylase

ß-Fluoromethylene-m-tyrosine is a substrate for dopa decarboxylase, an enzyme which resides in the presynaptic dopaminergic neuron (Palfreyman et al., 1985). Upon decarboxylation, it yields an irreversible MAO inhibitor which, upon reaction with MAO, becomes trapped in dopaminergic neurons. Studies are underway to assess the value of [^{18}F]fluoro-ß-fluoromethylene-m-tyrosine (**12**) for detecting and tracking neuronal degeneration (Murali et al., 1992; DeJesus, 1992). If this proves to be the case, then it may have a role in assessing the efficacy of inhibitors of dopa decarboxylase and for assessing the effectiveness of neuroprotective drugs. A similar argument may be made for 6-[^{18}F]fluoro-L-DOPA (**13**) which has been shown to be a tracer for L-aromatic amino acid decarboxylase (Firnau et al., 1986; Melega et al., 1991; Gjedde et al., 1991).

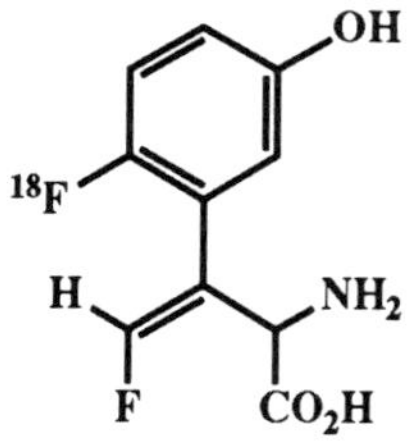

12 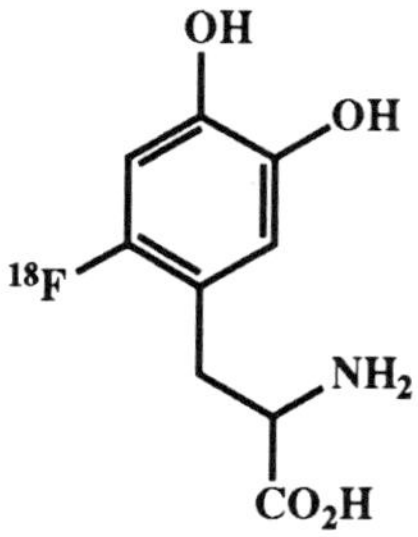 **13**

Diacylglycerol Kinase and Phosphoinositol Turnover

Neuronal activity is driven by second messenger systems. The possibility of imaging the receptor mediated phosphoinositol response with PET has been recently examined using sn-1,2-[^{11}C]diacylglycerol as a specific tracer for phosphoinositol turnover. Using this tracer and autoradiography in animals, discrete areas of Carbon-11 trapping are visualized in the neocortex suggesting that active neuronal processing occurs mediated by the phosphoinositol response (Imahori et al., 1991). Visualization of this pathway involves metabolic trapping of Carbon-11 via phosphorylation.

OTHER TRACERS

Fluconazole is a synthetic antifungal drug which has been labeled with Fluorine-18 (Livni et al., 1992). Its mechanism of action is the inhibition of fungal cytochrome P450 sterol alpha demethylation. It has been labeled in order to study its pharmacokinetics. The enzyme, urokinase, has been labeled with Fluorine-18 as a potential tracer for detecting deep vein thrombosis. This enzyme participates in clot formation and its use in labeled form may have potential for examining drugs which are used to treat clot formation and fibrinolysis (Muller-Platz et al., 1982).

OTHER APPROACHES: PROBING THE EFFECTS OF ENZYME INHIBITION INDIRECTLY BY PROBING CHANGES IN NEUROTRANSMITTER CONCENTRATION AND NEURO-TRANSMITTER INTERACTIONS

In the examples described above, the approach is to directly examine the interaction of the radiotracer with the enzyme. However, PET can also be used to examine the neurochemical effects of enzyme inhibitor drugs through the use of a radiotracer which probes changes in neurotransmitter concentration. For example, γ-vinyl-γ-aminobutyric acid (GVG, **14**) is an anticonvulsant drug. It is a suicide inactivator of the enzyme γ-aminobutyric acid (GABA) transaminase (Palfreyman et al., 1981). The mechanism for its anticonvulsant activity involves the facilitation of inhibitory GABAergic neurotransmission through the inhibition of the enzyme which metabolizes GABA. Although there are presently no PET tracers for probing GABA levels or for probing GABA transaminase levels, the well characterized inhibitory interaction of the GABAergic neuron on the dopaminergic neuron (Kubota et al.,1987) allows one to assess the effects of enzyme inhibition through the

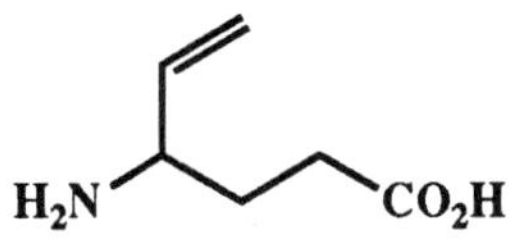

H_2N CO_2H

14

use of [^{11}C]raclopride (**15**) a selective, reversible inhibitor of the dopamine D2 receptor (Farde et al., 1986). GVG administration facilitates GABA thereby inhibiting dopamine release. This decrease in dopamine release is reflected by an increase in [^{11}C]raclopride binding (Dewey et al., 1992). The use of reversibly binding, D2 receptor selective tracers like [^{11}C]raclopride could also be envisioned to assess changes in synaptic dopamine arising from other enzyme-targeted drugs which enhance or decrease synaptic dopamine.

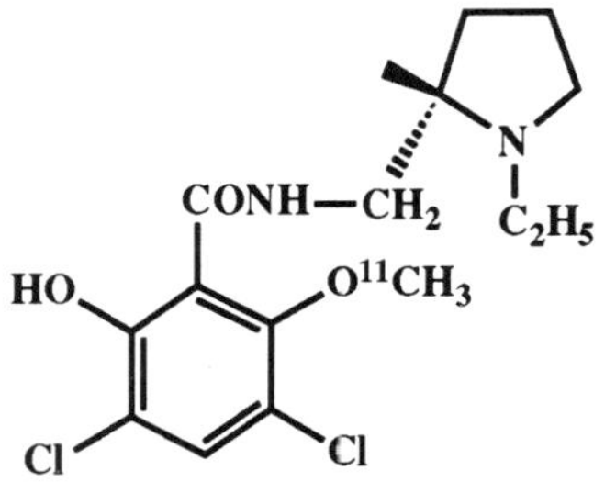

N
CONH—CH_2 C_2H_5
HO $O^{11}CH_3$
Cl Cl

15

CONCLUSION

In summary, PET is uniquely capable of providing insight into biochemical transformations in the living human and animal body. We have presented several examples of PET studies where the direct interaction of the tracer with the enzyme is used to provide information as well as indirect approaches where the effect of drug promoted alterations in enzyme activity are observed through neurotransmitter interactions resulting in an alteration in the behavior of another radiotracer. With the exception of FDG, to date there are only a few examples of the application of PET in drug research. In particular, in the development of drugs with enzyme targets, the use of this technique can be expected to increase in the future.

ACKNOWLEDGMENTS

This chapter was written at Brookhaven National Laboratory under contract DE-AC-76CH00016 with the U.S. Department of Energy and supported by its Office of Health and Environmental Research.

REFERENCES

Abeles RH and Maycock AL (1976): Suicide enzyme inactivators. *Acct Chem Res* 9: 313-319.

Adam MJ, McCarter JD and Withers SG (1992): Synthesis and evaluation of 2-deoxy-2-[^{18}F]-fluoro-ß-mannosyl fluoride as a mechanism-based imaging probe for glycosidase enzymes. *IXth International Symposium on "Radiopharmaceutical Chemistry"*, Paris, France, April 4-10: C9 (Abst).

Arnett CD, Fowler JS, MacGregor RR, Schlyer DJ and Wolf AP (1987): Turnover of brain monoamine oxidase measured *in vivo* by positron emission tomography using [^{11}C]-L-deprenyl. *J Neurochem* 49: 522-527.

Belleau B and Moran J (1963): Deuterium isotope effects in relation to the chemical mechanisms of monoamine oxidase. *Ann NY Acad Sci* 107: 822-839.

Bench CJ, Price GW, Lammertsma AA, Cremer JC, Luthra SK, Turton D, Dolan RJ, Kettler R, Dingemanse J, DaPrada M, Biziere K, McClelland GR, Jamieson VL, Wood ND and Frackowiak RSJ (1991): Measurement of human cerebral monoamine oxidase type B (MAO-B) activity with positron emission tomography (PET): A dose ranging study with the reversible inhibitor RO19-6327. *Eur J Clin Pharmacol* 10: 169-173.

Birkmayer W and Riederer P (1984): Deprenyl prolongs the therapeutic efficacy of combined L-DOPA in Parkinson's disease. *Adv Neurol* 40: 475-481.

Bonnot Lours S, Prenant C and Crouzel C (1992): Synthesis of [^{11}C]methylisocyanate. Application to the labelling of a potent acetylcholinesterase inhibitor: [^{11}C]physostigmine. *IXth International Symposium on "Radiopharmaceutical Chemistry"*, Paris, France, April 4-10: D4 (abst).

Brust P, Andersen BR, Diksic M and Gjedde A (1991): XVI-1. Monoamine oxidase B activity in human brain measured with L-[^{11}C]deprenyl. *J Cereb Blood Flow Metab* 11(Suppl 2): S227.

Cesura AM, Imhof R, Galva MD, Kettler R and DaPrada M (1988): Interactions of the novel inhibitors of MAO B RO 19-6327 and RO 16-6491 with the active site of the enzyme. *Pharmacol Res Commun* 20: 51-61.

Cohen G and Spina MB (1989): Deprenyl suppresses the oxidant stress associated with increased dopamine turnover. *Ann Neurol* 26: 689-690.

Crout JR (1989): Enzyme inhibitors as drugs. In: *Enzymes as Targets for Drug Design*, Palfreyman MG, McCann PP, Lovenberg W,

Temple JG Jr and Sjoerdsma A, eds. San Diego: Academic Press.

DeJesus OT, Holden JE, Endres C, Murali D, Oakes TR, Shelton S, Uno H, Houser D, Freund L, Perlman SB, Nickles RJ (1992): Visualization of dopamine nerve terminals by positron tomography using [^{18}F]-fluoro-ß-fluoromethylene-m-tyrosine. *Neuroscience J*: in press.

Dewey SL, Smith GS, Logan J, Brodie JD, Yu D-W, Ferrieri RA, King PT, MacGregor RR, Martin TP, Wolf AP, Volkow ND and Fowler JS (1992): GABAergic inhibition of endogenous dopamine release measured *in vivo* with ^{11}C-raclopride and positron emission tomography. *J Neuroscience*: in press.

Ding Y-S, Antoni G, Fowler JS, Wolf AP, Langstrom B (1989): Synthesis of L-[5-^{11}C]ornithine. *J Label Cmpds Radiopharm* 27: 1079-1090.

Douglas WW (1985): Polypeptides - Angiotensin, plasma kinins and others. In: *The Pharmacological Basis of Therapeutics*, Gilman AG and Goodman LS, eds. New York: Macmillan Publishing Co.

Elsinga PH, Daemen BJG and Vaalburg W (1989): Ornithine decarboxylase activity in tumor: A study with L-[1-^{14}C]- and L-[5-^{14}C]-Ornithine. *J Label Cmpds Radiopharm* 26: 481 (abst).

Farde L, Hall H, Ehrin E and Sedvall G (1986): Quantitative analysis of dopamine-D2 receptor binding in the living human brain by positron emission tomography. *Science* 231: 258-261.

Finberg JPM and Youdim MBH (1983): Selective MAO A and B inhibitors: Their mechanism of action and pharmacology. *Neuropharmacology* 22: 441-446.

Firnau G, Garnett ES, Chirakal R, Sood S, Nahmias C and Schrobilgen G (1986): [^{18}F]Fluoro-L-dopa for the *in vivo* study of intracerebral dopamine. *Appl Radiat Isot* 37: 669-675.

Fowler JS (1990): Enzyme activity: Monoamine oxidase. In: *Quantitative Imaging: Neuroreceptors, Neurotransmitters, and Enzymes*, Frost JJ and Wagner HN Jr, eds. New York: Raven Press.

Fowler CJ, Oreland L and Callingham BA (1981): The acetylenic monoamine oxidase inhibitors clorgyline, deprenyl, pargyline and J-508: Their properties and applications. *J Pharm Pharmacol* 33: 341-347.

Fowler CJ, Oreland L, Marcusson J and Winblad B (1980): Titration of human brain monoamine oxidase-A and -B by clorgyline and L-deprenyl. *Naunyn-Schmiedberg's Arch Pharmacol* 311: 263-272.

Fowler JS, MacGregor RR, Wolf AP, Arnett CD, Dewey SL, Schlyer D, Christman D, Logan J, Smith M, Sachs H, Aquilonius SM, Bjurling P, Halldin C, Hartwig P, Leenders KL, Lundquist H, Oreland L, Stalnacke C-G and Langstrom B (1987): Mapping human brain monoamine oxidase A and B with ^{11}C-suicide inactivators and positron emission tomography. *Science* 235: 481-485.

Fowler JS, Wolf AP and Volkow ND (1990): New directions in positron emission tomography-Part II. In: *Annual Reports in Medicinal Chemistry*, Bristol JA, ed. San Diego: Academic Press, Inc.

Fowler JS, Wolf AP, MacGregor RR, Dewey SL, Logan J, Schlyer DJ and Langstrom B (1988): Mechanistic positron emission tomography Studies: Demonstration of a deuterium isotope effect in the monoamine oxidase-catalyzed binding of [^{11}C]L-deprenyl in living baboon brain. *J Neurochem* 51: 1524-1534.

Friedhoff AJ (1990): Deprenyl and the progression of Parkinson's Disease. *Science* 249: 303.

Gallagher BM, Fowler JS, MacGregor RR and Wolf AP (1977): Evaluation of radiorespirometry for the determination of monoamine oxidase activity *in vivo* utilizing [^{11}C]octylamine as a substrate. *Biochem Pharmacol* 26: 1917-1922.

Gjedde A, Reith J, Dyve S, Leger G, Guttman M, Diksic M, Evans A and Kuwabara H (1991): Dopa decarboxylase activity of the living human brain. *Proc Natl Acad Sci* 88: 2721-2725.

Halldin C, Bjurling P, Stalnacke C-G, Jossan SS, Oreland L and Langstrom B (1989): ^{11}C-Labelling of dimethylphenethylamine in two different positions and biodistribution studies. *Appl Radiat Isot* 40: 557-560.

Hare MLC (1928): Tyramine oxidase. I. A new enzyme system in liver. *Biochem J* 22: 968-979.

Hashimoto K, Inoue O, Suzuki K, Yamasaki F and Kojima M (1986): Deuterium isotope effect of [^{11}C]N,N-dimethylphenethyl-amine-α,α-d_2; reduction in metabolic trapping rate in brain. *Nucl Med Biol* 13: 79-80.

Heikkila RE, Manzino L, Cabbat FS and Duvoisin RC (1984): Protection against the dopaminergic neurotoxicity of 1-methyl-4-phenyl-1,2,5,6-tetrahydropyridine by monoamine oxidase inhibitors. *Nature (London)* 311: 467-469.

Hwang D-R, Eckelman WC, Mathias CJ, Petrillo EW Jr, Lloyd J and Welch MJ (1991): Positron-labeled angiotensin-converting enzyme (ACE) inhibitor: Fluorine-18-fluorocaptopril. Probing the ACE activity *in vivo* by positron emission tomography. *J Nucl Med* 32: 1730-1737.

Imahori Y, Fujii R, Ueda S, Ido T, Nishino H, Moriyama Y, Yamamoto YL and Nakahashi H (1991): No-carrier-added carbon-11-labeled sn-1,2- and sn-1,3-diacylglycerols by [^{11}C]propyl ketene method. *J Nucl Med* 32: 1622-1626.

Inoue O, Tominaga T, Yamasaki T and Kinemuchi H (1985): Radioactive N,N-dimethylphenylethylamine: A selective radiotracer for *in vivo* measurement of monoamine oxidase-B activity in the brain. *J Neurochem* 44: 210-216.

Ishiwata K, Ido T, Yanai K, Kawashima K, Miura Y, Monma M, Watanuki S, Takahashi T and Iwata R (1985): Biodistribution of a positron-emitting suicide inactivator of monoamine oxidase, carbon-11 pargyline, in mice and a rabbit. *J Nucl Med* 26: 630-636.

Ishiwata K, Abe Y, Matsuzawa T and Ido T (1988): Tumor uptake studies of D,L-[5-^{14}C]ornithine and D,L-2-difluoromethyl[5-^{14}C]ornithine. *Int J Radiat Appl Instrum Part B* 15: 119.

Johnston JP (1968): Some observations upon a new inhibitor of monoamine oxidase in brain tissue. Biochem Pharmacol 17: 1285-1297.

Knoll J and Magyar K (1972): Some puzzling effects of monoamine oxidase inhibitors. *Adv Biochem Psychopharmacol* 5: 393-408.

Kofman OS (1991): Deprenyl: Protective vs. symptomatic effect. *Le Canadian J Neurol Sci* 18: 83-85.

Kubota Y, Inagaki S, Kito S and Wu JY (1987): Dopaminergic axons directly make synapses with GABAergic neurons in the rat neostriatum. *Brain Res* 406: 147-156.

Lammertsma AA, Bench CJ, Price GW, Cremer JE, Luthra SK, Turton D, Wood ND and Frackowiak RSJ (1991): Measurement of cerebral monoamine oxidase B activity using L-[^{11}C]deprenyl and dynamic positron emission tomography. *J Cereb Blood Flow Metab* 11: 545-556.

Langston JW (1990): Selegiline as neuroprotective therapy in Parkinson's disease. *Neurology* 40(Suppl 3): 61-66.

Livni E, Fischman AJ, Ray S, Sinclair I, Elmaleh DR, Alpert NM, Weiss S, Correia JA, Webb D, Dahl R, Robeson W, Margouleff D, Liss R, Strauss HW and Rubin RH (1992): Synthesis of ^{18}F-labeled fluconazole and positron emission tomography studies in rabbits. *Nucl Med Biol* 19: 191-199.

MacGregor RR, Halldin C, Fowler JS, Wolf AP, Arnett CD, Langstrom B and Alexoff D (1985): Selective, irreversible *in vivo* binding of [^{11}C]clorgyline and [^{11}C]-L-deprenyl in mice: Potential for measurement of functional monoamine oxidase activity in brain using positron emission tomography. *Biochem Pharmacol* 34: 3207-3210.

MacGregor RR, Fowler JS and Wolf AP (1988): Synthesis of suicide inhibitors of monoamine oxidase: Carbon-11 labeled clorgyline, L-deprenyl and D-deprenyl. *J Label Cmpds Radiopharm* 25: 1-9.

Maycock AL, Abeles RH, Salach JI and Singer TP (1976): The structure of the covalent adduct formed by the interaction of 3-dimethylamino-1-propyne and the flavine of mitochondrial amine oxidase. *Biochemistry* 15: 114-125.

Melega WP, Grafton ST, Huang S-C, Satyamurthy N, Phelps ME and Barrio JR (1991): L-6-[^{18}F]Fluoro-DOPA metabolism in monkeys and humans: Biochemical parameters for the formulation of tracer kinetic models with positron emission tomography. *J Cereb Blood Flow Metab* 11: 890-897.

Mulholland GK, Sugimoto H, Jewett DM and Kilbourn MR (1989): Simple preparation of a novel C-11 acetylcholinesterase inhibitor. *J Nucl Med* 30: 822.

Muller-Platz CM, Kloster G, Legler G and Stocklin G (1982): ^{18}F-Fluoroacetate: An agent for introducing no-carrier-added fluorine-18 into urokinase without loss of biological activity. *J Label Cmpds Radiopharm.* 19: 1645-1646.

Murali D, DeJesus OT, Sunderland JJ and Nickles RJ (1992): [^{18}F]-Fluoro-ß-fluoromethylene-m-tyrosine analogs, potential PET agents for presynaptic dopamine terminals: Synthesis and spectroscopic characterization. *Int J Radiat Appl Instrum Part A*: in press.

Murphy DL, Cohen RM, Siever LJ, Roy B, Karoum F, Wyatt RJ, Garrick NA and Linnoila M (1983): Clinical and laboratory studies with selective monoamine-oxidase-inhibiting drugs. *Mod Probl Pharmacopsychiat* 19: 287-303.

Oreland L (1991): Monoamine oxidase, dopamine and Parkinson's disease. *Acta Neurol Scand* 84(Suppl 136): 60-65.

Palfreyman MG, Schechter PJ, Buckett WR, Tell GP and Koch-Weser J (1981): The pharmacology of GABA-transaminase inhibitors. *Biochem Pharmacol* 30: 817-824.

Palfreyman MG, McDonald IA, Fozard JR, Mely Y, Sleight AJ, Zreika M, Wagner J, Bey P and Lewis PJ (1985): Inhibition of monoamine oxidase selectively in brain monoamine nerves using the bioprecursor (E)-ß-fluoromethylene-m-tyrosine (MDL 72394), a substrate for aromatic L-amino acid decarboxylase. *J Neurochem* 45: 1850-1860.

Patlak CS, Blasberg RG and Fenstermacher JD (1983): Graphical evaluation of blood-to-brain transfer constants from multiple-time uptake data. *J Cereb Blood Flow Metab* 3: 1-7.

Pare CMB (1972): Clinical implications of monoamine oxidase inhibition. In: *Monoamine Oxidase - New Vistas . Advances in Biochemical Psychopharmacology, Vol 5*, Costa E and Greengard P, eds. New York: Raven Press.

Reivich M, Kuhl D, Wolf AP, Greenberg J, Phelps M, Ido T, Casella V, Fowler JS, Hoffman E, Alavi A, Som P and Sokoloff L (1979): The

[18F]fluorodeoxyglucose method for the measurement of local cerebral glucose utilization in man. *Circ Res* 44: 127-137.

Robinson DS, Davis JM, Nies A, Colburn RW, Davis JN, Bourne HR, Bunney WE, Shaw DM and Coppen AJ (1972): Aging, monoamines, and monoamine oxidase levels. *Lancet* 1: 290-291.

Robinson JB (1985): Stereoselectivity and isoenzyme selectivity of monoamine oxidase inhibitors: Enantiomers of amphetamine, N-methylamphetamine and deprenyl. *Biochem Pharmacol* 34: 4105-4108.

Sammes PG (1990): Enzymes and other drug targets. In: *Comprehensive Medicinal Chemistry, The Rational Design, Mechanistic Study and Therapeutic Application of Chemical Compounds, Vol 2*, Hansch C, Sammes PG and Taylor JB, eds. Oxford: Pergamon Press.

Sarter M, Schneider HH and Stephens DN (1988): Treatment strategies for senile dementia: antagonist β-carbolines. *TINS* 11: 13-17.

Shinotoh H, Inoue O, Suzuki K, Yamasaki T, Iyo M, Hashimoto K, Tominaga T, Itoh T, Teteno Y and Ikehira H (1987): Kinetics of [11C]N,N-dimethylphenylethylamine in mice and humans: Potential for measurement of brain MAO-B activity. *J Nucl Med* 28: 1006-1011.

Tetrud JW and Langston JW (1989): The effect of deprenyl (selegiline) on the natural history of Parkinson's disease. *Science* 245: 519-522.

The Parkinson's Study Group (1989): Effect of deprenyl on the progression of disability in early Parkinson's disease. *N Engl J Med* 321: 1364-1371.

Waley SG (1985): Kinetics of suicide substrates. *Biochem J* 227: 843-849.

Walsh C (1982): Suicide substrates: Mechanism-based enzyme inactivators. *Tetrahedron* 38: 871-909.

Westlund KN, Denney RM, Kochersperger LM, Rose RM and Abell CW (1985): Distinct monoamine oxidase A and B populations in primate brain. *Science* 230: 181-183.

Withers SG, Street IP, Bird P and Dolphin DH (1987): 2-Deoxy-2-fluoroglucosides: A novel class of mechanism-based glucosidase inhibitors. J *Am Chem Soc* 109: 7530-7531.

Yu PH, Barclay S, Davis B and Boulton AA (1981): Deuterium isotope effects on the enzymatic oxidative deamination of trace amines. *Biochem Pharmacol* 30: 3089-3094.

Zeller EA, Barsky J and Berman ER (1955): Amine oxidases. XI. Inhibition of monoamine oxidase by 1-isonicotinyl-2-isopropylhydrazine. *J Biol Chem* 214: 267-274.

THE ROLE OF POSITRON EMISSION TOMOGRAPHY IN ASSESSING AND MONITORING DOPAMINE ACTIVE DRUGS

Dean F. Wong and Babington Yung
Johns Hopkins University School of Medicine
Department of Radiology, Division of Nuclear Medicine

INTRODUCTION

Neuroreceptors and neurotransmitters have been implicated in the pathophysiology, pathogenesis, and therapeutic mechanisms for a number of human disorders. The dopaminergic system is one of the most commonly implicated neuroreceptor neurotransmitter systems. Diseases such as Tourette's Syndrome, schizophrenia, and Parkinson's Disease are examples of such disorders where dopamine is suspected of playing an important role. Positron Emission Tomography (PET) and, more recently, Single Photon Emission Computed Tomography (SPECT) studies of neuroreceptors, re-uptake sites and neurotransmitter precursors as well as the indirect actions of endogenous neurotransmitters provide important methodological tools to investigate pathophysiology and design therapy. In this chapter we will describe how the PET studies can help design therapeutic stratagies through the monitoring and comparison of various drugs and toxins in the living human brain. Such measurements represent unique techniques that in many cases cannot be directly carried our by any other procedure.

Initial research on a potential receptor ligand which may be developed as a PET ligand includes tissue homogenate studies which are

Nuclear Imaging in Drug Discovery, Development, and Approval
H.D. Burns, R. Gibson, R. Dannals, Peter Siegl, Editors

typically performed to obtain binding parameters including receptor affinity and density. Autoradiographic studies on cryosections then provide information on receptor distribution in tissue. Ligand specificity is established by competition with drugs specific fo the receptor and demonstration of receptor mediated physiological or pharmacological effects.

Neuroreceptors exist in nanomolar concentrations in brain and therefore only minute quantities of radioactive ligands can be administered in order to avoid receptor saturation which would violate the tracer principle. Typically, natural neurotransmitters cannot be used to label receptors *in vivo*, owing to their low affinity, rapid metabolism, or inability to cross the blood-brain-barrier. Currently used receptor radiotracers are antagonists which have very high receptor affinities, with *in vitro* equilibrium dissociation constants (K_d) $\leq$ 2-3 nM (Sadzot and Frank, 1990). With PET technology, tracer concentration on the order of 0.1 nM can be detected, so receptor occupancy by a PET ligand can be sufficient low to avoid phamacological effects.

In order to image and quantitate a receptor population *in vivo* the contribution of non-specific binding of the radiotracer must be known. Non-specific binding is estimated by giving a saturating dose of an unlabeled competitor, using a lower affinity stereoisomer of the radiotracer, or by analysis of a region devoid of receptors. After subtracting count-rates associated with non-specific binding and with appropriate calibration, count rates can be converted into actual radiotracer concentrations. Multicompartmental modeling can be applied to describe the biological behavior of the drug after administration. The time-dependent changes in tracer concentration in each compartment are described by differential equations, which can be solved to obtain the rate constants defining the transfer of tracer between compartments. Using the dynamic method, kinetic analysis requires knowledge of the time-activity curve of radiotracer in the target structure, e.g., corpus striatum, and the input function which consists of time-activity curve of unmetabolized radiotracer in plasma. These can be fitted into the compartmental model to extract parameters of interest using non-linear regression analysis.

In high specific activity studies, the receptor occupancy is sufficiently low to avoid partial saturation of receptors which can lower the receptor specific signal intensity. These studies result in a combined term representing B_{max}/K_d (k_3/k_4, or binding potential). Receptor concentration and affinity cannot be separately estimated. In low specific activity studies, the receptor occupancy is not negligible. This allows separate estimation of B_{max} and K_d, and provides information that helps to decrease the number of unknown parameters. Only ligands with minimal pharmacological effects can be used for low specific activity studies.

PET receptor studies can be used to measure receptor density and affinity in normal conditions. This method can also be applied to study various diseases in their earlier stages, possibly before clinical manifestations. In competition experiments, in which unlabeled competitors are administered prior to radiotracer injection, the receptor occupancy of therapeutic drugs can be estimated. Receptor occupancy studies allow for: i) estimation of dose needed to attain a given occupancy, above which there is increased risk of side effects without therapeutic gain; ii) evaluation of access of the drug to receptor through which pharmacologic effect is produced; iii) study of rate of drug dissociation from receptors; and iv) study of the relationship between dose and plasma level, receptor occupancy, pharmacologic, behavioral, hormonal and electrophysiologic effects. For the purpose of measuring dose levels, simpler equipment such as a pair of scintillation detector probes can be used with tracer activity as low as 200-400 μCi. (Bice et al., 1986).

DOPAMINE RECEPTOR STUDIES

Dopamine Receptor Subtypes

In 1972, Kebabian et al. (1972) discovered a dopamine sensitive adenylate cyclase in rat striatum. The dopamine D_1 and D_2 receptor subtypes were later defined in 1979 by Kebabian et al. The D_1 receptor was defined initially as being coupled to adenylate cyclase (Kebabian and Caine, 1979), whereas the D_2 receptor was defined as either being uncoupled or inhibitory to this system. The two receptors systems may interact. For example, it has been shown that D_1 dopamine antagonists block the self-injurious behavior in rodent and baboon animal models of the Lesch-Nyhan syndrome, a disorder linked to the dopamine system (Goldstein et al., 1984; Goldstein et al., 1986; Breese, et al., 1984a; Breese et al., 1984b). Recently, Seeman et al. have shown that there is a fall with age for both dopamine D_1 and D_2 receptor subtypes (Seeman et al., 1987a). Abnormalities in the concentration of one or both subtypes have been found in Huntington's and Parkinson's disease postmortem striatal tissue (Seeman et al., 1987b). Additional dopamine receptor subtypes, e.g., D_3 (Sokoloff et al., 1990), D_4 (Van Tol et al., 1991), and D_5 (Sunahara et al., 1991), are being characterized.

Dopamine D_1 Receptor Ligands

A benzazepine (SCH 23390), developed by Iorio et al. in 1983 and found to have specific dopamine D_1 receptor binding by Hyttel (1983), led to a dramatic increase in the study of dopamine D_1 receptors. A saturation analysis study on human brain homogenates showed high densities of specific [^{3}H]SCH 23390 binding to D_1 receptors in basal ganglia, especially in lateral putamen and caudate nucleus (Farde et al., 1992). Binding in the neocortex was only 10-20% of binding in the caudate. A very low density of binding sites was found in cerebellar gray matter. No specific binding was demonstrated in white matter of the corpus callosum and cerebellum (Farde et al., 1992). Competitive analyses showed a 10 times lower affinity of SCH 23390 for serotonin S_2 receptors as compared to dopamine D_1 receptors.

Dopamine D_1 receptor binding by SCH 23390 is stereospecific. Only the active R-stereoenantiomer showed high accumulation in the basal ganglia and the neocortex (Halldin et al., 1986). *In vitro* experiments on human post mortem brain tissue lend support to serotonin S_2 receptor binding by [^{3}H]SCH 23390 as demonstrated by the effect of ketanserine on total brain binding (Hall et al., 1988). Dopamine D_1 receptors accounted for 85% of binding in the basal ganglia, but only 15% of the specific binding of [^{3}H]SCH 23390 in the temporal cortex, where the majority of binding was related to serotonin S_2 receptors.

Ravert et al. (1987) labeled SCH 23390 with Carbon-11 which led to baboon and human studies. [^{11}C]SCH 23390 has been used to study normal volunteers, schizophrenia patients (Farde et al., 1987) and Lesch-Nyhan syndrome (Wong et al., 1986a). With the availability of this D_1 antagonist and other D_2 receptor ligands, interactions between D_1 and D_2 receptors in health and disease can be investigated in the future. Preliminary *in vivo* studies of the D_1 and D_2 ratio have been performed on patients with Lesch-Nyhan Syndrome (Wong et al., 1986a) and further studies will be carried out on patients with Rett Syndrome. The balance between the two receptor systems may be a significant determinant of the therapeutic efficacy of certain drugs, such as clozapine, an atypical neuroleptic which possesses both D_1 and D_2 binding. Indeed, clozapine has recently been radiolabeled and used in PET imaging of striatum and frontal cortex in humans (Lundberg et al., 1989).

Dopamine D_2 Receptor Ligands

There are two main types of radioligands for dopamine D_2 receptor studies. The first group consists of spiperone and alkyl-derivatives of spiperone, such as [^{11}C]N-methylspiperone (NMSP). The second group consists of substituted benzamides, such as raclopride. There are some differences in receptor binding characteristics of NMSP and raclopride *in vitro*. [^{3}H]NMSP has 10 times higher affinity to D_2 receptors than [^{3}H]raclopride in rat striatum and human putamen (Hall et al., 1990). Endogenous dopamine was found to be more effective in inhibiting [^{3}H]raclopride than [^{3}H]NMSP binding to D_2 receptors (Hall et al., 1990). A study on Rhesus monkeys using NMSP and raclopride showed that NMSP binds to both serotonin receptors in frontal cortex and dopamine D_2 receptors in striatum. Raclopride binds selectively to dopamine D_2 receptors. NMSP has a lower dissociation constant (K_d) for D_2 receptors than raclopride (Hartvig et al., 1988).

Endogenous Dopamine Neurotransmitter Levels

A number of recent studies have attempted to use imaging procedures to examine the action of endogenous dopamine at the receptor site. Studies by Wong et al. (1986d) involving the use of intravenous cocaine did not demonstrate changes in NMSP binding using a simple caudate-to-cerebellar ratio method. However, it has been hypothesized that raclopride may be affected by endogenous dopamine. Seeman et al. (1989) directly demonstrated *in vitro* the ability to differentially affect endogenous dopamine levels and displace [^{3}H]raclopride as compared to [^{3}H]NMSP binding. Recent *in vivo* studies demonstrated dramatic changes in raclopride binding as compared to NMSP binding in the presence of intravenous amphetamine in rodent brain (Young et al., 1991). Baboon imaging studies, such as those by Innis et al. (1991), have shown the effect of amphetamine on other substituted benzamides such as [^{123}I]iodobenzamie (IBZM). It has been suggested in preliminary baboon studies by Dewey et al. (1991) that amphetamine may affect certain parameters of PET imaging with NMSP. A potential effect of endogenous dopamine that might be seen with PET imaging is internalization of the radioligand-receptor complex. Chugani et al. (1988) has suggested internalization of [^{3}H]spiperone in a rodent model.

Dopamine Transporter System

Recent developments have been made concerning the action of cocaine and the monitoring of 1-methyl-4-phenyl-1,2,3,5-tetrahydropyridine (MPTP) and other neurotoxins. Ritz et al. (1987) suggested that the dopamine transporter is the site of cocaine action. [^{11}C]cocaine has been imaged in living human brain by Fowler et al. (1989). Volkow et al. (1990) carried out a preliminary study examining the effects of chronic cocaine use on the binding of [^{18}F]NMSP to dopamine D_2 receptors. Pharmacokinetic measurements in blood and brain, as well as binding measurements at the presynaptic dopamine transporter site have been accomplished with these imaging studies. Prior to this, the dopamine uptake site was imaged with a more nonspecific PET radioligand, [^{11}C]nomifensine (Leenders et al., 1988). These studies have been employed in research on primates and human studies in Parkinson's disease.

Studies by Scheffel et al. (1989a) have demonstrated *in vivo* labeling of the dopamine transporter site with [^{11}C]WIN (35,428). This compound has a significantly higher target-to-background ratio and higher affinity than [^{11}C]cocaine. Recent imaging studies in baboons provide considerable promise for future measurements of the dopamine transporter using the single-photon emitting radiotracer [^{123}I]RTI-55 (Shaya et al., 1992). Studies with [^{18}F]GBR 13119 have also provided images of the dopamine transporter site (Kilbourn et al., 1988; Kilbourn et al.,1989). Thus, the availability of various ligands provides multiple probes for the study of the dopamine transporter site and the effects of drugs.

METHODOLOGY

Models used for receptor density and affinity quantification are basically divided into two types, dynamic non-equilibrium and equilibrium. For irreversibly-bound ligands, the "input function", consisting of tracer entry rate in receptor-rich regions, must be included in the kinetic modeling. Rate measurements of ligand delivery to receptors and ligand binding to receptors must be distinct in order to calculate receptor density (B_{max}) and affinity (K_d). Quasi-steady state may be attained (constant brain/blood ratio) even though full equilibrium is not reached during the PET imaging period. In this situation, the ratio of bound-to-free radioligand is constant (Wong et al., 1986b; Wong et al., 1986c). The amount of labeled metabolites and the partition coefficient between bound and free tracer can then be estimated using "kinetic" mathematical analysis.

The binding rate constants of D_2 receptor radiotracers in both unblocked and blocked states are calculated from the time course of radioactivity in regions of interest and the time course of radioactivity in plasma. For example, using [^{11}C]NMSP, a relatively irreversible ligand, receptor density can be calculated using the following equation (Wong et al, 1986c; Wong et al., 1986e):

$$B_{max} = \frac{D_w[H]}{[(1/k_3)_B-(1/k_3)_{UB}]}$$

where

$$D_w = \frac{\lambda}{(k'_{off})(MW)}$$

k'_{off} is the dissociation constant for the unlabeled inhibitor haloperidol, MW is the molecular weight of haloperidol, λ is the haloperidol partition coefficient of brain/plasma (bound and free), [H] is the haloperidol plasma concentration in ng/ml, and $(k_3)_{UB}$ and $(k_3)_B$ are the forward rate constants of binding (min^{-1}) of [^{11}C]NMSP to dopamine D_2 receptors in the unblocked and haloperidol blocked states, respectively. The k'_{off} and the partition coefficient are assumed to be a constant in normal and patient populations.

A number of studies have been carried out using other spiperone derivatives to estimate dopamine D_2 receptors with this dynamic (non-equilibrium) model (Mintun, et al., 1984; Huang et al., 1986; Logan et al., 1987; Perlmutter et al., 1987). In all these reports, equilibrium was not attained during the PET imaging period.

Lower affinity benzamides bind reversibly to receptors and reach an equilibrium within a short period of time. Scatchard analysis can be applied in such situations, such as the method reported by Farde et al, (1986). More commonly, a three or four compartment kinetic model is applied. By injecting a low specific activity radioligand and measuring the time course with high temporal resolution, separate estimation of B_{max} and K_d can be achieved. In this way, rate constants for tracer delivery into receptor rich regions and rate constants for receptor binding can be distinguished.

The above methods, employing partial occupancy by the same or related unlabeled ligands, have allowed the determination of B_{max} and K_d. This same principle of competitive inhibition is used in the measurement of receptor occupancy during drug monitoring.

Quantitative modeling studies may be too complex to carry out routinely. It may be acceptable to use approximations by some quantitative indices such as the ratio between counts in a receptor rich region to a region devoid of receptors. These simplified or semi-quantitative approaches are more practical for clinical studies. These approximations, however, have to be validated by extensive preliminary tracer kinetic modeling.

APPLICATIONS OF DOPAMINE SYSTEM IMAGING

Dopamine D_2 Receptors in Normal Physiology

A decrease in dopamine D_2 receptor concentration associated with aging has been established in normal subjects using increasingly more sophisticated methodology (Wong et al., 1984; Wong et al, 1988; Wong et al., 1989). This observation is consistent with most postmortem findings and that of other PET studies (Iyo et al., 1989). It may be possible to monitor normal hormonal effects on dopamine receptors such as the effect of the menstrual cycle (Wong et al., 1988).

Dopamine Neuronal Integrity and Dopamine D_2 Receptor Changes Following Drug Damage

Since the first imaging studies of human dopamine neurons using [^{18}F]fluro-L-DOPA (Garnett et al. 1983), there has been considerable interest in studying the effects of neurotoxins, such as MPTP, on the dopamine system. Recent quantitative methods such as those of Gjedde et al. (1991) suggest that the rate limiting dopa decarboxylase step is measurable in PET studies with [^{18}F]fluro-L-DOPA. The use of [^{18}F]fluro-L-DOPA has also been quite promising in describing MPTP induced damage as a measure of dopamine content and neuronal integrity. Studies by Calne et al. (1985) in MPTP patients have illustrated this point. More recently the use of dopamine transporter ligands, such as [^{123}I]RTI-55 have been employed to illustrate unilateral MPTP lesions in baboons (Shaya et al., 1992).

PET imaging of MPTP induced dopamine D_2 receptor damage has also been carried out with [^{18}F]spiperone in MPTP patients (Perlmutter et al., 1987; Perlmutter et al., 1989) and with [^{76}Br]bromospiperone in MPTP baboons (Hantraye et al., 1986).

Dopamine D_2 ligands with very high receptor affinity, such as [^{123}I]epidepride, have been employed to study extrastriatal (limbic and cortical) dopamine D_2 receptors in humans by SPECT (Kessler et al., 1991). Following further development, the method of imaging cortical D_2 receptors will have wide ranging implications in pathophysiological studies of neuropsychiatric disorders. For example, it would allow the study of the Pycock model of lesioning the cortical dopamine receptors (Pycock et al., 1980) which has been suggested to play a role in the pathogenesis of schizophrenia (Weinberger, 1987.)

Dopamine Receptor Studies in Schizophrenia

Carlsson et al. (1978) proposed a dopamine hypothesis for schizophrenia. Idiopathic psychoses may be caused by pathophysiologic overactivity of dopamine in the forebrain (Baldessarini, 1985). In the prevalence of dopamine hyperfunction hypothesis (McKenna, 1987), the question remains whether brain dopamine receptor density elevation in schizophrenics is independent of the increase expected as an adaptive response to long term neuroleptic treatment. Using [^{11}C]NMSP, Wong et al. (1986e) found a striking elevation in the B_{max} of caudate dopamine D_2 receptors in drug naive schizophrenic patients. Using [^{11}C]raclopride, Farde et al. (1987) found no difference in dopamine D_2 receptor density in basal ganglia in drug naive patients as compared to normal controls.

The above mentioned studies focused on striatal D_2 receptors. The role of extrastriatal (limbic and cortical) D_2 receptors remains to be evaluated. Limbic dopaminergic areas, such as nucleus accumbens, are difficult to resolve in PET studies because of proximity to the caudate-putamen complex (Waddington 1989).

Dopamine Receptor Studies in Drug Treatment Monitoring

Many antipsychotic agents appear to work by either stimulating or blocking neuroreceptors. Dopamine receptor blocking drugs help to decrease delusions and hallucinations in psychotic patients, and bring about cognitive improvement. Seeman et al. (1976) and Creese et al. (1976) both demonstrated the relationship between the *in vitro* binding affinity of various dopamine antagonists and their typical clinical daily dose used to treat schizophrenia.

Accurate assessment of dopamine D_2 receptor occupancy by various therapeutic drugs is certainly an important future application of PET imaging. Approximately 85 to 90 percent D_2 receptor occupancy in the putman by chemically distinct neuroleptics in therapeutic doses has been demonstrated (Farde et al., 1986; Farde et al., 1988). More recently, Wolkin et al. (1988) and Smith et al. (1988) used [^{18}F]NMSP to demonstrate an inverse relationship between striatal/cerebellar ratios and drug plasma levels. PET imaging can be utilized to monitor and correlate receptor occupancy with clinical status, thereby providing information for drug dosage adjustment. The relative *in vivo* drug affinities for receptors can also be compared using these methods.

Comparison of Dopamine D_2 Receptor Binding Affinity of Molindone with Haloperidol: A PET Study Involving Single Dose Receptor Occupancy as a Basis for Comparison of Relative Affinities In Vivo

As an example of the power of PET scanning to measure receptor occupancy, a small single blind crossover study comparing molindone and haloperidol is described. This study asks the question: what is the relative binding affinity and occupancy of two neuroleptics, haloperidol and molindone, when given in a clinically similar dose? Molindone, a second line neuroleptic, has a D_2 receptor binding affinity of approximately 1/30 that of haloperidol *in vitro*. However, clinical experience shows that similar antipsychotic action is achieved with an oral dose of molindone only five-fold higher than that of haloperidol. A further observation is that molindone has a much lower history of tardive dyskinesia. Thus the question is whether or not the apparent affinity *in vivo* is lower for molindone as compared to haloperidol, and where is the blocking site of molindone that relates to its antipsychotic action?

Thus, we carried out the first occupancy study at our center using the dopamine D_2 receptor radioligand, [^{11}C]NMSP. The study had two parts. The first part was to establish an appropriate single oral dose of haloperidol which could be measured by its competition with [^{11}C]NMSP. Three subjects obtained two PET scans each including a control study (in the absence of neuroleptic) followed by a PET scan which was preceded by an oral dose of haloperidol. The oral dose was established by doing a control 90 minute PET study following 20 mCi of [^{11}C]NMSP injected intravenously. Subjects then received either 5 or 7.5 mg oral haloperidol followed by a second PET scan several hours later. A marked decline in the binding was demonstrated with occupancy in the order of 60-80% following 7.5 mg haloperidol. The second part of the study consists of two segments which were carried out in 5 subjects. The first segment consisted of administering a dose of 37.5 mg of molindone, i.e., five-fold higher than that established for haloperidol, 1.5 hrs prior to the first PET scan with [^{11}C]NMSP. After a minimum of 1.5 wks, the second segment consisted of a second PET scan carried out with [^{11}C]NMSP preceded by 7.5 mg of oral haloperidol 4 hours prior to the [^{11}C]NMSP injection. Of the 5 in this series, 2 did not complete the second haloperidol PET scan.

The subjects consisted of normal, young, healthy males between 18 and 38 years of age. Each of the first 3 subjects underwent a localizing CT scan and multiple PET images as part of the control study. The analysis consisted of a comparison of the slopes of the caudate/cerebellar ratio versus time calculated for both the control and the subsequent haloperidol blocking studies. The slope has been previously established as an index of the rate of binding of [^{11}C]NMSP to D_2 receptors. A linear pattern over time was demonstrated as previously described (Wong et al., 1984). Thus, in establishing the percent blockade, a linear regression was employed for both unblocked and blocked cases. A typical blocking study is illustrated in Fig. 1 with the accompanying time course of the caudate/cerebellum ratio in Fig. 2. For this first group, the average caudate/cerebellar versus time slope was 0.07 ± 0.009/min for control and 0.02 ± 0.004/min for 3 subjects. Thus, there was a 60-70% blockade as measured by the slope method.

The second part of this study consisted of the molindone to haloperidol comparison in the same subjects. Again, the analysis consisted of the comparison of the slopes of the caudate/cerebellar versus time relationship for molindone and for haloperidol PET studies. The average comparison was approximately 0.019 ± 0.003/min for molindone versus 0.017 ± 0.006/min for the haloperidol PET scan averaged over all subjects. These similar slopes reflected similar rates of binding to the receptor and, by inference, similar occupancy for the dopamine D_2 receptor in as much as it had the same ability to compete for [^{11}C]NMSP as measured by PET imaging. A typical comparison between pre- and post-haloperidol PET scans for one of the subjects is given in Fig. 3. The caudate/cerebellum ratio as a function of time for molindone and haloperidol for one of the subjects is given in Fig. 4.

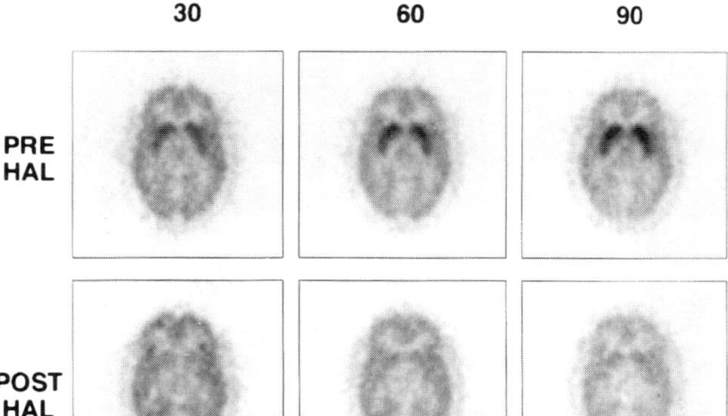

FIGURE 1. Establishment of haloperidol blocking dose PET scans demonstrating the control (upper row) and post haloperidol images (lower row) for the same individual. Both scans represent images acquired at 30, 60, and 90 minutes post injection of [^{11}C]NMSP.

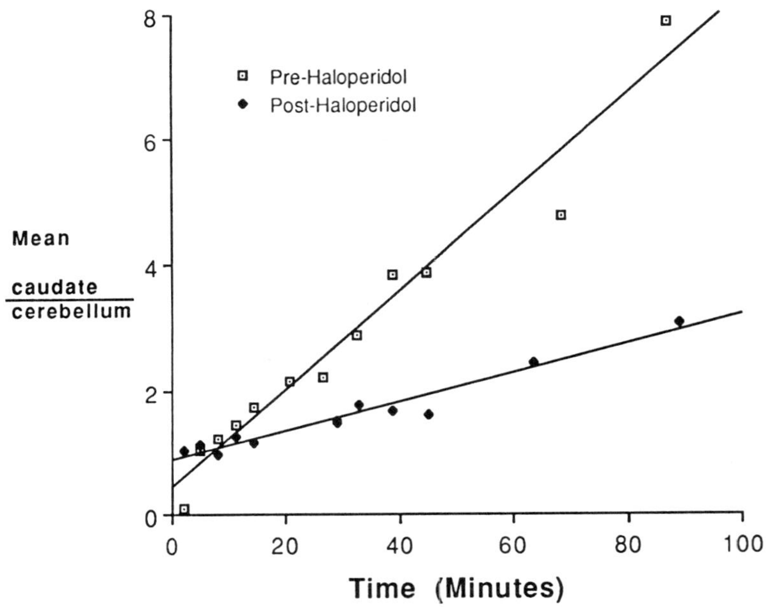

FIGURE 2. Caudate/cerebellum ratios plotted as a function of time before and after haloperidol administration for one subject in the study, demonstrating about 60-70% blockade due to haloperidol.

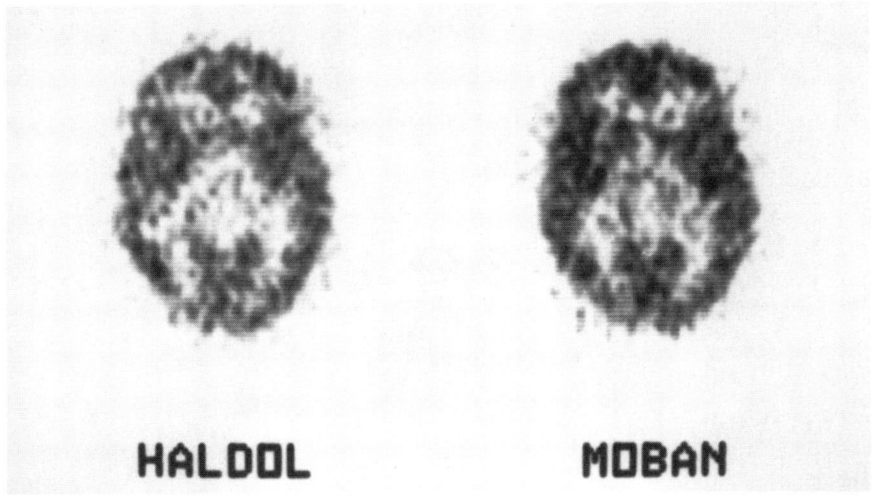

FIGURE 3. Comparison Haloperidol:Moban. PET scans imaged at same time post injection of [^{11}C]NMSP (approximately 42 to 48 minutes) for Haloperidol and Moban in the same subject.

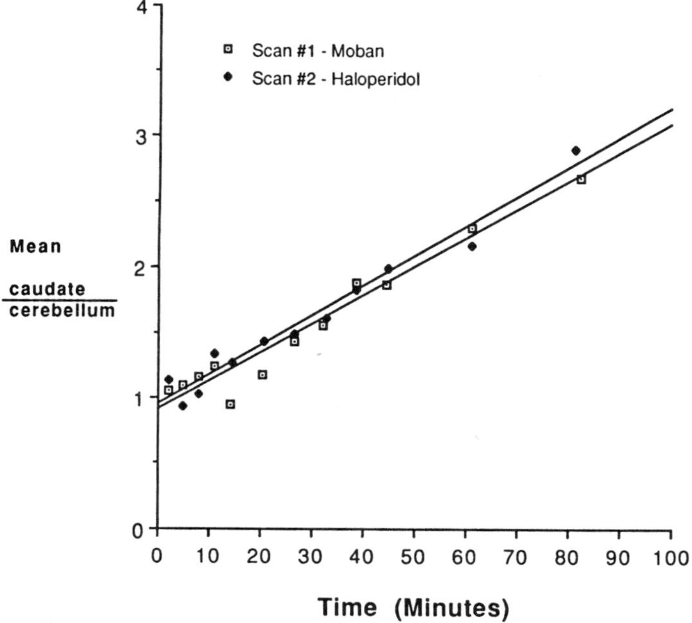

FIGURE 4. Caudate/cerebellum ratios plotted as a function of time after molindone and haloperidol administration for one subject in the study, indicating the similarity of blockade due to two "clinically efficacious" doses.

Subsequently, kinetic modeling was carried out measuring the forward binding rate, k_3, and employing multiple plasma samples and corrections for labeled metabolites using the method of Wong et al. (1986a). This method demonstrated insignificant differences in the k_3 for molindone and haloperidol (as illustrated in Wong et al., 1985).

Thus, the method of measuring *in vivo* competition of [^{11}C]NMSP with unlabeled neuroleptics can be used to detect the site and degree of neuroleptic blockade action *in vivo*. The study demonstrates that, in normal volunteers, molindone and haloperidol, given in a ratio of 5 to 1, respectively, on a milligram basis, were similar in both site and degree of blockade. This leads to the following conclusion: molindone probably blocks dopamine D_2 receptors in a site and manner similar to that of other neuroleptics such as haloperidol. Although this study cannot show all possible sites of action, it suggests that the dopamine blocking action is very similar to haloperidol except for the lower affinity. The apparent discrepancy between the *in vitro* affinity of molindone and haloperidol (molindone 1/30 of haloperidol) and the *in vivo* equivalency (approximately 5/1) may be due to differences in blood-brain-barrier penetration of other transport phenomena. An even more intriguing possibility is that the metabolites in molindone may also have active dopamine D_2 receptor properties which of course manifest themselves in an *in vivo* study and not in an *in vitro* measurement.

Dopamine D_2 receptor occupancy as a function of increasing neuroleptic dose

Another example of the role of PET imaging in monitoring drug therapy can be seen by doing multiple PET studies at increasing oral doses of haloperidol. Such a study in several schizophrenic patients was carried out at our center in collaboration with the Maryland Psychiatric Research Center. The subjects were diagnosed according to DSMIIIR criteria. Each was withdrawn from his usual neuroleptic for at least four weeks. Thereafter, PET scans with [^{11}C]NMSP were performed one week apart, four hours after a single oral dose of haloperidol. The data show a linear displacement of [^{11}C]NMSP binding by haloperidol for doses between 4 mg and 16 mg haloperidol (Fig. 5). Such a procedure

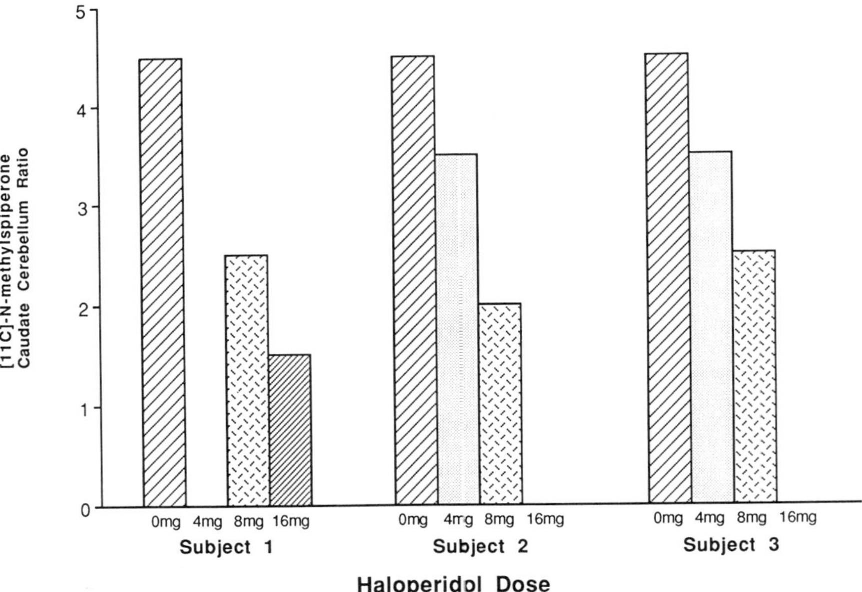

FIGURE 5. Ratio of ^{11}C-NMSP binding in caudate to the same measure in cerebellum in three schizophrenic subjects after single doses of placebo or increasing doses of haloperidol. Adapted with premission of American Psychiatric Press, Inc., Washington, D.C. from Tamminga et al. (1993)

demonstrates the dose response of typically therapeutic doses of haloperidol and suggests how PET imaging can be used to monitor drug therapy.

Dopamine Receptor Studies on Neuroleptic Nonresponders

There remains a group of schizophrenic patients who do not respond well to typical neuroleptics (dopamine antagonists). The key issue is whether or not the inadequate response is caused by failure to attain adequate receptor blockade, or does the inadequate response occur even with adequate receptor occupancy. Using [^{18}F]NMSP, the striatal/cerebellar uptake ratio was found to be linear for ≥ 4 hrs and highly correlated to plasma neuroleptic drug levels (Wolkin et al., 1989). In this study, nonresponders were defined as patients showing < 20% decrease in total Brief Psychiatric Rating Scale. Nonresponders were significantly older, and had more severe pretreatment positive symptoms than did responders. A ratio index was calculated as the slope of the regression line of striatal to cerebellar uptake ratio on time, multiplied by 100. The mean ratio index values for responders and nonresponders were similar, indicating comparable levels of dopamine blockade by haloperidol in both groups. There was also no significant difference in the percent changes in ratio indices pre- and post-treatment with haloperidol between the two groups.

Discrepancy Between Duration of Receptor Occupancy and Plasma Drug Washout

Both Farde et al. (1988) and Smith et al. (1988) showed persistent occupancy of dopamine D_2 receptors for at least 54 hrs after haloperidol withdrawal, and at least 27 hrs following sulpiride withdrawal, which are much longer than the decline in plasma drug levels. Possible explanations include slower dissociation of drug bound to receptors relative to the rate of reduction of free drug concentration or a curvilinear relationship between specific binding and free drug concentration. A saturation analysis of [^{11}C]raclopride binding on healthy volunteers has demonstrated a curvilinear relationship between specific binding and free drug concentration in brain (Farde et al., 1986). A similar relationship was found with sulpiride (Farde et al., 1988). The typical neuroleptic receptor occupancy of 65-85% may well fall on the horizontal part of a hyperbola, so that a change in free drug concentration will cause only minor change in receptor occupancy.

Study on Clozapine - an Atypical Neuroleptic

An atypical neuroleptic does not produce catalepsia in animals and has low incidence of extrapyramidal side effects. Clozapine, thioridazine, sulpiride and melperone belong to this category. Clozapine is chemically similar to imipramine. It has high antipsychotic potency, even in the most refractory cases (Kane et al., 1988). Clozapine was introduced in the late 1960s, and was withdrawn about 10 years later owing to causation of agranulocytosis (Hippus, 1989). It is now only licensed for use on psychotic patients not responding to typical antipsychotic therapy. Clozapine produces significantly greater symptomatic improvements in schizophrenic patients compared to haloperidol (Kane et al., 1988).

Studies in rats showed that prolonged treatment with clozapine does not reduce the number of spontaneously active neurons in the substantia nigra as observed with chronic haloperidol treatment (White and Wang, 1983). Haloperidol decreases basal dopamine release in both the striatum and nucleus accumbens, whereas clozapine only affects the accumbens and spares the striatum (Lane et al., 1988). The preserved release of dopamine from the striatal dopaminergic system in clozapine treatment has been related to the absence of extrapyramidal side effects (Imperato and Angelucci, 1989). PET studies are being performed to compare clozapine and haloperidol binding to D_2 receptors (Karbe et al., 1991).

Compared to typical neuroleptics, clozapine has a lower receptor occupancy of 40-65% for dopamine D_2 receptors, and a higher occupancy of 40-42% for dopamine D_1 receptors (Wiesel et al., 1990). Among the recently characterized dopamine receptor subtypes, the D_4 receptor has a high affinity for clozapine (Van Tol et al., 1991). Repeated administration of clozapine does not increase the abundance or sensitivity of dopamine receptors, although it may have such an effect on α_1-adrenergic receptors (Cohen et al., 1986). It has been shown to have anti-serotoninergic activity, especially at S_2 receptors (Meltzer et al., 1989). Clozapine has selective effects on neurons projecting to limbic areas (nucleus accumbens, amygdala and other sites) (Hand et al., 1987).

[^{123}I]IBZM was found to be a highly selective D_2 receptor ligand demonstrating saturable and reversible stereospecific binding (Kung et al., 1988). It has been used in SPECT studies in humans and demonstrated a curvilinear time/activity relation, with a plateau at 40-50 mins after intravenous injection (Alavi et al., 1989). Pilowsky et al. (1992) demonstrated with IBZM and SPECT imaging that clozapine treated patients showed significant clinical improvement in spite of a lower level of D_2 receptor blockade, whereas patients treated with typical antipsychotics showed poor therapeutic response despite a higher D_2 receptor blockade. This is in agreement with results obtained in PET studies.

Clozapine has good therapeutic efficacy in spite of only moderate dopamine D_2 receptor occupancy. It remains to be elucidated whether a moderate D_2 receptor occupancy is sufficient for antipsychotic effect or a combined D_1 and D_2 receptor blockade is needed for maximum potency. These issues are of increasing importance especially given recent studies suggesting high dopamine receptor occupancy despite therapy resistance (Coppers et al., 1991).

CONCLUSION

Dopamine receptor studies with PET methodology allow absolute quantification of receptor density and affinity. This provides a powerful tool for understanding human neurophysiologic states in various diseases as well as for studying therapeutic mechanisms and monitoring drug treatment efficacy. One of the most useful applications to drug development is the determination of receptor occupancy and site of action. For example, in our study on molindone, we showed that molindone provided equivalent receptor occupancy as haloperidol at equi-efficacious clinical dose. No other methodology is able to provide a direct measure of receptor occupancy.

ACKNOWLEDGMENTS

Special thanks to Carol Tamminga of the University of Maryland Psychiatric Center for permission to include our collaborative data and Figure 5 and Dr. Manley Paulos, Ph.D, for his collaborative role in the molindone/haloperidol studies. Special thanks to Solomon Snyder, M.D. and Joseph Coyle, M.D. for consultation concerning the molindone/haloperidol studies. Thanks to Catherine Chen, Boon Chan, and Victor Walker for technical assistance. Support was provided, in part, by the following grants: MH42821, HD24061, and Dupont Inc.

REFERENCES

Alavi A, Velchick MG, Kung HF, Kung MP, Chang W, Noto R, Pan S, Billings J, Sorgentoni K Rauch A and Reilly J (1989): Imaging the basal ganglia in the human brain with I-123-IBZM: A new CNS D_2 receptor agent. *J Nucl Med* 30:731.

Baldessarini RJ (1985): Antipsychotic agents. In: *Chemotherapy in Psychiatry: Principles and Practice*, Cambridge: Harvard University

Press.

Baldessarini RJ and Frankenburg FR (1991): Clozapine: A novel antipsychotic agent. *New Eng J Med* 324(11): 746-754.

Bice AN, Wagner HN Jr., Frost JJ, Natarajan TK, Lee MC, Wong DF, Dannals RL, Ravert HT, Wilson AA and Links JM (1986). Simplified detection system for neuroreceptor studies in human brain. *J Nucl Med* 27: 181-184.

Breese GR, Baumeister AA, McCown TJ, Emerick SG, Frye GD and Mueller RA (1984a): Neonatal-6-hydroxydopamine: Model of susceptibility for self-mutilation in the Lesch-Nyhan Syndrome. *Pharmacol Biochem Behav* 21: 459-461.

Breese GR, Baumeister AA, McCown TJ, Emerick SG, Frye GD, Crotty K and Mueller RA (1984b): Behavioral differences between neonatal and adult-6-hydroxydopamine-treated rats to dopamine agonists: Relevance to neurological symptoms in clinical syndromes with reduced brain dopamine. *J Pharmacol Exptl Ther* 231: 343-354

Calne DB, Langston JW, Martin WR, Stoessl AJ, Ruth TJ, Adam MJ, Pate BD and Schulzer M (1985): Positron emission tomography after MPTP: Observations relating to the cause of Parkinson's disease. *Nature* 317: 246-248.

Carlsson A (1978): Antipsychotic drugs, neurotransmitters, and schizophrenia. *Am J Psychiat* 135: 165-173.

Chugani DC, Ackermann RF and Phelps ME (1988): In vivo [^{3}H]spiperone binding: Evidence for accumulation in corpus striatum by agonist-mediated receptor internalization. *J Cereb Blood Flow Metab* 8: 291-303.

Cohen BM and Lipinski JF (1986): In vivo potencies of antipsychotic drugs in blocking alpha-1 noradrenergic and dopamine D_2 receptors: Implications for drug mechanisms of action. *Life Sci* 39: 2571-80.

Coppens HJ, Slooff CJ, Paans AMJ, Wiegman T, Vaalburg W and Korf J (1991): High central D_2-dopamine receptor occupancy as assessed with positron emission tomography in medicated but therapy-resistant schizophrenic patients. *Biol Psychiat* 29: 629-634.

Creese I, Burt DR and Snyder SH (1976): Dopamine receptor binding predicts clinical and pharmacological potencies of antischizophrenic drugs. *Science* 192: 481-483.

Dewey SL, Logan J, Wolf AP, Brodie JD, Angrist B, Fowler JS and Volkow ND (1991): Amphetamine induced decreases in (^{18}F)-N-Methylspiroperidol binding in the baboon brain using positron emission tomography (PET). *Synapse* 7: 324-327.

Farde L and Hall H (1992): Positron emission tomography - examination of chemical transmission in the living human brain: Development of radioligands. *Drug Res* 42: 260-264.

Farde L, Hall H, Ehrin E and Sedvall G (1986): Quantitative analysis of D_2 dopamine receptor binding in the living human brain by PET. *Science* 231: 258-261.

Farde L, Halldin C, Stone-Elander S and Sedvall G (1987): PET analysis of human dopamine receptor subtypes using ^{11}C-SCH 23390 and ^{11}C-raclopride. *Psychopharmacol* 92: 278-284.

Farde L, Wiesel F-A, Hall H, Halldin C, Stone-Elander S and Sedvall G (1987): No D_2 receptor increase in PET study of schizophrenia. *Arch Gen Psychiat* 44: 671-672.

Farde L, Wiesel FA, Halldin C and Sedvall G (1988): Central D_2-dopamine receptor occupancy in schizophrenic patients treated with antipsychotic drugs. *Arch Gen Psychiat* 45: 71-76.

Fowler JS, Volkow ND, Wolf AP, Dewey SL, Schlyer DJ, Macgregor RR, Hitzemann R, Logan J, Bendriem B and Gatley SJ (1989): Mapping cocaine binding sites in human and baboon brain *in vivo*. *Synapse* 4: 371-377.

Garnett RS, Firnau G and Nahmias C (1983): Dopamine visualized in the basal ganglia of living man. *Nature* 305: 137-138.

Gjedde A, Reith J, Dyve S, Leger G, Guttman M, Diksic M, Evans A and Kuwabara H (1991): Dopa decarboxylase activity of the living human brain. *Proc Nat Acad Sci* 88: 2721-2725.

Goldstein M and Kuga S (1984): Dopamine (DA) agonist induced compulsive biting (CB) behavior in monkeys: animal model for Lesch-Nyhan syndrome. *Soc Neurosci Abstr* 10: 239.

Goldstein M, Kuga S, Kusano N, Meller F, Dancis J and Schwarcz R (1986): Dopamine agonist induced self-mutilative biting behavior in monkeys with unilateral ventromedial tegmental lesions of the brain stem: Possible pharmacological model for Lesch-Nyhan syndrome. *Brain Res* 367: 14-20 .

Hall H, Farde L and Sedvall G (1988): Human dopamine receptor subtypes. In vitro binding analysis using [^{3}H]SCH23390 and [^{3}H]SCH raclopride. *J Neural Transm* 73: 7-21.

Hall H, Wedel I, Halldin C, Kopp J, Farde L, Ehrin E, Fasth KJ, Langstrom B and Sedvall G (1990): Comparison of the in vitro receptor binding properties of N-(^{3}H) raclopride to rat and human brain membranes. *J Neurochem* 55: 2048-2057.

Halldin C, Stone-Elander S, Farde L, Ehrin E, Fasth KJ, Langstrom B and Sedvall G (1986): Preparation of ^{11}C-labeled SCH23390 for the in vivo study of dopamine D-1 receptors using positron emission tomography. *Appl Radiat Isot* 37: 1039-1043.

Hand TH, Hu XT and Wang RY (1987): Differential effects of acute clozapine and haloperidol on the activity of ventral tegmental (A10) and nigrostriatal (A9) dopamine neurons. *Brain Res* 415: 257-69.

Hantraye P, Loch C, Tacke U, Riche D, Stulzaft O, Doudet D, Guibert G, Naquet R, Maziere B and Maziere M (1986): *In vivo* visualization by positron emission tomography of the progressive striatal dopamine receptor damage occurring in MPTP-intoxicated non-human primates. *Life Sci* 39: 1375-1382.

Hartivig P, Eckernas SA, Ekblom B, Lindstrom L, Lundqvist H, Axelsson S, Fasth KJ, Gullberg P and Langstrom B (1988): Receptor binding and selectivity of three ^{11}C-labeled dopamine receptor antagonists in the brain of Rhesus monkeys studied with positron emission tomography. *Acta Neurol Scand* 77: 314-321.

Hippus H (1989): The history of clozapine. *Psychopharmacol* 99: S3-S5.

Huang S-C, Barrio JR and Phelps ME (1986): Neuroreceptor assay with positron emission tomography: equilibrium versus dynamic approaches. *J Cereb Blood Flow Metab* 6 :515-52.1

Hyttel J (1983): SCH 23390 - The first selective D_1 antagonist. *Eur J Pharmacol* 91: 153-154.

Imperato A and Angelucci L (1989): The effects of clozapine and fluperlapine on the *in vivo* release and metabolism of dopamine in the striatum and in the prefrontal cortex of freely moving rats. *Psychopharmacol Bull* 25: 383-389.

Innis RB, Malison R, al-Tikriti M, Sybirska E, Zoghbi S, Baldwin RM, Smith E, Elsworth J and Kung H (1991): Amphetamine-induced dopamine release competes *in vivo* for I-123 IBZM binding to dopamine D_2 receptors in non-human primate brain. *J Nucl Med* 32:1007.

Iorio LC, Barnett A, Leitz FH, Houser VP and Korduba CA (1983): SCH 23390, a potential benzazepine antipsychotic with unique interactions on dopaminergic systems. *J Pharmacol Exptl Ther* 226: 462-468.

Iyo M, Yamasaki T, Fukuda H, Inoue O, Suzuki K, Shinotoh H, Ito T, Yonezawa H, Nishio M, Koseki A, Fukui S and Tateno Y (1989): Age-related decrease of ^{11}C-N-methylspiperone *in vivo* binding to human striatum detected by PET. *Jap J Nucl Med* 26: 213-220.

Kane J, Honigfeld G, Singer J and Meltzer H (1988): Clozapine for the treatment resistant schizophrenia. A double-blind comparison with chlorpromazine. *Arch Gen Psychiatry* 45: 789-796.

Karbe H, Wienhard K, Hamacher K, Huber M, Herholz K, Coenen HH, Stöcklin, Lövenich A and Heiss WD (1991): Positron emission tomography with (^{18}F)methylspiperone demonstrates D_2 dopamine receptor binding differences of clozapine and haloperidol. *J Neural Transm* 86: 163-173.

Kebabian JW and Greengard P (1972): Dopamine-sensitive adenyl cyclase: possible role in synaptic transmission. *Science* 174: 1346-

1349.

Kebabian JW and Calne DB (1979): Multiple receptors for dopamine. *Nature* 277: 93-96.

Kessler RM, Whetsell WO, Ansari MS, Votaw JR, de Paulis T, Bell RL, Schmidt D, Mason S and Manning RG (1991): Postmortem studies and *in vivo* aging of extrastriatal dopamine D2 receptors in man. *J Nucl Med* 32: 961.

Kilbourn MR, Carey JE, Koeppe RA, Haka MS, Hutchins GD, Sherman PS and Kuhl DE (1989): Biodistribution, dosimetry, metabolism and monkey PET studies of [^{18}F]GBR 13119. Imaging the dopamine uptake system *in vivo*. *Int J Rad Appl Instrument (B)* 16: 569-576.

Kilbourn MR and Haka MS (1988): *In vivo* binding of [18F]GBR13119, a presynaptic dopamine uptake antagonist. *Int J Rad Appl and Instrument (A)* 39: 279-82.

Kung HF, Kasliwal R, Pan S, Kung MP, Mach RH and Guo YZ (1988): Dopamine D_2 receptor imaging radiopharmaceuticals: Synthesis, radiolabeling and in vitro binding of (R)-(+)- and S-(-)-3-iodo-2-hydroxy-6-methoxy-N-((1-ethyl-2-pyrrolidinyl)-m-ethyl) benzamide. *J Med Chem* 31: 1039-43.

Lane RT, Blaha CD and Rivet JM (1988): Selective inhibition of mesolimbic dopamine release following chronic administration of clozapine, involvement of noradrenergic receptors demonstrated in vivo voltammetry. *Brain Res* 460: 398-401.

Leenders KL, Aquilonius SM, Bergstrom K, Bjurling P, Crossman AR, Eckernas S-A, Gee AG, Hartvig P, Lundqvist H, Langstrom B, Rimland A and Tedroff J (1988): Unilateral MPTP lesion in a rhesus monkey: Effects on the striatal dopaminergic system measured iin vivo with PET using various novel tracers. *Brain Res* 445: 61-67.

Logan J, Wolf AP, Shiue C-Y, and Fowler JS (1987): Kinetic modeling of receptor-ligand binding applied to positron emission tomographic studies with neuroleptic tracers. *J Neurochem* 48: 73-83.

Lundberg T, Lindstrom LH, Hartvig P, Eckerns SA, Ekblom B, Lundqvist H, Fasth KJ, Gullberg P and Langstrom B (1989): Striatal and frontal cortex binding of ^{11}C-labelled clozapine visualized by positron emission tomography (PET) in drug-free schizophrenics and healthy volunteers. *Psychopharmacol* 99: 8-12.

McKenna PJ (1987): Pathology, phenomenology and the dopamine hypothesis of schizophrenia. *Br J of Psych* 151: 288-301.

Meltzer HY, Matsubara S and Lee JC (1989): Classification of typical and atypical antipsychotic drugs on the basis of dopamine D-1, D-2 and serotonin2 pKi values. *J Pharmacol Exptl Ther* 252: 238-46.

Mintun MA, Raichle ME, Kilbourn MR, Wooten GF and Welch MJ (1984): A quantitative model for the *in vivo* assessment of drug binding sites with positron emission tomography. *Ann Neurol* 15: 217-227.

Perlmutter JS, Kilbourn MR, Raichle ME and Welch MJ (1987): MPTP-induced upregulation of *in vivo* dopaminergic radioligand-receptor binding in humans. *Neurol* 37: 1575-1579.

Perlmutter JS, Larson KB, Raichle ME, Markham J, Mintun MA, Kilbourn MR and Welch MJ (1986): Strategies for in vivo measurement of receptor binding using positron emission tomography. *J Cereb Blood Flow Metab* 6: 154-169.

Pilowsky LS, Costa DC, Ell PJ, Murray RM, Verhoeff NPLG and Kerwin RW (1992): Clozapine, single photon emission tomography, and the D_2 dopamine receptor blockade hypothesis of schizophrenia. *Lancet* 340: 199-202.

Pycock CJ, Kerwin RW and Carter CJ (1980): Effect of lesion of cortical dopamine terminals on subcortical dopamine receptors in rats. *Nature* 286: 74-77.

Ravert HT, Wilson AA, Dannals RF, Wong DF and Wagner HN Jr (1987): Radiosynthesis of a selective dopamine D-1 receptor antagonist: R(+)-7-Chloro-8-hydroxy-3[^{11}C]methyl-1-phenyl-2,3,4,5-tetrahydro-1H-3-benzazepine([^{11}C]SCH 23390). *Int J Appl Radiat Isot* 38: 305-306.

Ritz MC, Lamb RJ, Goldberg SR and Kuhar MJ (1987): Cocaine receptors on dopamine transporters are related to self-administration of cocaine. *Science* 237: 1219-1223.

Sadzot B and Franck G (1990): Non-invasive methods to study drug disposition: Positron emission tomography detection and quantification of brain receptors in man. *Eur J Drug Metab Pharmacokin* 15: 135-142.

Scheffel U, Boja JW and Kuhar MJ (1989a): Cocaine receptors: *in vivo* labeling with 3H-(-) cocaine, 3H-WIN 35, 065-2, and 3H-WIN 35,428. *Synapse* 4: 390-392.

Seeman P, Bzowej NH, Guan HC, Bergeron C, Becker LE, Reynolds GP, Bird ED, Riederer P, Jellinger K, Watanabe S and Tourtellotte WW (1987a): Human brain dopamine receptors in children and aging adults. *Synapse* 1: 399-404.

Seeman P, Bzowej NH, Guan HC, Bergeron C, Reynolds GP, Bird ED, Riederer P, Jellinger K and Tourtellotte WW (1987b): Human brain D_1 and D_2 dopamine receptors in Schizophrenia, Alzheimer's, Parkinson's, and Huntington's diseases. *Neuropsychopharmacol* 1: 5-15.

Seeman P, Guan H-C and Niznik HB (1989): Endogenous dopamine lowers the dopamine D_2 receptor density as measured by [^{3}H]raclopride: Implications for positron emission tomography of the human brain. *Synapse* 3: 96-97.

Seeman P, Lee T, Chau-Wong M and Wong K (1976): Antipsychotic drug doses and neuroleptic/dopamine receptors. *Nature* 261: 717-719.

Shaya EK, Scheffel U, Dannals RF, Ricaurte GA, Carroll FI, Wagner HN Jr, Kuhar MJ and Wong DF (1992): *In vivo* imaging of dopamine reuptake sites in the primate brain using single photon emission computed tomography (SPECT) and Iodine-123 labeled RTI-55. *Synapse* 10: 169-172.

Smith M, Wolf AP, Brodie JD, Arnett CD, Barouche F, Shiue CY, Fowler JS, Russell JAG, MacGregor RR, Wolkin A, Angrist B, Rotrosen J and Peselow E (1988): Serial [^{18}F]N-methylspiroperidol PET studies to measures changes in antipsychotic drug D-2 receptor occupancy in schizophrenic patients. *Biol Psychiat* 23: 653-663.

Sokoloff P, Giros B, Martres MP, Bouthenet ML and Schwartz JC (1990): Molecular cloning and characterization of a novel dopamine receptor (D_3) as a target for neuroleptics. *Nature* 347: 146-151.

Sunahara RK, Guan HC, O'Dowd BF, Seeman P, Laurier LG and Ng G (1991): Cloning of the gene for a human dopamine D_5 receptor with higher affinity for dopamine than D_1. *Nature* 350: 614-619.

Tamminga CA, Dannals RF, Frost JJ, Wong DF and Wagner HN Jr (1993): Neuroreceptor and neurochemistry studies with positron emission tomography in psychiatric illness: Promise and progress. In: *Reviews of Psychiatry, Vol 12*, Oldham JM, Tasman A and Reba MB, eds. Washington, D.C.: American Psychiatric Press, Inc.

Van Tol HHM, Bunzow JR, Guan HC, Sunahara RK, Seeman P and Niznik HB (1991): Cloning of the gene for a human dopamine D_4 receptor with high affinity for the antipsychotic clozapine. *Nature* 350: 610-614.

Volkow ND, Fowler JS, Wolf AP, Schlyer D, Shiue CY, Alpert R, Dewey S, Logan J, Bendriem B, Christman D, Hitzemann R and Henn F (1990): Effects of chronic cocaine abuse on postsynaptic dopamine receptors. *Am J Psychiat* 147: 719-724.

Waddington JL (1989): Sight and Insight: Brain dopamine receptor occupancy by neuroleptics visualised in living schizophrenic patients by positron emission tomography. *Brit J Pscychiat* 154: 433-436.

Weinberger DR (1987): Implications of normal brain development for the pathogenesis of schizophrenia. *Arch Gen Psychiat* 44: 660-669.

White JF and Wang RY (1983): Differential effects of classical and atypical antipsychotic drugs on A9 and A10 dopamine neurons. *Science* 221: 1054-1057.

Wolkin A, Barouche F, Wolf AP, Rotrosen J, Fowler JS, Shiue CY, Cooper TB and Brodie JD (1989): Dopamine blockade and clinical response: Evidence for two biological subgroups of schizophrenia. *Am J Psychiat* 146: 905-908.

Wolkin A, Brodie JD, Barouche F, Rotrosen J, Wolf AP, Smith M, Fowler J and Cooper TB (1989): Dopamine receptor occupancy and plasma haloperidol levels. *Arch Gen Psychiat* 46: 482-484.

Wong DF, Braestrup C, Harris J, Dannals RF, Ravert H, Wilson A, Williams J, Links J, Scheffel U, O'Tuama L, Fanaras G, Moser H, Naidu S, Nyhan W, Wagner HN Jr and Gjedde A (1986a): Assessment of D_1 and D_2 dopamine receptors in Lesch-Nyhan syndrome by positron emission tomography. *J Nucl Med* 27: 1027.

Wong DF, Brousselle E, Wand G, Villemagne V, Dannals RF, Links JM, Zacur HA, Harris J, Naidu S, Braestrup C, Wagner HN Jr and Gjedde A (1988): *In vivo* measurement of dopamine receptors in human brain by positron emission tomography: age and sex differences. In: *Central Determinants of Age-Related Declines in Motor Function*, Joseph JA, ed. New York: Annals New York Academy of Sciences.

Wong DF, Gjedde A and Wagner HN Jr (1986b): Quantification of neuroreceptors in the living human brain. Part I. Association rate of irreversibly bound ligands. *J Cereb Blood Flow Metab* 6: 137-14.6

Wong DF, Gjedde A, Wagner HN Jr, Dannals RF, Douglass KH, Links JM and Kuhar MJ (1986c): Quantification of neuroreceptors in the living human brain. Part II. Inhibition studies of receptor density and affinity. *J Cereb Blood Flow Metab* 6: 147-153.

Wong DF, Ross C, Wagner HN Jr, Pearlson G, Links JM, Brousolle E, Fanaras G, Fischman M, Danashvar D, Wilson A, Ravert H and Dannals RF (1986d): The effect of IV cocaine on the kinetics of [^{11}C]-3-n-methylspiperone binding in the human caudate. *J Nucl Med* 27: 1074.

Wong DF, Wagner HN Jr, Coyle J, Snyder S, LaFrance N, Bice A, Dannals R, Links J, Paulos M, Hoehn-Saric R, Tune L and Kuhar MJ (1985): Assessment of dopamine receptor blockade by neuroleptic drugs in the living human brain. *J Nucl Med* 26: 52.

Wong DF, Wagner HN Jr, Dannals RF, Links JM, Frost JJ, Ravert HT, Wilson AA, Rosenbaum AE, Gjedde A, Douglass KH, Petronis JD, Folstein MF, Toung JKT, Burns HD and Kuhar MJ (1984): Effects of age on dopamine and serotonin receptors measured by positron tomography in the living human brain. *Science* 226: 1393-1396.

Wong DF, Wagner HN Jr, Tune LE, Dannals RF, Pearlson GD, Links JM, Tamminga CA, Broussolle EP, Ravert HT, Wilson AA, Toung JKT, Malat J, Williams JA, O'Tuama LA, Snyder SH, Kuhar MJ and Gjedde A (1986e): Positron emission tomography reveals elevated D_2 dopamine receptors in drug-naive Schizophrenics. *Science* 234:

1558-1563.

Wong DF, Young LT, Pearlson G, Singer H, Tune L, Ross C, Dannals RF, Wilson AA, Ravert HT, Links J, Wagner HN Jr and Gjedde, A (1989a): D_2 dopamine receptor densities measured by PET are elevated in several neuropsychiatric disorders. *J Nucl Med* 30: 731.

Wong DF, Gjedde A, Young D, Tune L, Pearlson G, Shaya E, Young T, Chan B, Burkhardt D, Wilson PD, Dannals, RF, Wilson AA, Ravert HT, Natarajan TK and Wagner HN Jr (1991b): PET demonstrates reduced NMSP receptor affinity in psychosis. *J Cereb Blood Flow Metab* (suppl): S650.

Wuesel F-A, Farde L, Nordstrom AL and Sedvall G (1990): Central D_1- and D_2-receptor occupancy during antipsychotic drug treatment. *Prog Neuro-Psychopharmacol Biol Psychiat* 14: 759-767.

Young LT, Wong DF, Goldman S, Minkin E, Chen C, Matsumura K, Scheffel U and Wagner HN Jr (1991): Effects of endogenous dopamine on kinetics of [^{3}H]N-Methylspiperone and [^{3}H]Raclopride binding in rat brain. *Synapse* 9: 188-194.

IODINATED DOPAMINE D_1 AND D_2 RECEPTOR IMAGING AGENTS FOR SPECT

Hank F. Kung
Departments of Radiology and Pharmacology
University of Pennsylvania

INTRODUCTION

In vivo imaging of the dopaminergic system in the mammalian brain has been the subject of extensive studies in recent years. Not only is this part of the central nervous system (CNS) very important for normal brain function, it is also a possible primary action site for antiparkinsonian agents or neuroleptic drugs for schizophrenia. Addiction to cocaine, which is an acute social and economic problem, is related to the CNS reward system involving dopaminergic function. Binding of cocaine to the dopamine reuptake sites, which leaves more dopamine in the synaptic cleft for binding to the postsynaptic neuron, is the primary mechanism for stimulation. There is a wealth of pharmacological information about the dopamine receptor (Creese and Fraser, 1987; Beaulieu, 1987; Waddington, 1986; Waddington, 1989; Stoof and Kebabian, 1984; Waddington and O'Boyle, 1987; Waddington and O'Boyle, 1989; Sidhu et al., 1991).

Initial classification of dopamine receptor subtypes is based on the ability of agonists and antagonists to discriminate between two different and distinct dopamine receptors, which are designated as D_1 and D_2 (Waddington and O'Boyle, 1987; Kebabian and Calne, 1979). Recent advances in the molecular biology of the dopamine receptor have

Nuclear Imaging in Drug Discovery, Development, and Approval
H.D. Burns, R. Gibson, R. Dannals, Peter Siegl, Editors

significantly modified and complicated the functional aspects and classification of receptor subtypes. Cloning of D_1 and D_2 receptors clearly demonstrates that these two receptor subtypes are structurally different (Monsma et al., 1990a; Monsma et al., 1990b; Fremeau et al., 1991; Grandy et al., 1989; Weiner et al., 1991; Dearry et al., 1990a; Dearry et al., 1990b; Albert et al., 1990). Discovery of isoforms, long and short forms, of D_1 and D_2 receptors suggests that the subtypes may not be uniform (Sunahara et al., 1990a; Sunahara et al., 1990b; Zhou et al., 1991; Bunzow et al., 1988; Mansour et al., 1990a; Mansour et al., 1990b; Monsma et al., 1989). In addition, D_3, D_4 and D_5 receptor subtypes have been cloned (Snyder, 1990; Sunahara et al., 1991; Van Tol et al., 1991; Sokoloff et al., 1990). One of the most important findings on the "post-D_2" receptor subtypes is the implication of different mechanisms for atypical neuroleptics. The D_3 receptor appears to localize in the mesolimbic area (Sokoloff, 1990), while the extra-striatal D_4 receptor displays high binding affinity to clozapine, an atypical neuroleptic (Van Tol et al., 1991). Since a majority of the D_2 receptor selective ligands invariably show equally high potency in D_3 and D_4 receptor binding, and they display homology of a large portion of the amino acid sequences, it is possible that these subtypes have the same evolutionary origin.

Imaging studies of CNS dopamine D_2 receptors in humans with [^{11}C]raclopride by Positron Emission Tomography (PET) have been reported (Farde et al., 1985; Farde et al., 1986; Farde et al., 1987a; Farde et al., 1987b; Farde and von Bahr, 1990; Farde et al., 1990). A high ratio of specific striatal to nonspecific cerebellar binding in living human brain was observed. Using an equilibrium model and Scatchard plots, the binding parameters (K_d = 7.1 nM, B_{max} = 15 pmole/ml) in living human brain was measured. The values for dopamine D_2 receptor density were comparable to those determined earlier with a different imaging agent: [^{11}C]N-methylspiperone (NMSP, K_d = 0.097 nM, B_{max} = 16.6 pmole/g) (Wagner et al., 1983; Wong et al., 1985; Wong et al., 1986). In two separate studies, different receptor concentrations were reported based on quantitative *in vivo* imaging studies of schizophrenic patients with [^{11}C]NMSP (Wong et al., 1986) and [^{11}C]raclopride (Farde et al., 1990). [^{11}C]NMSP showed a higher D_2 dopamine concentration and [^{11}C]raclopride displayed no change in the dopamine density. The discrepancy may be a result of the differential effects of endogenous dopamine concentrations between raclopride and NMSP (Young et al., 1991). In addition, differences due to methodology and/or patient population have also been suggested (Waddington, 1989; Waddington and O'Boyle, 1989; Swart and Korf, 1989; Wong et al., 1990).

However, despite the attractive features associated with PET studies, this technique is still limited to a handful of major university medical centers and national laboratories, where an abundance of technical and financial resources are available. For the majority of nuclear medicine clinics, Single Photon Emission Computed Tomography (SPECT) and planar imaging studies are the only procedures currently available. The resolution of SPECT (full width half maximum, FWHM; 7-15 mm) is generally lower than that of PET (FWHM 4-7 mm), but due to the lower cost, the SPECT technique may be the method of choice for widespread application.

IODINATED DOPAMINE D_2 RECEPTOR IMAGING AGENTS

Three types of iodinated compounds have been reported as dopamine D_2 receptor SPECT imaging agents: iodobenzamides, iodospiperone and iodolisuride.

Iodobenzamides

A variety of substituted benzamide derivatives (see Fig. 1, next page) possessing antidopaminergic properties have been reported (Bishop et al., 1991; Ögren et al., 1984; de Paulis et al., 1985; Hall and Wedel, 1986). Of these, IBZM (Kung et al., 1988a; Kung et al., 1990a; Verhoeff et al., 1991; Verhoeff, 1991), ioxipride (NCQ 298) (Hall et al., 1990), epidepride (Kessler et al., 1991a; Kessler et al., 1991b) and IBF (**1**, K_d = 0.106 nM)(Kung et al., 1990b; Murphy et al., 1990) show specific D_2 receptor binding activity and high affinity (lower K_d) in rat striatum tissue preparations and low nonspecific binding. All of these iodinated agents are potentially useful SPECT imaging agents.

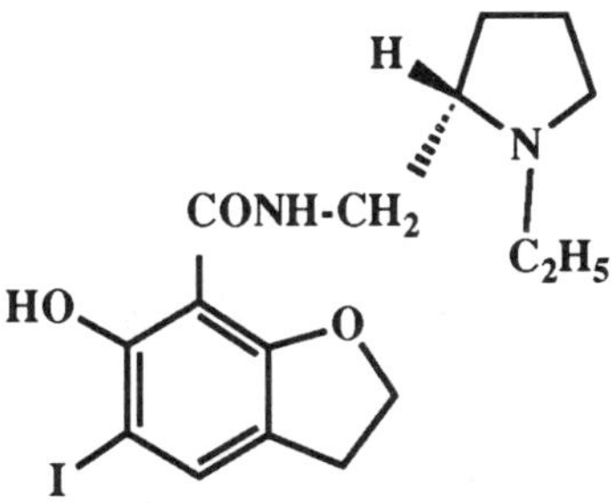

1

Imaging studies in humans with [^{123}I]IBZM demonstrate that this agent, as expected, localizes in the basal ganglia area (Kung et al.,

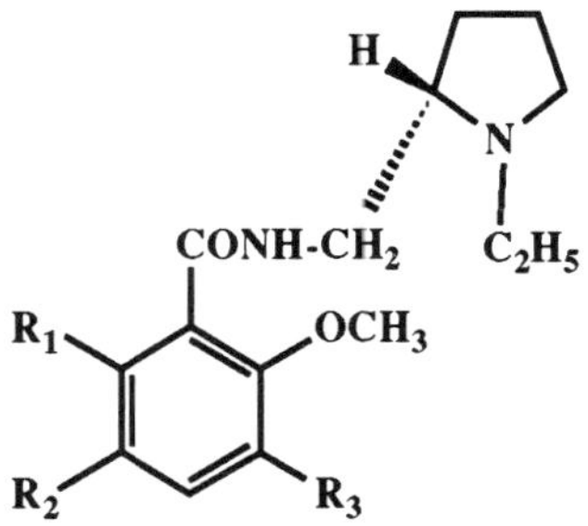

	R_1	R_2	R_3	K_d (nM)
Iodosulpiride	I	SO_2NH_2	H	1.5
Raclopride	HO	Cl	Cl	1.1
Eticopride	HO	Et	Cl	0.17
IBZM	HO	I	H	0.43
Iodopride	H	I	H	3.0
Epidepride	H	I	CH_3O	0.025
Ioxipride	HO	I	CH_3O	0.019

FIGURE 1. Chemical structures of benzamides and their affinity to dopamine D_2 receptors.

1990a; Verhoeff et al., 1991; Verhoeff, 1991). SPECT imaging of the normal human brain displayed a pattern which clearly indicated the location of the basal ganglia in the brain. This agent is commercially available in Europe. Initial studies indicate that the agent is useful in characterization of dopamine D_2 receptor in patients (Cordes et al., 1991; Costa et al., 1990; Brucke et al., 1991).

In order to investigate the displacement effect of various agents on D_2 receptor binding and to further characterize these D_2 receptor SPECT imaging agents by an *in vivo* imaging technique, a series of imaging studies in monkeys was carried out. SPECT imaging with [^{123}I]IBZM, [^{123}I]IBF and [^{123}I]NCQ 298 in monkeys all showed high selectivity and localization in a D_2 receptor-rich region, i.e. basal ganglia (Figs. 2, 3 and 4). Monkeys were injected with 5-6 mCi of [^{123}I]IBZM, [^{123}I]IBF or [^{123}I]NCQ 298 and images were collected with a Picker T3000 triple-head camera fitted with a fan beam collimator. In each study, Haldol was injected intravenously 80 min post-injection to examine the ease with

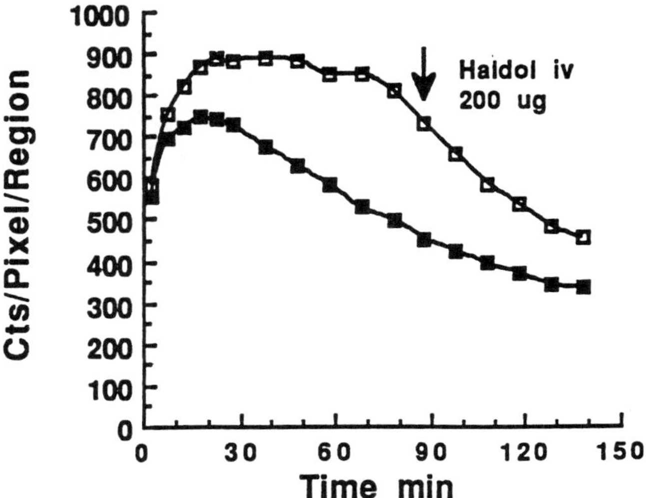

FIGURE 2. Kinetics of uptake of [^{123}I]IBZM in monkey basal ganglia (□) and cerebellum (■). Haldol clearly competes with [^{123}I]IBZM binding to the dopamine D_2 receptors in the basal ganglia and increases the washout rate.

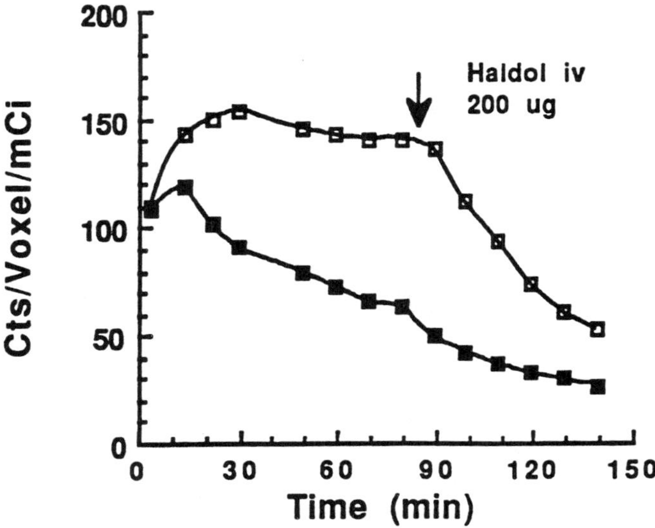

Figure 3. Kinetics of uptake of [^{123}I]IBF in monkey basal ganglia (□) and cerebellum (■). The Haldol clearly competes with [^{123}I]IBF binding to the dopamine D_2 receptor in the basal ganglia and increases the washout rate.

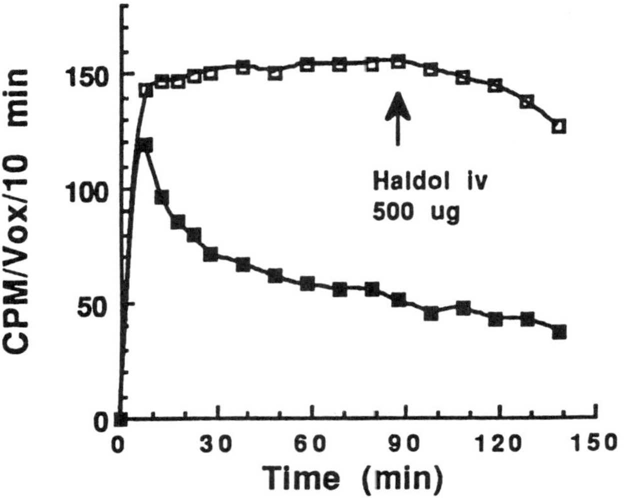

FIGURE 4. Kinetics of uptake of [^{123}I]NCQ 298 (Ioxipride) in monkey basal ganglia (□) and cerebellum (■). Although Haldol shows competition with radiotracer binding to the dopamine D_2 receptor in the basal ganglia, the washout rate does not significantly increase as with the other two D_2 agents.

which radiotracer binding can be reversed. The uptake in the cerebellum region, i.e., a region low in dopamine D_2 receptors, displayed lower localization and faster washout. After the D_2 receptor imaging agent reached a relatively stable distribution in basal ganglia, an intravenous (iv) dose of Haldol displaced the selective uptake and significantly increased the rate of washout from the target region. The data strongly suggest that Haldol competes with the same binding sites in the brain. Despite the similarity of these three agents in the blocking studies, there is a significant difference in the ability of Haldol to displace the NCQ 298 binding. Even at a dose of 500 μg Haldol (iv), the inhibition is less (Fig. 3) than that observed for IBZM and IBF (at a dose of 200 μg, iv). This observation is consistent with the *in vitro* binding affinity data (obtained using rat striatal membrane preparation) that NCQ 298 (K_d = 0.019 nM) has an affinity five times higher than IBF (K_d = 0.106 nM). Based on this data it is reasonable to conclude that, for drug competition studies, either for new neuroleptics or new agents affecting postsynaptic dopamine binding, it is better to use the D_2 receptor ligands, IBZM and IBF, which show high selectivity but lower affinity.

Iodospiperone

Preliminary studies of an iodinated 2'-iodospiperone (**2**; 2'-ISP) indicate that the spiperone analogue displays excellent D_2 receptor specificity (K_d = 0.25 nM in rat striatum) and *in vivo* stability as compared to the 4-iodo-spiperone reported earlier (Saji et al., 1987; Nakatzuka et al., 1987). *In vitro* binding data for 2'-ISP appears to show a higher nonspecific binding (>40%) than that observed with [^{125}I]IBZM (5%).

Iodolisuride

Lisuride is a dopamine agonist and it is currently being used in Europe as an antiparkinson agent. Iodolisuride (**3**) is an antagonist of the dopamine D_2 receptor and shows promise as a SPECT imaging agent.

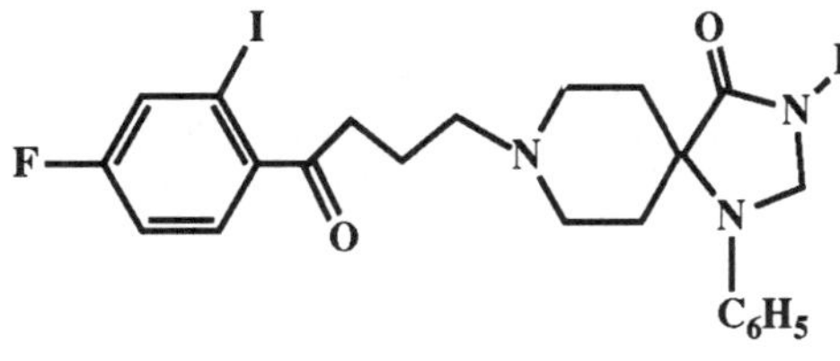

3

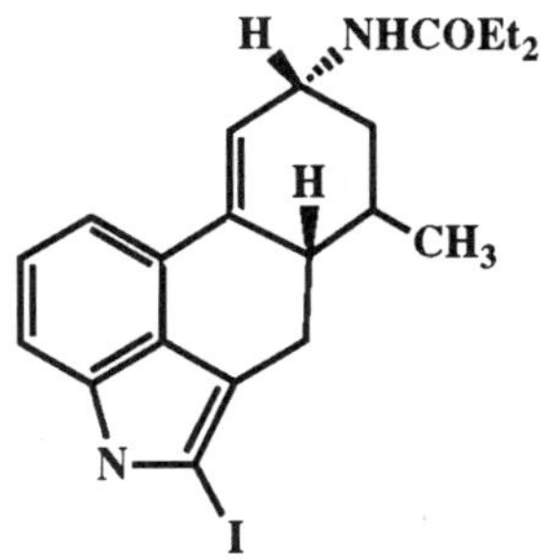

2

Initial clinical study indicated that the agent, as expected, concentrated in the region of the basal ganglia in the brain (Maziere and Maziere, 1990). However, this agent appears to have a slower washout rate in the basal ganglia and other brain regions as compared to those of the benzamides.

IODINATED DOPAMINE D_1 RECEPTOR IMAGING AGENTS

PET studies in humans and nonhuman primates with [^{11}C]SCH 23390 (Farde et al., 1987a; Farde et al., 1987b) and its analogues have demonstrated the feasibility of imaging the D_1 receptor *in vivo*. Several iodinated analogues of SCH 23390 as D_1 receptor ligands have been investigated. When labeled with Iodine-123 ($T_{1/2}$ = 13 hr; 159 keV), they can be used in conjunction with widely available SPECT techniques to study the CNS dopamine D_1 receptor in the living monkey and human brain.

[^{125}I]SCH 23892 (IBZP or SKF-103108a) (Kung et al., 1988b; Manik et al., 1988; Sidhu et al., 1986; Moerlein et al., 1990; Thonoor et al., 1988) has a K_d of 0.7 nM in the rat striatum and is a commonly used commercial agent for D_1 receptor autoradiography and *in vitro* binding studies. Although the Iodine-123 labeled agent has shown high uptake and selectivity in the early phases of its administration to primates (Alavi and Kung, 1989), significant non-specific binding has been observed at later time points. The low target-to-nontarget ratio is probably due to an *in vivo* instability leading to the presence of free iodide in the blood circulation. In order to improve the *in vivo* stability of the iodinated ligand, a 4'-iodinated SCH 23390 was prepared (FISCH, an acronym for 4-iodo-SCH 23390). FISCH was found to be more stable *in vivo*, and (±)-[^{125}I]FISCH had a high binding affinity for D_1 receptors (K_d = 1.4 nM, rat striatum preparation) (Alavi and Kung, 1989; Chumpradit et al., 1989a; Chumpradit et al., 1989b; Billings et al., 1989). However, the *in vivo* biodistribution of (±)-[^{125}I]FISCH in monkeys indicated that FISCH has low striatum/cortex and striatum/cerebellum ratios. For quantitative SPECT imaging, a high ratio is necessary for determining small changes in D_1 receptor densities. These small changes can occur in many neurologic disease states (Joyce et al., 1988). A novel derivative of SCH 23390 was prepared: 7-chloro-8-hydroxy-(3'-iodophenyl)-3-methyl-2,3,4,5-tetrahydro-1H-3-benzazepine (**3**, TISCH) (Chumpradit et al., 1990; Chumpradit et al., 1991; Billings et al., 1991). This D_1 receptor ligand has the stereoselective preference for the R isomer of the 1-phenyl benzazepine compounds (Neumeyer et al., 1990; McQuade et al., 1988a; McQuade et al., 1988b). The binding affinity of R-(+)-[^{123}I]TISCH using

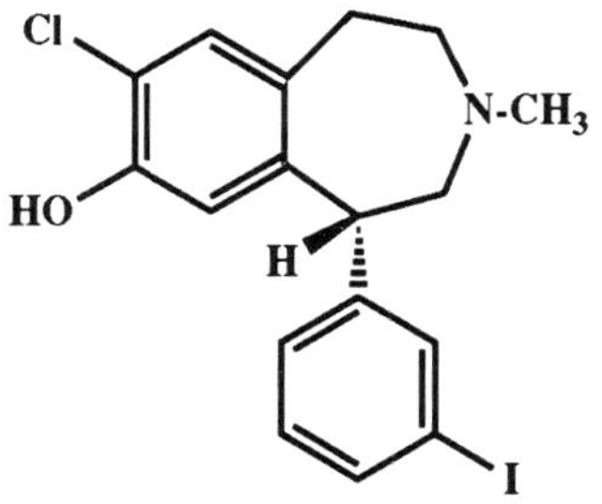

Cl
N-CH_3
HO
H
I

3

rat striatal membrane preparations showed a K_d = 0.13 nM and a pharmacological profile consistent with a selective D_1 receptor ligand. Initial SPECT imaging studies showed superior contrast in basal ganglia vs. cerebellum (Billings et al., 1991).

Using the same type of SPECT imaging techniques as described for Fig. 2, 3 and 4 and R-(+)-[^{123}I]TISCH, the effect of various drugs on the dopamine D_1 receptor binding (*in vivo*) were investigated (Fig. 5). The

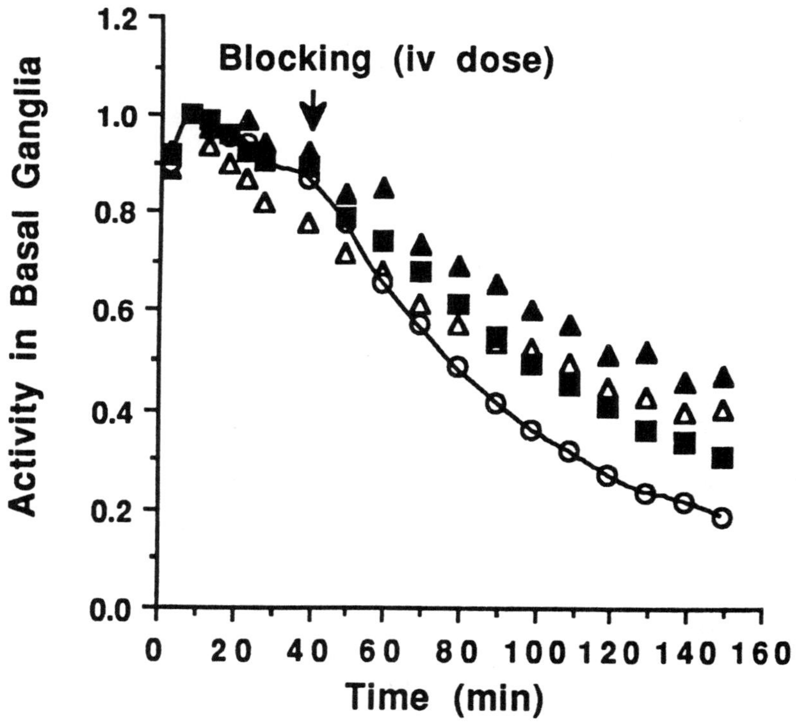

FIGURE 5. Kinetics of uptake of [^{123}I]TISCH in monkey basal ganglia (▲, Δ). At 40 min post-injection either SCH 23390 (3.6 mg) or clozapine (15.6 mg) was injected (iv). The SCH 23390 (O), a selective D_1 agent, clearly competed with TISCH binding to the dopamine D_1 binding sites in the basal ganglia and increased the washout rate, while the atypical neuroleptic, clozapine (■), displayed little effect on the washout of TISCH.

results suggest that the selective ligand SCH 23390 clearly competes with the binding of TISCH in basal ganglia, while the atypical neuroleptic, clozapine, does not show any significant effect in washout of TISCH binding. The affinity of new neuroleptics to the D_1 dopamine receptor *in vivo* can be evaluated with this imaging technique.

CONCLUSION

Significant progress has been made in the past few years in developing new iodinated SPECT imaging agents for *in vivo* dopamine receptor imaging. However, only limited clinical studies with SPECT have been reported in the literature. A more extensive clinical evaluation may be required before the full potential of this type of imaging study can be applied on a routine basis. There are several obstacles remaining to be resolved before a widespread application of this imaging method in humans and monkeys can be achieved: i) the supply of Iodine-123 is still limited; ii) the cost of this isotope is relatively high (compared to Technetium-99m); iii) the imaging procedure may require the use of dedicated SPECT devices to increase the sensitivity and quantitation of dopamine receptors; and iv) B_{max} measurement using kinetic data from SPECT has not yet been fully investigated. For a relative and non-quantitative study of *in vivo* binding of dopamine D_1 and D_2 receptors using regions of interest, comparison may provide useful information for elucidation of mechanisms of action of new drugs. The same technique may also be very effective in planning the dose schedules for clinical trials of new drugs based on the *in vivo* binding and competition studies at the preclinical and/or phase I and phase II stages of a new drug development.

ACKNOWLEDGEMENT:

Financial support from PHS (NS-24538). The author thanks Catherine Cartwright for her editorial assistance.

REFERENCES

Alavi A and Kung HF (1989): Imaging of the brain with SPECT: Is it coming of age? In: *Yearbook of Nuclear Medicine*, Hoffer PB, ed. Chicago: Year Book Medical Publishers.

Albert PR, Neve KA, Bunzow JR and Civelli O (1990): Coupling of a cloned rat dopamine-D_2 receptor to inhibition of adenylyl cyclase and prolactin secretion. *J Biol Chem* 265: 2098-2104.

Beaulieu M (1987): Clinical importance of D_1 and D_2 receptors. *Can J*

Neurol Sci 14 (3 Suppl): 402-406.

Billings J, Kung M-P, Chumpradit S, Pan S and Kung HF (1989): [^{125}I](±)FISCH: A new CNS D_1 dopamine receptor imaging ligand. *Life Sci* 45: 711-718.

Billings JJ, Kung M-P, Chumpradit S, Mozley D, Alavi A and Kung HF (1991): Characterization of radioiodinated TISCH: A high affinity and selective ligand for mapping CNS dopamine D_1 receptor. *J Neurochem* 58: 227-236.

Bishop JE, Mathis CA, Gerdes JM, Whitney JM, Eaton AM and Mailman RB (1991): Synthesis and *in vitro* evaluation of 2,3-dimethoxy-5-(fluoroalkyl)-substituted benzamides: high affinity ligands for CNS doapmine D_2 receptors. *J Med Chem* 34: 1612-1624.

Brucke T, Podreka I, Angelberger P, Wenger S, Topitz A, Kufferle B, Muller C and Deecke L (1991): Dopamine D_2 receptor imaging with SPECT: Studies in different neuropsychiatric disorders. *J Cerb Blood Flow Met* 11: 220-228.

Bunzow JR, Van Tol HHM, Grandy DK, Albert P, Salon J, Christie M, Machida CA, Neve KA and Civelli O (1988): Cloning and expression of a rat D_2 dopamine receptor cDNA. *Nature* 336: 783-787.

Chumpradit S, Kung HF, Billings J, Kung M-P and Pan S (1989a): (±)-7-Chloro-8-hydroxy-1-(4'-[^{125}I]iodophenyl)-3-methyl-2,3,4,5-tetrahydro-1H-3-benzazepine: A potential CNS D_1 dopamine receptor imaging agent. *J Med Chem* 32: 1431-1435.

Chumpradit S, Billings J, Kung M-P, Pan S and Kung HF (1989b): An improved CNS D_1 dopamine receptor imaging ligand: [^{123}I](±)FISCH. *J Nucl Med* 30: 803.

Chumpradit S, Billings JJ, Kung M-P and Kung HF (1990): R-(+) and S-(-) TISCH: New CNS D_1 dopamine receptor ligands. *J Nucl Med* 31: 899.

Chumpradit S, Kung MP and Kung HF (1991): New Derivatives of SCH-23390 as CNS D_1 Dopamine Receptor Ligands. *J Labelled Compd Radioparm* 30: 347-348.

Cordes M, Henkes H, Laudahn D, Bräu H, Kramp W, Girke W, Hierholzer J, Eichstädt H and Felix R (1991): Initial experience with SPECT examinations using [^{123}I]IBZM as a D_2-Dopamine receptor antagonist in Parkinson's Disease. *Eur J Radiol* 12: 182-186.

Costa DC, Verhoeff NPLG, Cullum ID, Ell PJ, Syed GMS, Barrett J, Palazidou E, Toone B, Van Royen E and Bobeldijk M (1990): *In Vivo* characterization of 3-Iodo-6-methoxybenzamide ^{123}I in humans. *Eur J Nucl Med* 16: 813-816.

Creese I and Fraser CM (1987): Dopamine receptors. In: *Receptor*

Biochemistry and Methodology, Venter JC and Harrison LC, eds. New York: Alan Liss, Inc.
Dearry A, Gingrich JA, Falardeau P, Fremeau RT Jr, Bates MD and Caron MG (1990a): Molecular cloning and expression of the gene for a human D_1 Dopamine receptor. *Nature* 347: 72-76.
Dearry A, Edelman JL, Miller S and Burnside B (1990b): Dopamine induces light-adaptive retinomotor movements in bullfrog cones via D_2 receptors and in retinal pigment epithelium via D_1 receptors. *J Neurochem* 54: 1367-1378.
Farde L, Ehrin E, Eriksson L, Greitz T, Hall H, Hedström C-G, Litton J-E and Sedvall G (1985): Substituted benzamides as ligands for visualization of dopamine receptor binding in the human brain by Positron Emission Tomography. *Proc Natl Acad Sci* 82: 3863-3867.
Farde L, Hall H, Ehrin E and Sedvall G (1986): Quantitative analysis of D_2 dopamine receptor binding in the living human brain by PET. *Science* 231: 258-261.
Farde L, Wiesel F-A, Hall H, Halldin C, Stone-Elander S and Sedvall G (1987a): No D_2 receptor increase in PET study of schizophrenia. *Arch Gen Psychiat* 44: 671-672.
Farde L, Halldin C, Stone-Elander S and Sedvall G (1987b): PET analysis of human dopamine receptor subtypes using [^{11}C]-SCH 23390 and [^{11}C]-Raclopride. *Psychopharmacol* 92: 278-284.
Farde L and von Bahr C (1990): Distribution of remoxipride to the human brain and central D_2-Dopamine receptor binding examined *in vivo* by PET. *Acta Psychiatr Scand* 358(Suppl): 67-71.
Farde L, Weisel F-A, Stone-Elander S, Halldin C, Nordström A-L, Hall H and Sedvall G (1990): D_2 dopamine receptors in neuroleptic-naive schizophrenic patients. A PET study with [^{11}C]raclopride. *Arch Gen Psychiat* 47: 213-219.
Fremeau RT Jr, Duncan GE, Fornaretto MG, Dearry A, Gingrich JA, Breese GR and Caron MG (1991): Localization of D_1 dopamine receptor mRNA in brain supports a role in cognitive, affective, and neuroendocrine aspects of dopaminergic neurotransmission. *Proc Natl Acad Sci* 88: 3772-3776.
Grandy DK, Marchionni MA, Makam H, Stofko RE, Alfano M, Frothingham L, Fischer JB, Burke-Howie KJ, Bunzow JR, Server AC and Civelli O (1989): Cloning of the cDNA and gene for a human D_2 dopamine receptor. *Proc Natl Acad Sci* 86: 9762-9766.
Hall H and Wedel I (1986): Comparisons between the *in vitro* binding of two substituted benzamides and two butyrophenones to dopamine-D_2 receptors in the rat striatum. *Acta Pharmacol et Toxicol* 58: 368-373.
Hall H, Högberg T, Halldin C, Köhler C, Ström P, Ross SB, Larsson SA and Farde L (1990): NCQ 298, a new selective iodinated

salicylamide ligand for the labelling of dopamine D_2 receptors. *Psychopharmacol* 103: 16-18.

Joyce JN, Lexow N, Bird E and Winokur A (1988): Organization of D_1 and D_2 receptors in human striatum: receptor autoradiographic studies in Huntington's disease and schizophrenia. *Synapse* 2: 546-557.

Kebabian JW and Calne DB (1979): Multiple receptors for dopamine. *Nature* 277: 93-96.

Kessler RM, Ansari MS, de Paulis T, Schmidt DE, Clanton JA, Smith HE, Manning RG, Gillespie D and Ebert MH (1991a): High affinity dopamine D_2 receptor radioligands. 1. regional rat brain distribution of iodinated benzamides. *J Nucl Med* 32: 1593-1600.

Kessler RM, Ansari MS, Schmidt DE, de Paulis T, Clanton JA, Innis R, Al-Tikriti M, Manning RG and Gillespie D (1991b): High affinity dopamine D_2 receptor radioligands. 2. [^{125}I]Epidepride, a potent and specific radioligand for the characterization of striatal and extrastriatal dopamine D_2 receptors. *Life Sci* 49: 617-628.

Kung HF, Guo Y-Z, Billings J, Xu X, Mach RH, Blau M and Ackerhalt RE (1988a): Preparation and biodistribution of [^{125}I]IBZM: a potential CNS D_2 dopamine receptor imaging agent. *Nucl Med Biol* 15: 195-201.

Kung HF, Billings J, Guo Y-Z, Blau M and Ackerhalt RE (1988b): Preparation and biodistribution of [^{125}I]IBZP: a potential CNS D_2 dopamine receptor imaging agent. *Nucl Med Biol* 15: 187-193.

Kung HF, Alavi A, Chang W, Kung M-P, Keyes JW Jr, Velchik MG, Billings J, Pan S, Noto R, Rausch A and Reilly J (1990a): *In vivo* SPECT imaging of CNS D_2 dopamine receptors: Initial studies with [^{123}I]IBZM in humans. *J Nucl Med* 31: 573-579.

Kung M-P, Kung HF, Billings J, Yang Y, Murphy RA and Alavi A (1990b): The characterization of IBF as a new selective dopamine D_2 receptor imaging agent. *J Nucl Med* 31: 648-654.

Manik CP, Molinoff PB and McGonigle P (1988): Comparison of ^{125}I-SCH 23982 and [^{3}H]SCH 23390 as ligands for the D_1 dopamine receptor. *J Neurochem* 51: 391-397.

Mansour A, Meador-Woodruff JH, Bunzow JR, Civelli O, Akil H and Watson SJ (1990a): Localization of dopamine D_2 receptor mRNA and D_1 and D_2 receptor binding in the rat brain and pituitary: An *in situ* hybridization-receptor autoradiographic analysis. *J Neurosci* 10: 2587-2600.

Mansour A, Meador-Woodruff JH, Camp DM, Robinson TE, Bunzow J, Van Tol HH, Civelli O, Akil H and Watson SJ (1990b): The effects of nigrostriatal 6-hydroxydopamine lesions on dopamine D_2 receptor mRNA and opioid systems. *Prog Clin Biol Res* 328: 227-230.

McQuade RD, Chipkin R, Amlaiky N, Caron M, Iorio L and Barnett A (1988a): Characterization of the radioiodinated analogue of SCH 23390: *In vitro* and *In vivo* D_1 dopamine receptor binding studies. *Life Sci* 43: 1151-1160.

McQuade RD, Ford D, Duffy RA, Chipkin RE, Iorio LC and Barnett A (1988b): Serotonergic component of SCH 23390: *In vitro* and *In vivo* binding analyses. *Life Sci* 43: 1861-1869.

Maziere B and Maziere M (1990): Where have we got to with neuroreceptor mapping of the human brain? *Eur J Nucl Med* 16: 817-835.

Monsma FJ Jr, McVittie LD, Gerfen CR, Mahan LC and Sibley DR (1989): Multiple D_2 dopamine receptors produced by alternative RNA splicing. *Nature* 342: 926-929.

Monsma FJ Jr, Mahan LC, McVittie LD and Gerfen CR (1990a): Molecular cloning and expression of a D_1 dopamine receptor linked to adenylyl cyclase activation. *Proc Natl Acad Sci* 87: 6723-6727.

Monsma FJ Jr, Barton AC and Sibley DR (1990b): Expression of functional D_2 dopamine receptors following differentiation of y-79 human retinoblastoma cells. *J Neurochem* 54: 1200-1207.

Moerlein SM, Parkinson D and Welch MJ (1990): Radiosynthesis of high effective specific-activity [^{123}I]SCH 23982 for dopamine D_1 receptor-based SPECT imaging. *Appl Radiat Isot* 41: 381-385.

Murphy RA, Kung HF, Kung M-P and Billings J (1990): Synthesis and characterization of iodobenzamide analogues: Potential D_2 dopamine receptor imaging agents. *J Med Chem* 33: 171-178.

Nakatzuka I, Saji H, Shiba K, Shimizu H, Okuno M, Yoshitake A and Yokoyama A (1987): *In vitro* evaluation of radioiodinated butyrophenones as radiotracer for dopamine receptor study. *Life Sci* 41: 1989-1997.

Neumeyer JL, Baindur N, Yuan J, Booth G, Seeman P and Niznik HB (1990): Development of a high affinity and stereoselective photoaffinity label for the D_1 dopamine receptor: Synthesis and resolution of 7-[^{125}I]Iodo-8-hydroxy-3-methyl-1-(4'-azidophenyl)-2,3,4,5-tetrahydro-1H-3-benzazepine. *J Med Chem* 33: 521-526.

Ögren SO, Hall H, Köhler C, Magnusson O, Lindbom L-O, Ängeby K and Florvall L (1984): Remoxipride, a new potential antipsychotic compound with selective antidopaminergic actions in the rat brain. *Eur J Pharmacol* 102: 459-474.

de Paulis T, Kumar Y, Johansson L, Rämsby S, Florvall L, Hall H, Ängeby-Möller K and Ögren SO (1985): Potential neuroleptic agents. 3. Chemistry and antidopaminergic properties of substituted 6-Methoxysalicylamides. *J Med Chem* 28: 1263-1269.

Saji H, Nakatzuka I, Shiba K, Tokui T, Horiuchi K, Yoshitake A,

Torizuka K and Yokoyama A (1987): Radioiodinated 2'-iodospiperone: A new radioligand for *in vivo* dopamine receptor study. *Life Sci* 41: 1999-2006.

Sidhu A, van Oene JC, Dandridge P, Kaiser C and Kebabian JW (1986): [^{125}I]SCH 23982: The ligand of choice for identifying the D_1 dopamine receptor. *Eur J Pharmacol* 128: 213-220.

Sidhu A, Sullivan M, Kohout T, Balen P and Fishman PH (1991): D_1 dopamine receptors can interact with both stimulatory and inhibitory guanine nucleotide binding proteins. *J Neurochem* 57: 1445-1451.

Snyder SH (1990): Pharmacology. The dopamine connection. *Nature* 347: 121-112.

Sokoloff P, Giros B, Martres MP, Bouthenet ML and Schwartz JC (1990): Molecular cloning and characterization of a novel dopamine receptor (D_3) as a target for neuroleptics. *Nature* 347: 146-151.

Stoof JC and Kebabian JW (1984): Two dopamine receptors: Biochemistry, physiology, and pharmacology. *Life Sci* 35: 2281-2296.

Swart JAA and Korf J (1989): Quantification of dopamine D_2 receptors by irreversible tracer binding in the living human brain: The model-dependent correction of metabolites. *J Cereb Blood Flow Metab* 9: 906-908.

Sunahara RK, Niznik HB, Weiner DM, Stormann TM, Brann MR, Kennedy JL, Gelernter JE, Rozmahel R, Yang YL, Israel Y, Seeman P and O'Dowd BF (1990a): Human dopamine D_1 receptor encoded by an intronless gene on chromosome 5. *Nature* 347: 80-83.

Sunahara RK, Murphy BJ and Tuana BS (1990b): A sensitive and rapid method for identification and characterization of low abundance receptors. *Anal Biochem* 185: 143-146.

Sunahara RK, Guan HC, O'Dowd BF, Seeman P, Laurier LG, Ng G, George SR, Torchia J, Van Tol HH and Niznik HB (1991): Cloning of the gene for a human dopamine D_5 receptor with higher affinity for dopamine than D_1. *Nature* 350: 614-619.

Thonoor CM, Couch MW, Greer DM, Thomas KD and Williams CM (1988): biodistribution and radiation dosimetry of radioiodinated-SCH 23982, a potential dopamine D_1 receptor imaging agent. *J Nucl Med* 29: 1668-1674.

Van Tol HH, Bunzow JR, Guan HC, Sunahara RK, Seeman P, Niznik HB and Civelli O (1991b): cloning of the gene for a human dopamine D_4 receptor with high affinity for the antipsychotic clozapine. *Nature* 350: 610-614.

Verhoeff NPLG, Bobeldijk M, Feenstra MGP, Boer GJ, Maas MAW,

Erdtsieck-Ernste E, de Bruin K, and van Royen EA (1991): *In vitro* and *in vivo* D_2-dopamine receptor binding with [^{123}I]S(–)iodobenzamide) [^{123}I]IBZM) in rats and human brain. *Nucl Med Biol* 18: 837-846.

Verhoeff NPLG (1991): Pharmacological implication for neuroreceptor imaging. *Eur J Nucl Med* 18: 482-502.

Wagner HN Jr, Burns HD, Dannals RF, Wong DF, Langstrom B, Duelfer T, Frost JJ, Ravert HT, Links JM, Rosenbloom SB, Lukas SE, Kramer AV and Kuhar MJ (1983): Imaging dopamine receptors in the human brain by positron tomography. *Science* 221: 1264-1266.

Waddington JL (1986): Behavioural correlates of the action of selective D_1 dopamine receptor antagonists: Impact of SCH 23390 and SKF 83566, and functionally interactive D_1:D_2 receptor systems. *Biochem Pharmacol* 35: 3661-3667.

Waddington JL and O'Boyle KM (1987): The D_1 dopamine receptor and the search for its functional role: From neurochemistry to behaviour. *Rev Neurosci* 1: 157-184.

Waddington JL (1989): Sight and insight: brain dopamine receptor occupancy by neuroleptics visualised in living schizophrenic patients by Positron Emission Tomography. *Br J Psychiat* 154: 433-436.

Waddington JL and O'Boyle KM (1989): Drugs acting on brain dopamine receptors: a conceptual re-evaluation five years after the first selective D_1 antagonist. *Pharmac Ther* 43: 1-52.

Weiner DM, Levey AI, Sunahara RK, Niznik HB, O'Dowd BF, Seeman P and Brann MR (1991): D_1 and D_2 dopamine receptor mRNA in rat brain. *Proc Natl Acad Sci* 88: 1859-1863.

Wong DF, Wagner HN Jr, Pearlson G, Dannals RF, Links JM, Ravert HT, Wilson AA, Suneja S, Bjorvvinsen E, Kuhar MJ and Tune L (1985): Dopamine receptor binding of C-11-3-N-methylspiperone in the caudate in schizophrenia and bipolar disorder: A preliminary report. *Psychopharmacol Bull* 21: 595-598.

Wong DF, Wagner HN Jr, Tune LE, Dannals RF, Pearlson GD, Links JM, Tamminga CA, Broussole EP, Ravert HT, Wilson AA, Toung JKT, Malat J, Williams JA, O'Tuama LA, Snyder SH, Kuhar MJ and Gjedde A (1986): Positron Emission Tomography reveals elevated D_2 dopamine receptors in drug-naive schizophrenics. *Science* 234: 1558-1563.

Wong DF, Minkin E, Wilson AA, Young LT, Dannals RF, Ravert HT and Wagner HN Jr (1990): *In vivo* studies of D_2 dopamine receptors with C-11 and I-125 Iodobenzamide. *J Nucl Med* 31: 882.

Young LT, Wong DF, Goldman S, Minkin E, Chen C,

Matsumura K, Scheffel U and Wagner HN Jr (1991): Effects of endogenous dopamine on kinetics of 3H N-methylspiperone and 3Hraclopride binding in the rat brain. *Synapse* 9: 188-94.

Zhou GZ, Katki AG, Schwarz S, Munson PJ and Rodbard D (1991): Quantitative characterization of multiple binding sites for phencyclidine and n-allylnormetazocine in membranes from rat and guinea pig brain. *Neuropharmacol* 30: 775-786.

RADIOLABELED ATRIAL NATRIURETIC PEPTIDE AND SOMATOSTATIN FOR *IN VIVO* IMAGING OF RECEPTORS

Richard J. Flanagan
Merck Frosst

INTRODUCTION

For many years peptides and proteins labeled with Iodine-125 on tyrosine residues have been standard tools of biochemistry for the *in vitro* mapping of receptors. Such methods provide microscopic information on a cellular level but little overall information about the *in vivo* biodistribution of the receptors and their dynamic relationship with their endogenous ligands. The gathering of this *in vivo* information by techniques such as Nuclear Imaging has been limited by the rapid catabolism of radiolabeled peptides in circulation. This is not as a result of labeling but is due to the extremely short half-life of peptides *in vivo*. The normal fate of peptides in circulation is to be partitioned between a large pool of low affinity non-specific peptidases and a small quantity of specific, high affinity receptors. Under such circumstances the labeled peptide will preferentially bind to the high affinity receptors but injection of amounts greater than the capacity of the receptors will cause overflow into the non-specific peptidase pool, with corresponding loss of information. Furthermore, it is important that receptor occupancy by the labeled peptide be limited, perhaps to less than 10% of receptor mass, in order that the ligand/receptor interactions observed are true tracer phenomena and not due to perturbation of the natural equilibrium by the

Nuclear Imaging in Drug Discovery, Development, and Approval
H.D. Burns, R. Gibson, R. Dannals, Peter Siegl, Editors

injected material. Since many of these peptides have physiological thresholds at the microgram level, indicative of very low receptor populations, it is essential that very small quantities of peptide be injected. This sets severe limitations on the nature of the labeled material and explains why there have been few reports of nuclear imaging of receptors using labeled peptides. Recent advances in radiolabeling technique and the use of shorter lived isotopes have allowed the development of peptide tracers in such high specific activity that the amount of material injected is within these limitations.

The term best suited for describing these limits on mass is specific activity, which is measured in Curie/μmole (Ci/μmole). It now appears that specific activities in the range 1 to 10 Curie/μmole are required for the successful use of labeled peptides for *in vivo* imaging of many receptors. For a typical imaging procedure involving 2 mCi of injected radioactivity, this would correspond to 2 to 0.2 nanomoles of injected peptide. With Iodine-125 and Iodine-131, specific activities of 1 Ci/μmole is just achievable but values of 10 Ci/μmole are beyond the limit for these isotopes. The theoretical specific activities and the percentage of this theoretical value achievable for the commonly used isotopes are shown in Table 1.

Isotope	Half-Life	Maximum Specific Activity (Ci/μmole)	% of Maximum Specific Activity Attainable
^{123}I(p,2n)	13 hours	241.04	70
^{123}I(p,5n)	13 hours	241.04	<1
^{125}I	60 days	2.17	80
^{131}I	8 days	16.23	10
^{111}In	2.8 days	46.63	>50
^{99m}Tc	6 hours	522.24	Unknown

TABLE 1. Theoretical and attainable specific activities for common isotopes.

It can be seen that ^{123}I(p,2n), Indium-111 and Technetium-99m, the isotopes of choice for receptor imaging, offer higher attainable specific activities than the other isotopes. ^{123}I(p,5n) is of comparable specific activity to Iodine-125 and Iodine-131 and is thus only suitable for imaging less demanding receptor systems. It should be noted that Indium-111 is often contaminated with other metals and purification by High Performance Liquid Chromatography (HPLC) after labeling avoids lowered specific activity but not other problems (*vide infra*).

For each receptor/ligand family there is a cutoff value for the specific activity above which successful *in vivo* imaging of receptors is possible and below which only non-specific blood pool imaging is observed. Such a cutoff can be mathematically modelled but an empirical approach would be to inject sufficient material to saturate 10% of the receptors in the target organ. For target organs at some distance from the site of injection (e.g., kidney via a femoral vein injection) this figure can be substantially increased, especially if there is competing extraction from other upstream organs. The important restriction is to avoid overflow from receptor mediated uptake onto non-specific peptidases. It can be anticipated that material of specific activity close to the cutoff value will demonstrate great variability in distribution since doubling the amount of injected material will result in 50% non-specific uptake and probable total loss of imaging.

ATRIAL NATRIURETIC PEPTIDE

Atrial natriuretic peptide (ANP) is an important peptide hormone produced in the cardiac atrium which acts upon the kidney, playing an important role in fluid, electrolyte and blood pressure homeostasis (De Bold, 1985; Genest and Cantin, 1987; Wildey et al., 1992). *In vitro* studies have demonstrated a high density of specific ANP receptors in the kidneys (Napier et al., 1984), adrenals (De Lean et al., 1985) and lungs (Olins et al., 1986). In the kidney, these receptors are most numerous in the glomeruli (Healy and Fanestil, 1986).

Three different types of ANP receptors have now been reported, A- and B-receptors, which contain particulate guanylate cyclase in their intracellular domain (Chang et al., 1989) and the more abundant C-receptor which is not coupled to guanylate cyclase (Fuller et al., 1988), but has ≈30% homology in the extracellular compartment with A- and B-receptors. The biological responses to ANP have been correlated with elevated levels of cGMP (Hamet et al., 1986) stimulated by guanylate cyclase, and A- and B-receptors containing this enzyme seem to express the peptide's biological actions. The function of the C-receptor is not yet fully known, but a clearing or buffering role has been postulated (Maack et al., 1987).

ANP is a twenty eight amino acid peptide (commonly numbered 99 to 126) containing one tyrosine at the C-terminus and no lysine groups. There are two forms commonly used, a human form ($_{H}ANP^{99-126}$) which has a methionine at position 110 and the rat form ($_{R}ANP^{99-126}$) which has an isoleucine at this position. The rat form is preferable for most studies as it has an equivalent affinity to the human form but is not as susceptible to oxidative damage (Fig. 1). In all of our studies we have used $_{R}ANP^{101-126}$ which is a shortened form of rat ANP lacking the serine at 99 and the leucine at 100.

In humans the half-life of ANP is reported to be between 1.7 and 4.5 minutes (Schwartz et al., 1990), a figure which seems very short compared to the localization times of most radiopharmaceuticals. Yet, it is clear this hormone must bind to the receptor within this time frame to exert its physiological effect. Of greater importance for imaging is the site and nature of the destruction of ANP. ANP in plasma is reported to be relatively stable. The major loci of degradation are renal brush border membranes, where a number of endopeptidases have been identified as being responsible for its inactivation (Kenny and Stevenson, 1988). The major sites of cleavage are the peptide bonds at 123-124, 105-106 and 110-111, with the site at 123-124 being the most significant (Fig. 1).

A final consideration that must be taken into account when labeling peptides is the active site of the molecule. The minimum configuration which will still demonstrate biological activity is a peptide fragment comprising residues 105 to 114, with the active site being residues 109 to 113 (Bovy, 1988). The position of radiolabeling must be at some

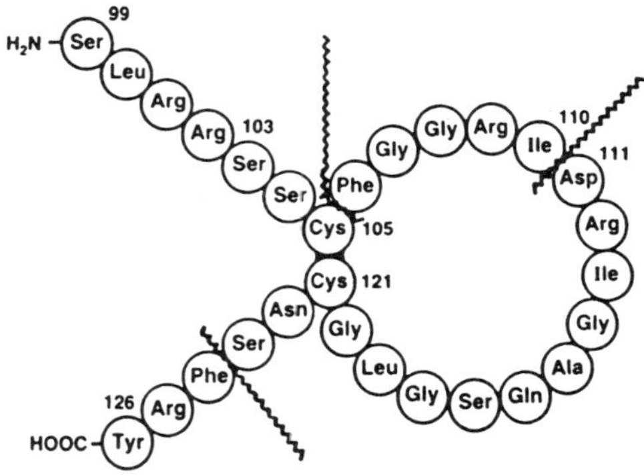

FIGURE 1. Primary sequence of rat ANP including major sites of enzymatic cleavage.

distance from these sites and with as little alteration to the conformation as possible (Fig. 2).

Strategies for radiolabeling ANP should also involve the placement of a radiolabel on the correct side of a cleavage site, so that cleavage results in a biologically active labeled fragment. Thus, the placement of an iodine on the C-terminal tyrosine would be expected to produce inactive radiolabeled fragments due to cleavage at 123-124. This is unfortunate considering the excellent positioning of the radiolabel at residue 128 with respect to the active site. The attachment of a chelator for indium or technetium at the N-terminal amino acid or the reduction of the disulphide bridge with concomitant insertion of a technetium atom, followed by ring closure, are more attractive possibilities since they satisfy both the requirements with respect to distance from the active site and cleavage sites.

Chelated versions of ANP can be prepared by the reaction of the dianhydride of diethylene-triaminepentacetic acid (DTPA) or N,N'-bis(2'-hydroxybenzyl)ethylene-diaminediacetic acid (HBED) with the N-terminal amino group of ANP under mildly basic conditions. It is important that the arginine residues be converted to their hydrochloride salts to prevent their participation in the reaction. The chemistry is surprisingly easy and the chelates can be isolated by HPLC.

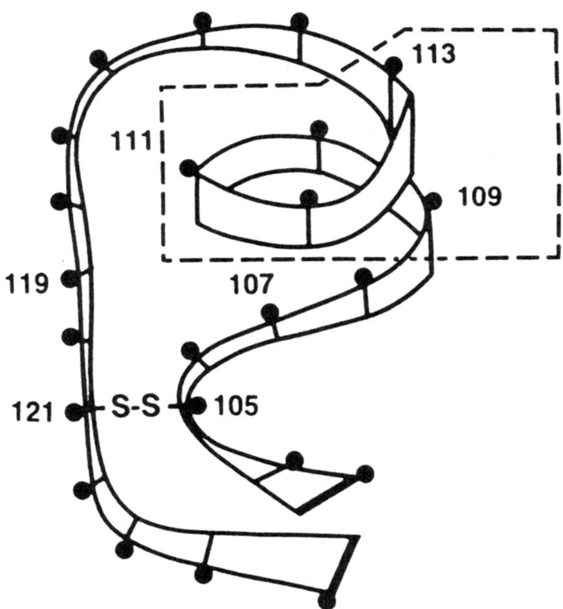

FIGURE 2. Proposed active conformation of ANP.

METHODS

Radioiodination of ANP

$_R$ANP$^{101\text{-}126}$ is easily iodinated with ^{123}I(p,2n) using the chloramine-T method; 10 µg of ANP$^{101\text{-}126}$ is dissolved in 40 µL of pH 7.8 borate buffer and 20 mCi of [^{123}I]NaI is added, followed by 5 µg of chloramine-T in 5 µl of water. The solution is vortexed at 0° C for 2 minutes and the reaction then quenched with cysteine. Monoiodinated ANP is isolated using HPLC on a C_{18} µ-Bondapak column by elution with a linear gradient of 20-40% acetonitrile/water (0.1% TFA) at 0.5%/min and 1 ml/min. The specific activity of [^{123}I]ANP is typically ≥ 50 Ci/µmole, this being the upper limit of measurement by HPLC/UV spectrometry at 214 nm.

This method produces only mono-iodo-ANP in about 40% yield after isolation by HPLC which is in sharp contrast to the nearly equal amounts of mono-iodinated and di-iodinated derivatives produced by iodination of ANP with [^{131}I]NaI under similar conditions. This can be explained by the much lower specific activity of Iodine-131, where further iodination of the mono-iodinated species can take place.

Preparation of DTPA-ANP and [^{111}In]DTPA-ANP

$_R$ANP$^{99\text{-}126}$ (0.79mg) is added to a 1 mL vial equipped with a small magnetic stirring bar. A solution of DTPA-anhydride (0.233mg) in DMSO (47.5 µL containing 4.91 µg/µL) is added, followed by a solution of HOBT (3,4-dihydro-3-hydroxy-4-oxo-1,2,3-benzotriazine) in pyridine (65.4 µL containing 0.653 µg/µL). The reaction mixture is stirred at room temperature for 22 hours and then H_2O (20 µL) is added. The solvent is removed *in vacuo* and the residue redissolved in water. Purification of the reaction mixture by HPLC (Partisil 5 ODS(3) Phenomenex) using a gradient system (Solvent A: 0.1% TFA in acetonitrile, Solvent B: 0.1% TFA in water, T = 0 min 20% A, 80% B, T = 5min 20%A, 80%B, T = 50 min 40% A, 60% B) affords pure DTPA-ANP. Unreacted ANP elutes at 21.77 min and the DTPA-ANP elutes at 24.09 min. The product was identified by mass spectrometry: MS(Electrospray, Hypermass) 810.25 (z = 4), 1080.00 (z = 3), 1619.5 (z = 2); Calculated compound mass = 3237.6, and measured compound mass = 3237.31.

In a 0.3 mL React-Vial is placed [^{111}In]InCl$_3$ solution (25 µL Millex water >18 MOhms, 25 µL [^{111}In]InCl$_3$), citrate buffer (50 µL, 0.1M,pH 4.5) and DTPA-ANP (16 µg in 10 µL of Millex water). The mixture is allowed to stand for 15 minutes and the progress of the reaction monitored by ITLC (Gelman-SG, 0.1M citrate as solvent). The crude reaction mixture is chromatographed on HPLC (Partisil 5 ODS(3), Phenomenex) using a gradient of 10% - 40% of 0.1% TFA in acetonitrile against 0.1% TFA in water. The desired labeled chelate elutes at 27.48 and 28.32 min post-injection. There appears to be two forms of [^{111}In]-DTPA-ANP as detected by HPLC, which we have called Peak-1 and Peak-2. Both of these peaks were isolated and studied. It seems likely that these two peaks represent diastereomers of DTPA-ANP.

Preparation of HBED-ANP and [^{111}In]HBED-ANP

To a 5 mL round bottomed flask is added $_R$ANP$^{101\text{-}126}$ (5.0 mg), bicarbonate/phosphate buffer (0.2M, pH 8.5, 400 µL), DMSO (500 µL) and HBED-NCS (12.1 mg) in DMSO (700 µL). The reaction mixture is stirred at room temperature for 20 hours followed by 6 hours at 38°C. The solvent is removed *in vacuo* and the residue dissolved in 10% acetic acid in water. Purification of the reaction mixture by HPLC (Partisil 5 ODS(3) Phenomenex) using a gradient system (Solvent A: 0.1%TFA in acetonitrile, Solvent B: 0.1% TFA in water, T = 0 min 20% A, 80% B, T = 40 min 50% A, 50% B) affords pure ANP-HBED. Unreacted ANP elutes at 10.5 min and the HBED-ANP elutes at 12.78 min. The product was identified by mass spectrometry: MS(Electrospray, Hypermass) 828.0 (z = 4), 1103.4 (z = 3), 1654.8 (z = 2), Calculated compound mass = 3307.38, and meassured compound mass = 3307.58.

In a 0.3 mL React-Vial is placed [^{111}In]$InCl_3$ solution (10 µL Millex water >18 Mohms, 1 µL [^{111}In]$InCl_3$), citrate buffer (20 µL, 0.01M, pH 7.6) and ANP-HBED (5 µg in 5 µL of Millex water). The mixture is incubated for 30 minutes and the progress of the reaction monitored by ITLC (Gelman-SG, 0.1M citrate as solvent). The crude reaction mixture is purified by chromatography using a PRP-1 solid-phase extractor (Hamilton & Co.). Elution of the PRP column with acetonitrile afforded pure [^{111}In]HBED-ANP. [^{111}In]HBED-ANP is unstable at the conditions required for HPLC (the chelate is unstable at low pH, the peptide will not elute at neutral pH) the material was analyzed by ITLC-SG. A single peak at the origin after development with 0.1M citrate indicates [^{111}In]HBED-ANP free from unchelated indium.

The insertion of metals into high affinity chelates such as DTPA and HBED with [^{111}In]$InCl_3$ proceeds with ease. The chelate is dissolved in a solution of the metal and reaction is rapid and kinetically controlled with DTPA chelation being somewhat faster than HBED because of its more flexible backbone. Nevertheless, both reactions are very fast, chelation with DTPA is complete in minutes while chelation with HBED requires approximately 10 minutes. It is important that all solutions be free from contaminating transition metals as these interfere with the kinetics of chelate formation and cause lower yields.

RESULTS

Biochemical Properties of Radiolabeled ANP Derivatives

Because of the short half-life of Iodine-123, the binding affinity of [^{123}I]ANP was not measured, but it should be identical to that of [^{125}I]ANP. Iodinated ANP is reported to be less active than ANP itself, with an IC_{50} of 1.6 pM compared to 0.5 pM for ANP in the rat aorta relaxation assay (Brady et al., 1986).

The affinity of [^{111}In]DTPA-ANP (Peak-1) was determined using a competition assay on solubilized bovine adrenal membranes. The chelate had a displaceable affinity for both the A and C receptor but did not distinguish between the two. The measured affinity of DTPA-ANP is lower than that of ANP by one order of magnitude (Tremblay, 1991).

[^{111}In]HBED-ANP was also found to be approximately one order of magnitude less potent than ANP and indistinguishable from DTPA-ANP, which is reasonable since both chelators are similar in size (Tremblay, 1991).

Biological Properties of Radiolabeled ANP Derivatives

[^{123}I]ANP$^{101-126}$, when injected into rats, dogs and monkeys shows rapid clearance from blood with uptake in receptor bearing tissues such as lung, kidney and liver. The biodistributions reveal interspecies differences in receptor population, with uptake in the monkey and rat being predominantly in the kidney, while for the dog uptake is mostly in the lung. A quantitative biodistribution study in rats at different specific activities was carried out by diluting high specific activity [^{123}I]ANP with unlabeled ANP (Fig. 3). This study illustrates the effect that specific activity has on uptake and allows the cutoff between specific and non-specific biodistribution to be observed, which in the rat is between 0.3 and 3.0 Ci/μmole.

The biological behaviour of Peak 1 and Peak 2 of [^{111}In]DTPA-ANP and [^{111}In]HBED-ANP in male rats at 60 minutes post injection is shown in the Table 2. It can be seen that the blood clearance of Peak-1 is much faster (i.e., better) than that of Peak-2 (3.2% vs 7.3% at 1 hr post-injection). Peak-1 also gives better uptake in the kidney at 1 hr post-injection (11.2 vs. 9.4). Peak-2 shows slightly higher non-specific uptake as evidenced by muscle uptake (6.8 vs. 4.6). All subsequent

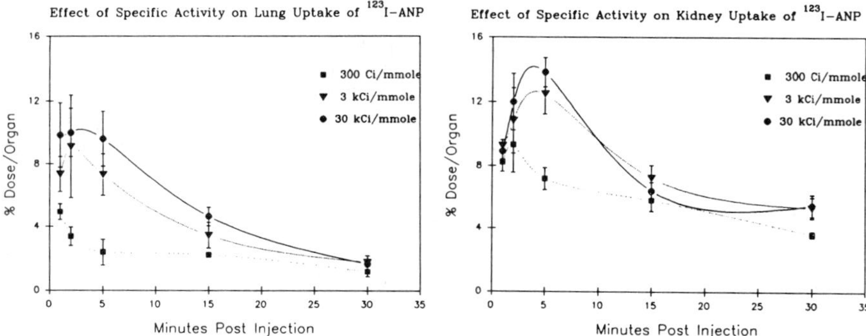

FIGURE 3. Effect of specific activity on the biodistribution of [^{123}I]ANP in Rats.

studies were carried out with Peak-1 only. The biological behavior of [^{111}In]HBED-ANP is also very similar to that of [^{111}In]DTPA-ANP in keeping with their similar binding affinities.

	Percent Injected Dose/Organ		
	[^{111}In]DTPA-ANP		[^{111}In]HBED-ANP
Organ	Peak 1	Peak 2	
Blood	3.2	7.3	5.3
Liver	5.3	4.4	7.2
Kidneys	11.2	9.4	10.9
Stomach	0.5	0.5	0.7
Lungs	1.8	1.6	1.12
Spleen	0.1	0.1	0.1
Muscle	1.8	6.8	5.1
Heart	0.4	0.4	0.4

TABLE 2. Biodistribution of Chelated ANP in Rats at 60 min post-injection.

Imaging with Labeled ANP Derivatives

Perhaps the most interesting experiments that can be performed with labeled ANP derivatives are imaging studies in live animals (Willenbrock et al., 1992). These permit the non-invasive observation of the passage of the ANP via the circulation to target organs, the processing of the ANP by the target organs and the subsequent excretion or redistribution of the ligand. These experiments offer the best glimpse of the true biological fate of peptides *in vivo* and it is impressive to watch the efficiency and speed with which they bind to their receptors. We have studied the fate of labeled ANP in rats, dogs and monkeys. In all cases the labeled ligand clears the circulation in less than 1 minute. At high specific activity the target organs are lungs and kidney with excretion via the kidneys; at lower specific activities there is noticeable uptake in liver with excretion via the bile into the intestine. Such a biodistribution pattern suggests that liver excretion is receptor mediated and not a function of lipophilicity since non-specific peptides of similar molecular weight show no liver excretion (see [^{111}In]DTPA-MK-678).

Figure 4A shows the biodistribution of [^{123}I]ANP$^{101\text{-}126}$ (specific activity ≥ 50 Ci/µmole) in green vervet monkeys at 0-5 minutes, 30-35 minutes and 60-65 minutes post injection. Figure 4B shows the biodistribution of [^{123}I]ANP$^{101\text{-}126}$ at the same intervals but with the co-injection of 20 µgram of unlabeled ANP. The dramatic differences in the biodistribution indicate that uptake in 4A is receptor mediated. It is important to note that 20 µg of ANP is close to the therapeutic threshold for the green vervet monkey but in this case was not sufficient to initiate natriuresis/diuresis. The material was injected via a catheter inserted into the inferior vena cava and so had to pass through the lungs to get to the kidney. The fact that the majority of the ligand is bound to the kidney confirms that the major site of action of ANP in the monkey is the kidney.

Figure 5A shows a baseline image of [^{111}In]DTPA-ANP$^{101\text{-}126}$ in a normotensive dog. The dog was anaesthetized with Ketamine/Rompun and injected via the femoral vein with 3.0 mCi [^{111}In]DTPA-ANP$^{101\text{-}126}$ in 1.5 mL of normal saline and 10 mg of albumin as carrier. The image shows the rapid clearance of the radiopharmaceutical from the blood with corresponding uptake in the lung and kidney. There is no evidence of stomach uptake, liver uptake (i.e., metabolism of ANP), or significant smooth muscle uptake. The image is consistent with rapid receptor mediated uptake in those organs known to possess either A-receptors (e.g., kidney) or C-receptors (e.g., lung and kidney). The image is quite unusual in that it represents successful receptor imaging *in vivo* using a chelated radiopharmaceutical.

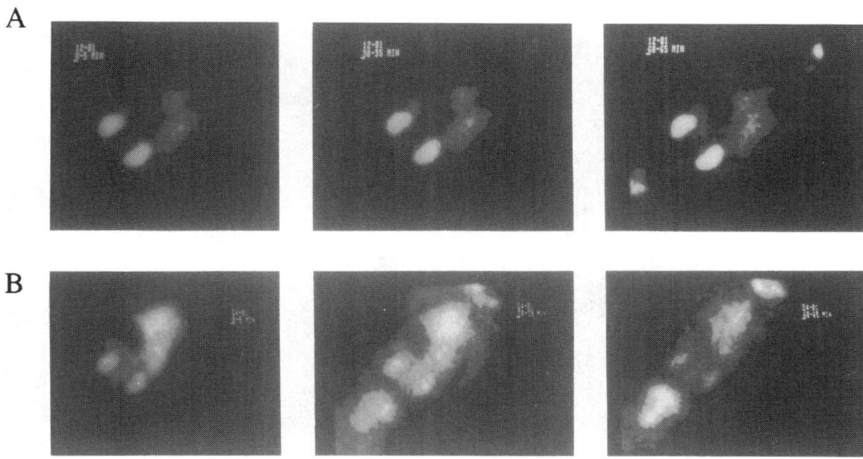

FIGURE 4. (A) Biodistribution of [^{123}I]ANP$^{101-126}$ in green vervet monkeys at 0-5 min, 30-35 min and 60-65 min post injection. (B) Biodistribution of [^{123}I]ANP$^{101-126}$ at the same intervals but with the co-injection of 20 μg of unlabeled ANP .

To confirm that the image in Fig. 5A is receptor mediated uptake, the same animal was imaged on a subsequent day using a prior injection of C-ANP, a selective C-receptor antagonist (Maack et al., 1987). This material would be expected to alter uptake of ANP in those organs possessing mainly C-receptors such as the lungs. Figure 5B shows the biological distribution at 5 min post injection in the presence of selective C-receptor antagonism. Lung uptake is diminished but there still is uptake in the kidney, probably due to A-receptors and there is now a focal site of uptake in the heart.

Figure 6 shows a similar experiment to that of Fig. 5B except that the animal was given a prior injection of ANP in an attempt to saturate both the A and C receptors. This is apparent in the diminished uptake in both the lung and kidney. Liver is now evident and the major focal uptake in the heart, found in Fig. 5B is still visible. The nature of the uptake in the heart must be receptor mediated and one can conclude that these heart receptors are not inhibited by ANP or C-ANP and can still bind the radiopharmaceutical. This is unusual but is perhaps due to a large population of low affinity receptors.

It is interesting to note that co-injection of unlabeled ANP in the monkey produced a non-specific biodistribution, while in the dog it results in an alternative specific biodistribution. Clearly, interspecies

A

BASELINE
0-5 MIN.

B

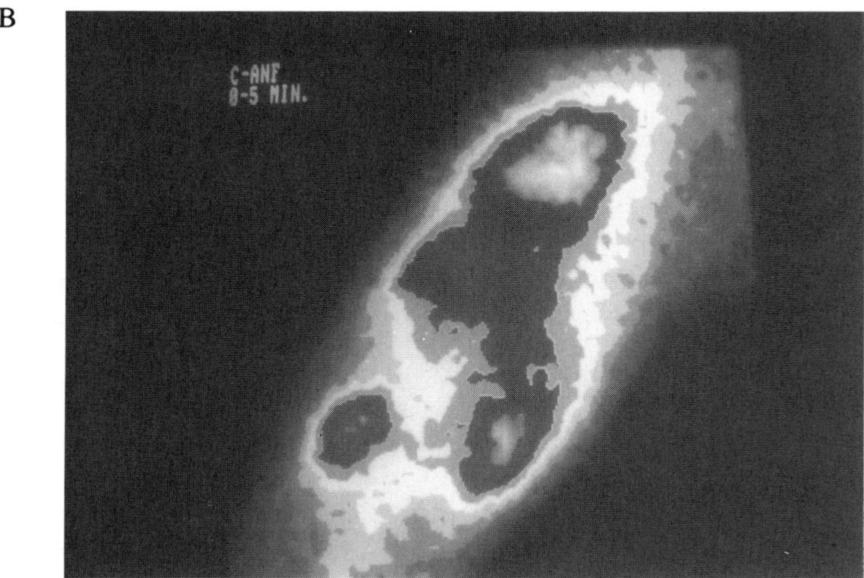

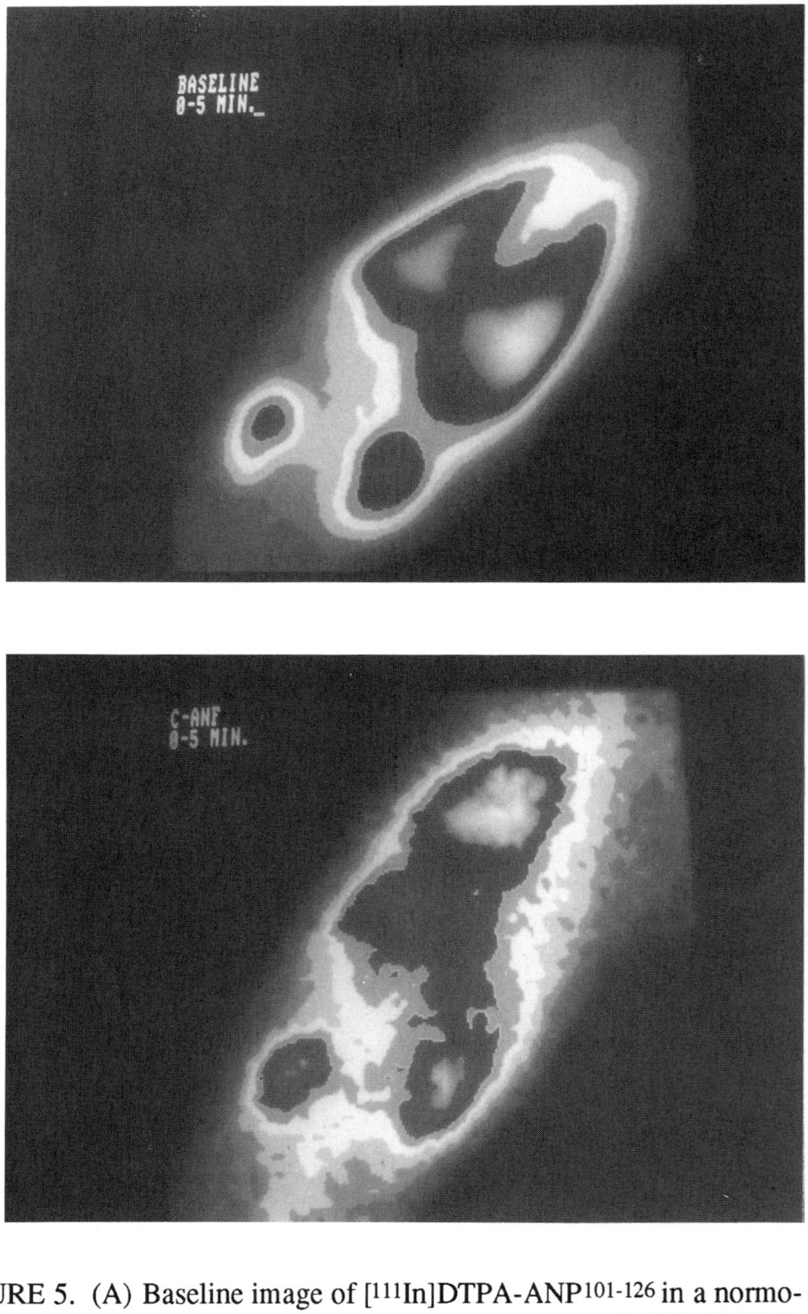

FIGURE 5. (A) Baseline image of [^{111}In]DTPA-ANP$^{101\text{-}126}$ in a normotensive dog at 0-5 min post-injection. (B) Image of [^{111}In]DTPA-ANP$^{101\text{-}126}$ in the same dog at 0-5 min post-injection in the presence of selective C-receptor antagonism.

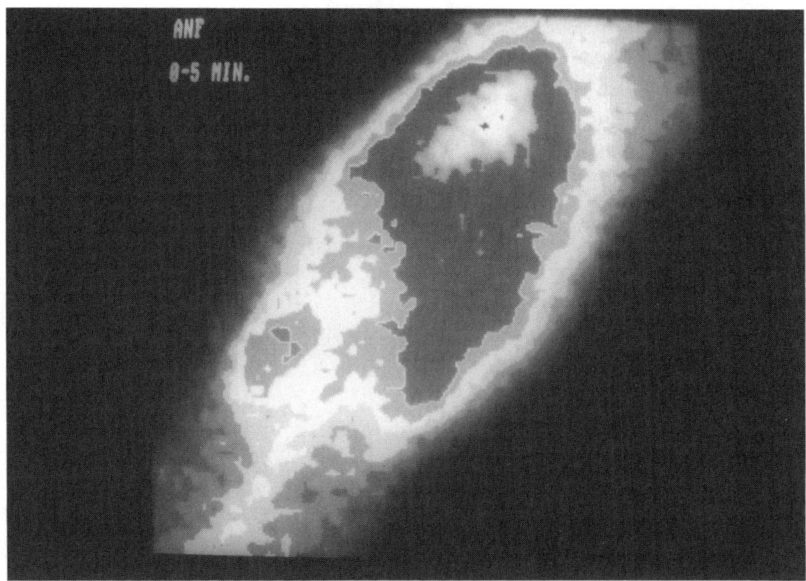

FIGURE 6, Image of [^{111}In]DTPA-ANP$^{101-126}$ in the same dog as Fig. 5 at 0-5 min post-injection in the presence of ANP in an attempt to saturate both the A and C receptors.

differences are very significant in the study of receptors.

Figure 7 is an *in vivo* autoradiographic study in a dog where the kidney was removed 30 min post injection of [^{111}In]HBED-ANP$^{101-126}$, sectioned and used for autoradiography. The image shows that the radiolabeled material is bound to glomeruli in the kidney; indeed the individual glomeruli can be counted. It is hard to imagine a more definitive proof that the biodistribution of this radiolabeled chelated ANP derivative is receptor mediated.

SOMATOSTATIN

Somatostatin (SRIF, Somatotropin Releasing Inhibiting Factor) is a tetradecapeptide originally isolated from the hypothalmus and named for its ability to inhibit the secretion of growth hormone (Brazeau et al., 1972). It is widely distributed in other tissues and inhibits insulin and glucagon secretion (Rechlin, 1983). The structure-activity-relationship of somatostatin is well defined and it is known that the active site resides in a four amino sequence comprising Phe-Trp-Lys-Thr (Fig. 8). Sandostatin is a synthetic octapeptide derivative of somatostatin

FIGURE 7. *In vivo* autoradiographic study in a dog where the kidney was removed 30 min post-inection of [^{111}In]HBED-ANP$^{101-126}$, sectioned and used for autoradiography.

containing this active sequence in a reduced cysteine bridged ring (Kvols et al., 1986). Interest in the use of labeled somatostatin derivatives for imaging was engendered by reports of large numbers of high affinity

Somatostatin:	A-G-C-K-N-F-F-W-K-T-F-T-S-C
Sandostatin:	F-C-F-W-K-T-C
Tyr-Sandostatin:	F-C-Y-W-K-T-C
MK-678:	Me-A-Y-W-K-V-F

FIGURE 8. Structure of somatostatin and peptide analogues.

somatostatin receptors in endocrine tumors, pituitary adenomas, meningiomas, breast tumors, neuroblastomas and medulloblastomas (Reubi et al., 1987a; Reubi et al., 1987b; Reubi et al., 1987c). In work by Krenning et al. (1989), a derivative of Sandostatin, in which the phenylalanine was replaced by a tyrosine, showed excellent uptake in SRIF positive tumors when labeled with Iodine-123 or Iodine-131 (Krenning et al., 1989). More recent work with an [^{111}In]DTPA derivative of Sandostatin also shows great promise (Krenning et al., 1992).

An even more minimalist derivative of SRIF is MK-678, a cyclic hexapeptide in which the active sequence is Tyr-Try-Lys-Val (Veber et al., 1984). This corresponds to the substitution of tyrosine for phenylalanine and valine for threonine, a net transfer of a hydroxyl from threonine to phenylalanine. The alanine group is methylated on the amide nitrogen to increase the half-life *in vivo* by reducing susceptibility to metabolism by proteases. In fact, MK-678 is so stable that it is excreted via the liver and bile into the intestine and can be recovered intact in the faeces. Since MK-678 has no free amino group it cannot be labeled via a chelator without further modification, so we describe herein initial results obtained with the iodinated version.

METHODS

High Specific Activity Radioiodination of MK-678

MK-678 can be labeled with iodine using the chloramine-T method in such a way that either a mono-iodinated or di-iodinated derivative can be produced. To produce a di-iodinated derivative using Iodine-123 requires the addition of cold Iodine-127. The use of a doubly labeled di-[^{131}I]MK-678 allows an effective doubling of specific activity.

To 26 μg of MK-678 in 10 μL of 2.0 M phosphate buffer (pH 7.5) is added 23.1 mCi of [^{131}I]NaI in 10 μL. Chloramine-T (64 μg in 4 μL of water) is added and the solution vortexed for 5 min at 0° C. The reaction is promptly quenched with the addition of 1 mg of sodium thiosulphate and purified by direct injection on a HPLC PRP-1 column using water (0.1% TFA)/acetonitrile as solvent and a gradient of 80:20 to 40:60 over 40 min. The MK-678 elutes at 22 min, the [^{131}I]MK-678 at 26.2 min and the di-[^{131}I]MK-678 elutes at 30.1 min. [^{123}I]MK-678 is prepared in a similar fashion but in order to produce the di-[^{123}I]MK-678, cold $^{127}I_2$ must be added (3.0 ng/mCi) to the [^{123}I]NaI.

RESULTS

Biological Properties of MK-678

The biodistribution of various labeled forms of MK-678 is shown in Table 3. The results for [^{123}I]MK-678 are essentially identical to [^{131}I]MK-678, inferring that, for the somatostatin receptor family, the cutoff for specific/non-specific biodistribution is less than 1.0 Ci/μmole; thus, there are substantially more receptors for SRIF than ANP. It is also interesting to note that the biodistribution pattern for di-[^{131}I]MK-678 is very similar to that of the mono-iodinated derivative which indicates that disubstitution of the tyrosine does not reduce affinity for the receptor.

Since the lysine of somatostatin is part of the active site, one would expect that derivatization of the lysine amino by DTPA would destroy the biological potency of the molecule. [^{111}In]DTPA-MK-678 shows an altered biodistribution which appears to be entirely non-specific. Since [^{111}In]DTPA-MK-678 shows no liver excretion, the elimination of SRIF via the liver into the intestine must be a receptor mediated pathway.

	Percent Injected Dose/Organ		
Organ	[^{131}I]MK-678	di-[^{131}I]MK-678	[^{111}In]DTPA-MK-678
Blood	1.3	2.0	6.9
Liver	4.2	4.9	2.0
Kidneys	1.0	0.7	4.0
Intestines	81.0	81.7	7.7
Gallbladder	-	-	0.93
Lungs	0.3	-	0.78
Spleen	0.1	-	0.1
Muscle	4.4	8.6	10.9
Heart	0.4	0.1	1.0

TABLE 3. Biodistribution of labeled MK-678 at 30 min post injection.

Imaging with [^{123}I]MK-678

Figure 9 is a 48-hour image of [^{123}I]MK-678 in a 62-year old male with Type I Multiple Endocrine Neoplasia. There are sites of uptake in the lung and liver, which are visible as early as 8 minutes post injection in the case of the lung mass. Since there are somatostatin receptors present in normal liver, the abnormal, non-excreting tumor receptors only become visible after the normal receptors have cleared via the bowel.

CONCLUSION

Labeled peptides used in nuclear imaging offer a simple method for the *in vivo* study of receptors for Nuclear Medicine or drug discovery. They provide information which at present cannot be acquired by any other method. Chelated peptides allow the prior synthesis of precursors, which can then be radiolabeled just before use. It is surprising how

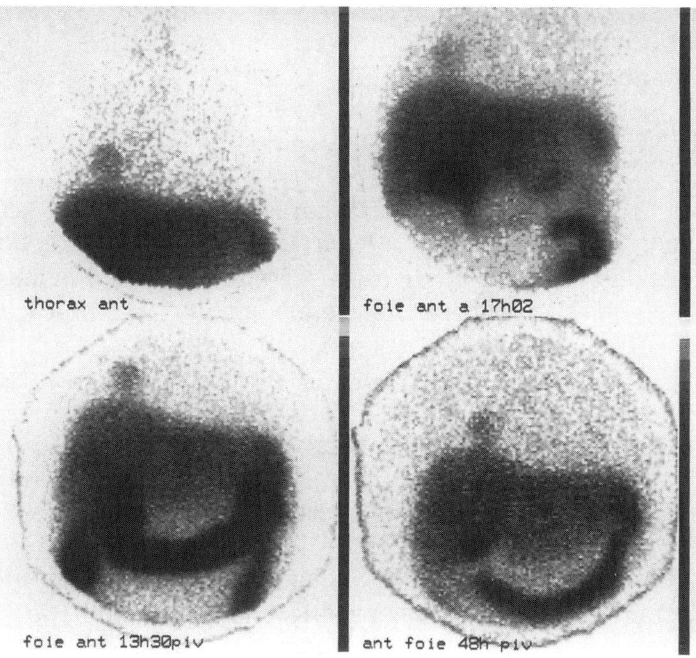

FIGURE 9. Image obtained 48 hr after injection of [^{123}I]MK-678 in a 62-year old male with Type I Multiple Endocrine Neoplasia.

tolerant small peptides, such as ANP and SRIF, are to the introduction of large chelate residues, modifications that would seem to contradict most existing peptide structure-activity-relationships.

Since ANP is implicated in hypertension, diabetic nephropathy, renal artery stenosis and congestive heart failure it seems likely that labeled ANP will provide a new tool for the non-invasive imaging of these pathological conditions. Similarly, if more tumors are found to express somatostatin receptors, radiolabeled SRIF will become a general purpose tumor imaging probe.

ACKNOWLEDGEMENTS

The studies on ANF were conducted in collaboration with F.P. Charleson, J-M. Dufour, K. Hogan, R. Limoges and D. Tartaglia (Merck Frosst, Montreal, Canada), J. Ethier, P. Hamet, Y. Langlois, J. Tremblay and R. Willenbrock (Centre de Recherche, Hôtel-Dieu Hospital, Montreal, Canada) and D. Lalonde and R. Lambert (Department of Nuclear Medicine, Hôtel-Dieu Hospital, Montreal, Canada). The studies on MK-678 were conducted in collaboration with F.P. Charleson, J-M. Dufour, K. Hogan, R. Limoges and D. Tartaglia (Merck Frosst, Montreal, Canada), and M. Picard (Department of Nuclear Medicine, Hôpital St Luc., Montreal, Canada).

REFERENCES

Bovy PR, O'Neal JM, Olins GM, Patton DR, Mahta PP, McMahon EG, Palomo M, Schuh J and Blehm D (1988): A synthetic linear decapeptide binds to atrial natriuretic receptors and demonstrates cyclase activation and vasorelaxation activity. *J Biol Chem* 264: 20309-20313.

Brady SF, Napier MA, Nutt RF and Vandlen RF (1986): Cardiac atrial peptides. *US Patent No: 4,609,725.*

Brazeau P, Vale W, Burgus R, Ling N, Butcher M, Rivier J and Guillemin R (1972): Hypothalmic polypeptide that inhibits the secretion of immunoreactive pituitary growth hormone. *Science* 129: 77-79.

Chang M-s, Lowe DG, Lewis M, Hellmiss R, Chen E and Goeddel DV (1989): Differential activation by atrial and brain natriuretic factor of two different guanylate cyclases. *Nature* 341: 68-72.

De Bold AJ (1985): Atrial Natriuretic Factor: a hormone produced by the heart. *Science* 230: 676-770.

De Lean A, Vinay P and Cantin M (1985): Distribution of Atrial Natriuretic Factor receptors in dog kidney fractions. *FEBS Lett* 193:

239-242.

Fuller F, Porter G, Arfsten AE, Miller J, Schilling JW, Scarborough RM, Lewicki JA and Schenk DB (1988): Atrial natriuretic peptide clearance receptor. *J Biol Chem* 269: 9395-9401.

Genest J and Cantin M (1987): Atrial Natriuretic Factor. *Circulation* 75 (Suppl.1): 1118-1124.

Hamet P, Tremblay J, Pang SC, Skuherska R, Schiffrin EL, Garcia R, Catnin M, Genest J, Palmour R and Ervin ER (1986): Cyclic GMP as mediator and biological marker of atrial natriuretic factor. *J Hypertens* 4 (Suppl 2): S49-56.

Healy DP and Fanestil DD (1986): Localization of atrial natriuretic peptide binding sites within the rat kidney. *Am J Physiol* 250: F573-F578.

Kenny AJ and Stephenson SL (1988): Role of endopeptidase-24.11 in the inactivation of atrial natriuretic peptide. *FEBS Lett* 232: 1-8.

Krenning EP, Bakker WH, Breeman WA, Koper JW, Kooij PP, Ausema L, Lameris JS, Reubi JC and Lamberts SW (1989): Localization of endocrine-related tumours with radioiodinated analogue of somatostatin. *Lancet* 1: 242-244.

Krenning EP, Bakker JW, Kooij PP, Breeman WA, Oei HY, de Jong M, Reubi JC, Visser TJ, Bruns C and Kwekkeboom DJ (1992): Somatostatin receptor scintigraphy with ^{111}In-DTPA-D-Phe-1-Octreotide in man - metabolism, dosimetry and comparison with ^{123}I-Tyr-3-Octreotide. *J Nuc Med* 33: 652-658.

Kvols LK, Moertel CG, O'Connell MJ, Schuth AJ, Rubin J and Hahn RG (1986): Treatment of the malignant carcinoid syndrome: evaluation of a long-acting somatostatin analogue. *N Eng J Med* 315: 663-666.

Maack T, Suzuki M, Almeida FA, Nussenzveig D, Scarborough RM, McEnroe GA and Lewicki JA (1987): Physiological role of silent receptors of atrial natriuretic factor. *Science* 238: 675-678.

Napier MA, Vandlen RL, Albers-Schönberg G, Nutt RF, Brady S, Lyle T, Winquist R, Faison EP, Heinel LA and Blaine EH (1984): Specific receptors for atrial natriuretic factor in renal and vascular tissue. *Proc Nat Acad Sci* 81: 5946-5950.

Olins GM, Patton DR, Tjoeng FS and Blehm DJ (1986): Specific Receptors for Atriopeptin III in rabbit lung. *Biochem Biophys Res Commun* 140: 302-307.

Rechlin S (1983): Somatostatin. *N Eng J Med* 309: 1495-1501.

Reubi J-C, Häcke WH and Amberts SWJ (1987a): Hormone producing gastrointestinal tumours contain a high density of somatostatin receptors. *J Clin Endocrinol Metab* 65: 1127-1134.

Reubi J-C, Maurer R, von Werder K, Torhorst J, Klijn GM and Lamberts

SWJ (1987b): Somatostatin receptors in human endocrine tumours. *Cancer Res.* 47: 551-558.

Reubi J-C, Lang W, Maurer R, Koper JW and Lamberts SWJ (1987c): Distribution and biochemical characteristics of somatostatin receptors in human endocrine tumours. *Cancer Res* 47: 5758-5764.

Schwartz J-C, Gros C, Lecomte JM and Bralet J (1990): Enkephalinase (EC 3.4.24.11) Inhibitors: Protection of endogenous ANF against inactivation and potential therapeutic applications. *Life Sci* 47: 1279-1297.

Tremblay J and Hamet P (1991): Personal Communication.

Veber DF, Saperstein R, Nutt RF, Freidinger RM, Brady SF, Curey P, Perlow DS, Paleveda WJ, Colton CD, Zacchei AG, Tocco DJ, Hoff DR, Vandlen RL, Gerich JE, Hall L, Mandarino L, Cordes EH, Anderson PS and Hirschmann R (1984): A super active cyclic hexapeptide analogue of somatostatin. *Life Sci* 34: 1371-1378.

Wildey GM, Misano KS and Graham KM (1992): Atrial Natriuretic Factor, Biosynthesis and Mechanism of Action. In : *The Heart and Cardiovascular System*, Fozzard HA, Haber E, Jennings RB, Katz AM and Morgan HE, eds. New York: Raven Press.

Willenbrock R, Lambert R, Tremblay J, Bavaria G, Langlois Y, Léveille J, Flanagan R and Hamet P (1992): *In vivo* measurement of Atrial Natriuretic Peptide receptors using nuclear imaging, *Am J Hypertens* in press.

USE OF RADIONUCLIDES IN EXPERIMENTAL VASCULAR THROMBOSIS

Linda W. Schaffer, John T. Davidson, and Peter K.S. Siegl
Departments of Pharmacology and Radiopharmacology
Merck Research Laboratories

INTRODUCTION

Vaso-occlusive thrombosis resulting in ischemic cardiovascular, cerebrovascular, and peripheral vascular disease is a leading cause of mortality in the world (Collen and Gold, 1990). Arterial high shear conditions promote the formation of thrombus composed primarily of platelets, while low shear venous conditions promote the formation of thrombus composed primarily of fibrin and red blood cells (Cadroy and Harker, 1990). The critical role of platelets and fibrin in the pathogenesis of vascular thrombosis make these two components of the hemostatic system ideal targets for radiolabeling to facilitate study of the effects of pharmacological intervention on experimental thrombosis. In this chapter, we describe the systemic application of nuclear medicine technology to assess the antithrombotic efficacy of putative therapeutic agents in a baboon model of platelet-dependent vascular graft thrombosis.

Nuclear Imaging in Drug Discovery, Development, and Approval
H.D. Burns, R. Gibson, R. Dannals, Peter Siegl, Editors

ANIMAL MODELS

Several animal species and numerous techniques have been used to investigate the pathophysiology of vascular thrombosis and to evaluate pharmacological means of preventing its occurrence (Didisheim, 1972; Bush and Shebuski, 1990; Harker et al., 1991). The baboon model of vascular graft thrombosis described in this chapter is one of several developed by Harker and Hanson and has been used extensively to evaluate antithrombotic agents (Harker and Hanson, 1979; Hanson et al., 1985a; Hanson et al., 1985b; Cadroy et al., 1989a; Gruber et al., 1990; Schaffer et al., 1991). The baboon was selected based upon primate vascular anatomy, size, and acceptance of a chronic arteriovenous shunt. Furthermore, hemostatic mechanisms in the baboon closely resemble those of humans as assessed by concentrations of coagulation factors and plasminogen, bleeding times, platelet count, platelet volume, platelet morphology, platelet aggregation and release reaction *in vitro,* plasma concentration of platelet factor 4, B-thromboglobulin, fibrinopeptide A, factor V, protein C and S, von Willebrand factor, factor VIII:C, and tissue plasminogen activator determined by human radioimmune assays (Hanson et al., 1985a).

In this baboon model, thrombogenesis is initiated by inserting an uncrimped, knitted Dacron vascular graft (5.0 cm in length, 4.0 mm i.d.) into the femoral arteriovenous shunt as shown in Fig. 1. The resultant occlusive thrombus is composed primarily of platelets, with fibrin strands and entrapped red blood cells. The composition of this experimental thrombus, which is formed under high shear conditions, is similar to acute arterial thrombi in man. The Dacron graft segment elicits reproducible platelet deposition and occlusive thrombosis over a sixty minute period. Additionally, this vascular graft model may have clinical relevance as vascular implants are widely used in man.

We have modified the method originally described by Harker and Hanson (1979) to circumvent problems associated with maintaining chronic vascular catheters which include: catheter patency, septicemia, and restrictive suits for the baboon to prevent manipulation of the arteriovenous shunt. With our technique, the baboon is sedated, intubated, and maintained on 2% isoflurane for the duration of the study. This method permits acute aseptic placement of the femoral arteriovenous shunt on the day of study, but requires that the animal remain anesthetized. It is possible that the anesthetic agent itself, or its effect on cardiovascular parameters may potentiate the antithrombotic efficacy of some agents; however, we have successfully used this technique and have not identified contrasting effects in anesthetized versus conscious baboons.

RADIOPHARMACEUTICALS

A number of radiotracers have been evaluated as agents for imaging vascular thrombi and the properties of these compounds have been recently reviewed (Koblik et al., 1989; Knight, 1990). The tracers used in our laboratory are detailed in the following sections.

Indium-111 Labeled Platelets

Under normal conditions, platelets circulate in a discoid shape at a concentration of 250,000-400,000/μl and have a life span of 9-12 days in

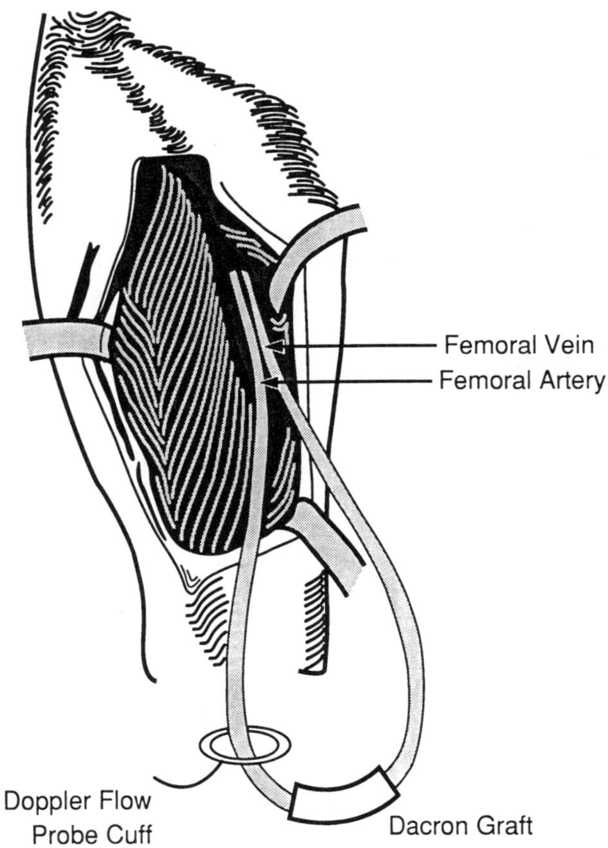

FIGURE 1. Schematic representation of the acutely placed femoral arteriovenous shunt in an anesthetized baboon.

humans and 6-7 days in baboons (Lotter, 1985). Approximately one third of the total platelet population remains pooled in the spleen in a state of dynamic equilibrium with the circulating platelets. Following severe vascular injury, exposure of subendothelial collagen stimulates platelet adhesion mediated by the platelet membrane glycoprotein receptor GPIb, plasma von Willebrand factor and fibronectin (Fressinaid et al., 1987). After adhesion, platelets form pseudopods and spread over the damaged surface. The interaction between collagen and platelets results in the release of adenosine diphosphate (ADP) and arachidonic acid which leads to the formation of thromboxane A2, a potent vasoconstrictor and platelet activator. Thrombin, which is produced locally at the site of tissue injury, can directly induce platelet aggregation and cause release of ADP, thromboxane A2, and serotonin. The final obligatory step in platelet aggregation, independent of the agonist, is binding of the platelet-membrane complex GPIIb/IIIa with fibrinogen (Plow et al., 1987). The pivotal role of platelets in arterial thrombosis has been well established in several animal models using selective antagonists of the GPIIb/IIIa fibrinogen receptor (Cadroy et al., 1989a; Mickelson et al., 1989; Coller et al., 1989; Haskel et al., 1989).

Since the introduction of Indium-111 oxine-labeled blood products in 1976, Indium-111 has been used extensively as a tracer for platelets in animal models of vascular disease (Callow et al., 1980; Hanson et al., 1985a; Lam et al., 1986; Heraset et al., 1988; Mickelson et al., 1989). Indium-111 has a half-life of 2.8 days and has efficient gamma photons (173 keV 84% and 247 kev 94%) that permit excellent images with doses of less than 500 μCi using scintillation equipment. Furthermore, Indium-111 is incorporated within the platelet without altering the morphology, biochemistry, or physiologic function of the platelet. Due to the high affinity of plasma transferrin for Indium-111, platelets must be separated from whole blood through differential centrifugation and incubated *in vitro* with the cells suspended in a nonplasma medium (Lotter, 1985). Numerous methods for platelet labeling have been described in the literature. We have used a modified method of the technique described by Bauman and Landry (1986) because this procedure routinely yields viable platelet populations with labeling efficiencies of greater than 95% and requires approximately 90 minutes to complete. A diagrammatic representation of the labeling procedure is shown in Fig. 2. The optimal amount of radioactivity used to label the cells depends upon the method of detection subsequently used. The labeled platelets are injected 1 hr before initiation of the study to permit equilibration of the labeled and unlabeled platelet population between the splenic pool and the systemic circulation.

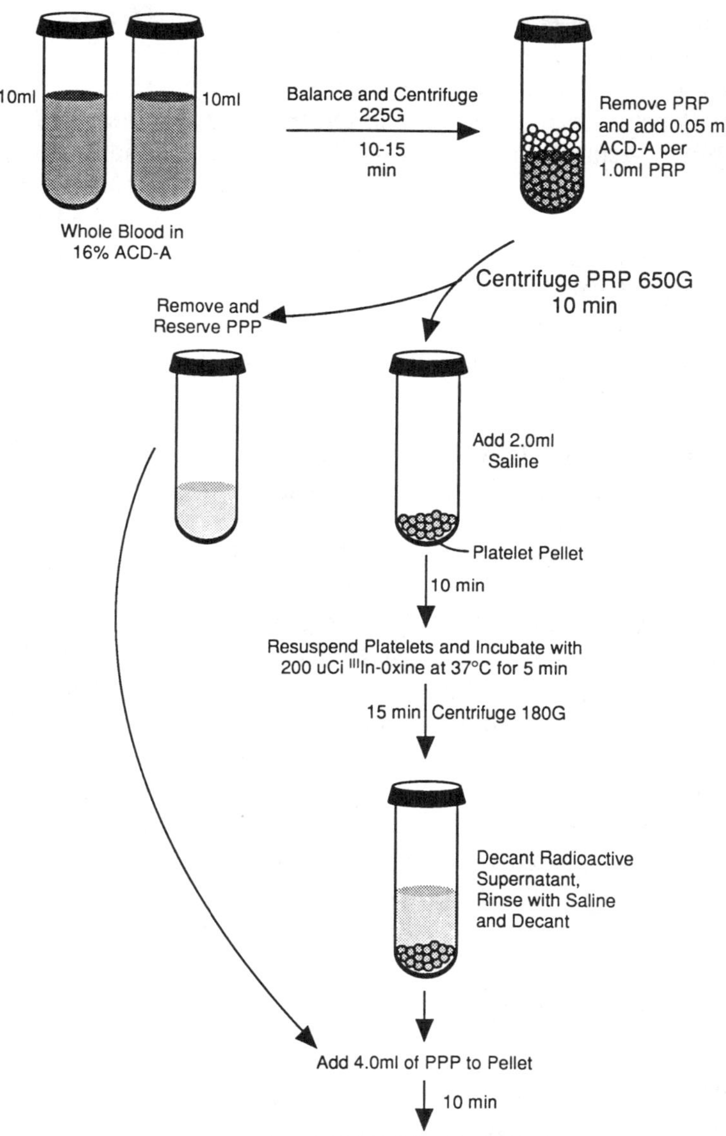

FIGURE 2. Schematic representation of the procedure used to label autologous baboon platelets with Indium-111 oxine. PPP = platelet poor plasma; PRP = platelet rich plasma; ACD-A = acid-citrate dextrose solution A.

Iodine-125 Labeled Fibrinogen

Fibrinogen is a high molecular weight protein (340,000) occurring in the plasma at a concentration of 1-7 mg/ml and in platelet alpha granules (George, 1990). Thrombin, a multifunctional serine protease, cleaves four small peptides from the amino termini of the α and β chains of fibrinogen to yield fibrin. The fibrin molecules are covalently linked by factor XIIIa to form a stable clot (Owen et al., 1983). In animal studies, preferential accretion of $[^{131}I]$fibrinogen was demonstrated in 1960 (Hobb and Davies, 1960). The disadvantages of Iodine-131 as a label for fibrinogen led to its replacement by Iodine-125 (Atkins and Hawkins, 1965). The labeled fibrinogen is incorporated into actively propagating intravenous thrombi and exhibits the same biological behavior as native fibrinogen. Iodine-125 has a half-life of 60 days which permits monitoring of thrombi over extended periods (Beckerman, 1984). Fibrinogen is commercially available, or homologous fibrinogen can be isolated from citrated plasma using the β-alanine precipitation technique (Jakobsen and Kierulf, 1973).

There are numerous methods available for labeling fibrinogen. We have used the Iodogen technique using 1-2 atoms of iodine per molecule of fibrinogen corresponding to approximately 1 mg of fibrinogen labeled with 100 mCi of Iodine-125 (Knight et al., 1981). The free Iodine-125 is removed and the tagged fibrinogen is injected one hr prior to initiation of the experiment. The amount of radioactivity injected depends on the method of subsequent detection. The clottability of the labeled fibrinogen was greater than 85% in all studies using this method. The low energy gamma radiation (35 keV) and high tissue absorbance of Iodine-125 limits detectable surface counts, making gamma camera imaging impractical with this isotope; however, a NaI probe can be used to dynamically detect Iodine-125 labeled fibrinogen in the exteriorized, shielded arteriovenous shunt of the baboon. However, $[^{123}I]$labeled fibrinogen can be used to image thrombi in experimental models and in man (DeNardo et al., 1974). The use of Iodine-123 for labeling fibrinogen in our studies is limited because Iodine-123 is not readily accessible from commercial sources, has a short half-life (13.1 hrs) necessitating frequent labeling, and is expensive. Recently, fragments of monoclonal antibodies to fibrin and fragment E_1, a plasmin degradation product of human cross-linked fibrin which binds specifically to polymers of fibrin, have been labeled with Technetium-99m (Cerqueira et al., 1992; Knight et al., 1992). Experimental studies with these Technetium-99m labeled proteins have produced diagnostic quality images of venous thrombosis and may be useful in the future.

[^{111}In]PLATELET AND [^{125}I]LABELED FIBRINOGEN DETECTION

There are two options for quantitative measurement of thrombus formation in the graft segment, gamma scintillation counter and NaI probe. The principles of instrumentation and data processing are reviewed by Links in Chapter 2. The gamma camera was invented by Hal Anger in 1958 and is the most commonly used imaging device in Nuclear Medicine today. The image quality improves when the 173 and the 247 keV gamma photons of Indium-111 are simultaneously detected using a dual spectrometer, but if the camera has only one window, it is best to use the 247 keV peak. The collimator must be a medium energy high-sensitivity parallel hole collimator because a low energy collimator is penetrated by the high energy Indium-photons resulting in a loss of resolution even if only the 173 keV photopeak is used (Lotter, 1985). A dedicated computer system is used to process the data and permits rapid time-sequence image acquisition. Time-activity curves can be generated from selected regions of interest in the dynamic study. As previously stated, it is not possible to use the gamma camera to image Iodine-125 labeled fibrinogen; however, fibrinogen is not a major component of the experimental thrombus formed in this model.

We have successfully used a NaI probe to monitor dynamic [^{111}In]platelet and [^{125}I]fibrinogen deposition onto the Dacron vascular graft in the baboon model. A NaI probe has numerous advantages over the gamma camera including expense (approximately 30 times less expensive), small size facilitating excellent shielding of the probe to limit background and scatter radiation, as well as ability to detect Iodine-125 gamma photons (35 keV) simultaneously with Indium-111 gamma photons since the spilldown from the Indium-111 photopeaks can be calculated and subtracted from the 35 keV window of the [^{125}I]fibrinogen for each data acquisition.

CHARACTERIZATION OF THE MODEL AND RESPONSES TO ANTITHROMBOTIC THERAPY.

As previously stated, the thrombus elicited by insertion of the thrombogenic Dacron graft segment into the femoral arteriovenous shunt of the baboon is composed primarily of platelets and is similar in morphology to acute arterial thrombi in man. The reproducible rate and extent of platelet deposition and predictable occlusive thrombus formation within a short period are characteristics which permit the use of this model to evaluate potential antithrombotic agents. A typical experimental protocol which can be modified by altering the route or duration of administration of the test compound is shown in Fig. 3. In

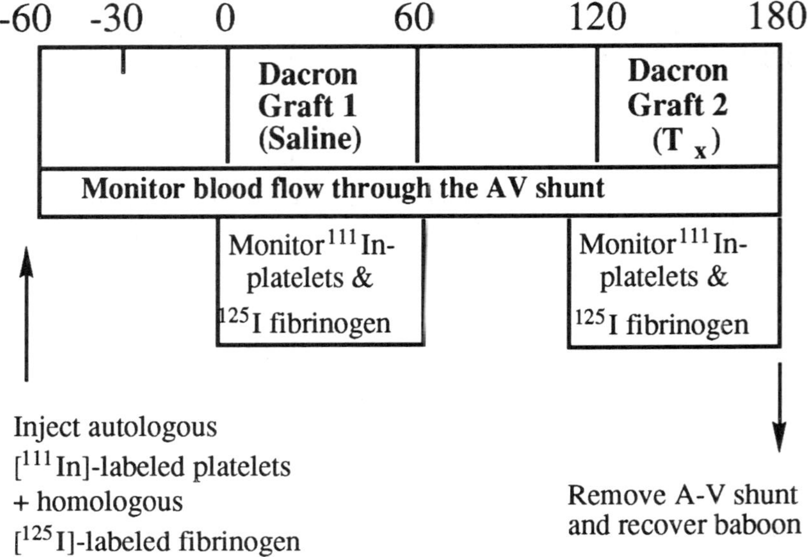

FIGURE 3. Typical experimental protocol used to evaluate antithrombotic efficacy of test compounds. Baboons are rested 4 to 6 weeks between studies.

studies completed in our laboratory, saline was infused during exposure to the first Dacron graft segment to establish the control thrombotic response in individual baboons. Following occlusion of the initial graft, the segment was removed and the arteriovenous shunt was reestablished. The appropriate therapy is administered prior to insertion of the second graft segment. In control animals, saline was infused during exposure to the first and second Dacron grafts. As shown in Fig. 4, platelet and fibrinogen deposition (panel A), and time to occlusion, as assessed by the cessation of detectable blood flow (panel B), were not different between the sequentially placed Dacron segments. Therefore, the response to the initial control graft can be used to assess the antithrombotic effect of the test compound on the second graft within individual animals.

Antithrombotic Therapy

The development of safe and efficacious therapies for the treatment and prevention of vascular thrombosis is an area of active research and

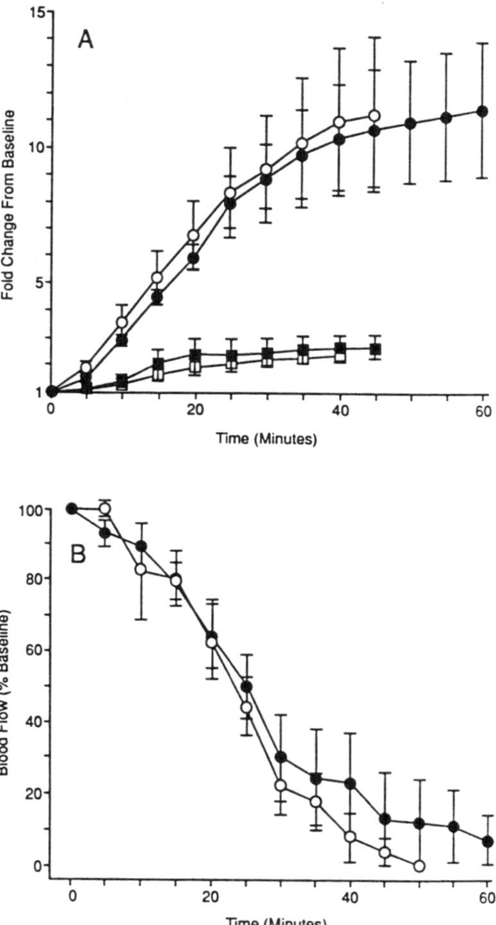

FIGURE 4. Effect of saline infusion on [^{111}In]-labeled platelet and [^{125}I]labeled fibrin(ogen) deposition onto sequentially placed graft segments shown in panel A. Fold increase in platelet deposition in graft 1 (○, n=4) and graft 2 (●, n=4) and fibrin(ogen) deposition in graft 1 (□, n=3) and graft 2 (■, n=4) represent mean ± standard error. The effect of saline infusion on blood flow during graft 1 (○, n=4) and graft 2 (●, n=4) exposure is shown in panel B and represents mean ± standard error. Reprinted with permission of the American Heart Association, Inc., from Schaffer et al. (1991).

the reader is directed to several thorough reviews of this topic (Cadroy and Harker, 1990; Fuster et al., 1990; Collen and Gold, 1990). In this section, the effects of selective agents will be discussed to provide a pharmacological profile of the baboon model.

Clinically, aspirin is the most commonly used antiplatelet agent. Aspirin exerts its antiplatelet effect by irreversibly acetylating cyclooxygenase for the life span of the platelet. Aspirin has been inconsistent in its efficacy in experimental animal models (Hanson and Harker, 1985; Haskel et al., 1989; Bush and Shebuski, 1990; Muller et al., 1992). In our laboratory, aspirin dose-dependently delayed occlusion of the arteriovenous shunt, but did not effect the rate of platelet deposition. Recently, a new class of antiplatelet agents that block the platelet membrane glycoprotein IIb/IIIa (GPIIb/IIIa) receptor and inhibit fibrinogen binding has been described. Monoclonal antibodies against the GPIIb/IIIa peptides and mimetics containing the Arg-Gly-Asp (RGD) sequence prevent experimental arterial thrombosis in numerous animal models (Gold et al., 1988; Hanson et al., 1988; ; Mickelson et al., 1989; Shebuski et al., 1990; Bates et al., 1991). The experimental results of an RGD-containing fibrinogen receptor antagonist in the baboon model is shown in Fig. 5. Infusion of the compound dose-dependently inhibited platelet deposition (panel A) and prevented occlusion of the arteriovenous shunt (panel B) concurrent with marked elevation in bleeding times.

Standard unfractionated heparin is the most commonly used antithrombotic agent. Heparin has multiple actions and can accelerate inhibition of thrombin and factors IXa, Xa, and XIa by antithrombin III (Heras et al., 1990). The doses of heparin which are effective in arterial thrombosis exceed those used to successfully treat venous thrombosis. Bleeding complications associated with heparin are dose dependent and require monitoring by coagulation tests concurrent with therapy. In experimental studies, heparin inhibits arterial thrombus formation in some models (Heras et al., 1988, Cadroy et al., 1989b; Eidt et al., 1989) but not in others (Kelly et al., 1991; Schaffer et al., 1991). The explanation for the differential effects of heparin may be related to the variable severity of the thrombogenic challenge or dose of heparin used in different models, as well as the inability of heparin to access fibrin-bound thrombin (Weitz et al., 1990).

Although the efficacy of heparin, an indirect inhibitor of thrombin, has been disappointing in the baboon model, potent synthetic and natural inhibitors of thrombin have dramatic antithrombotic effects. Hirudin, a polypeptide obtained from the medicinal leech *Hirudo medicinalis*, dose-dependently inhibits Dacron vascular graft thrombosis (Kelly et al., 1991). A synthetic peptide, chloromethyl ketone inhibitor of thrombin (PPACK) also interrupts platelet-dependent thrombosis in this model

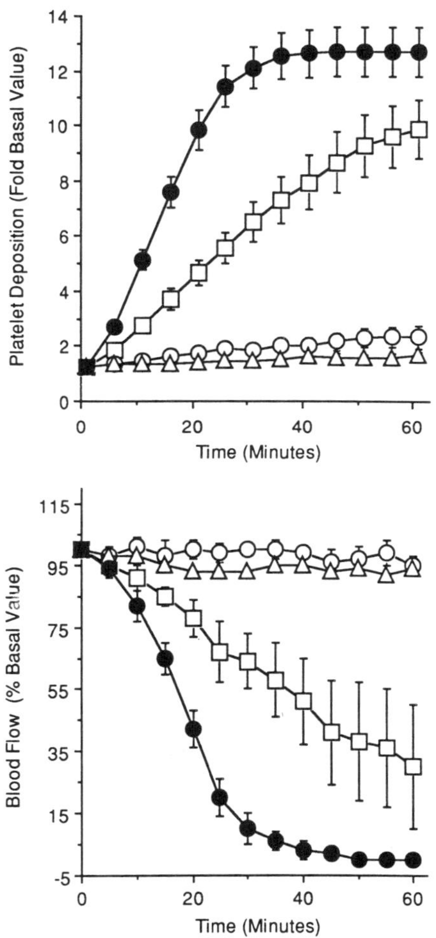

FIGURE 5. Effect of RGD-based peptide infusion on [^{111}In]-labeled platelet deposition onto a thrombogenic Dacron graft segment shown in panel A. Values for platelet deposition (Panel A) and blood flow (Panel B) during systemic intravenous infusion of saline (●, n=13), or 0.3 µg/kg/min (□, n=4), 1.0 µg/kg/min (○, n=4), and 3.0 µg/kg/min (Δ, n=4) of the fibrinogen receptor antagonist represent mean ± standard error.

(Hanson et al., 1988). The promising antithombotic effects of hirudin have been reported in numerous animal models of platelet-dependent arterial thrombosis (Heras et al., 1989; Kelly et al., 1991; Jang et al., 1990). In most studies, efficacy in platelet-dependent thrombosis models

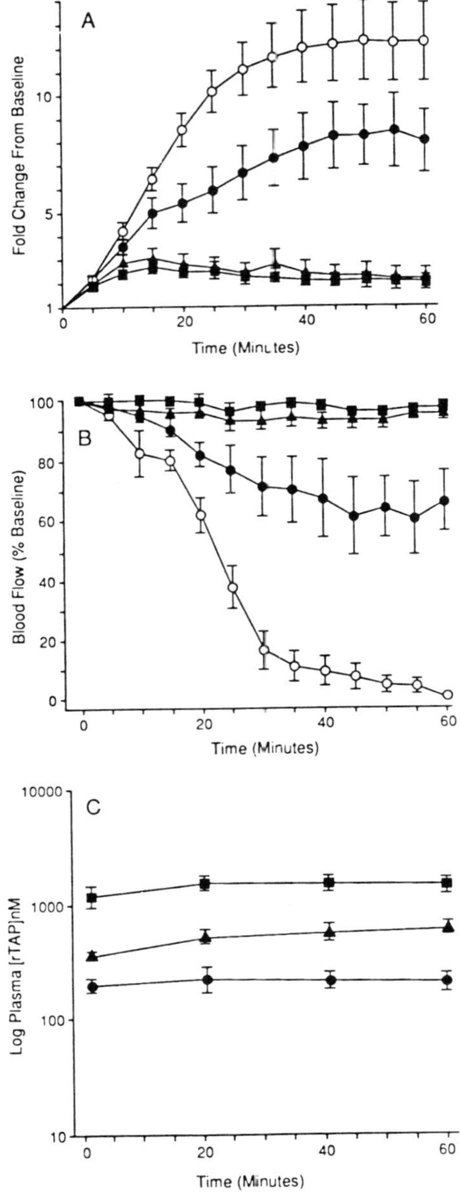

FIGURE 6. Effect of 6.25 µg/kg/min (●, n=4), 12.5 µg/kg/min (▲, n=4), or 25.0 µg/kg/min (■, n=4) rTAP or saline (○, n=12) on [^{111}In]labeled platelet deposition (panel A), blood flow (panel B), and plasma rTAP concentration (panel C) during graft replacement. All values represent mean ± standard error. Reprinted with permission from the American Heart Association, Inc., from Schaffer et al. (1991).

necessitates doses of the thrombin inhibitors that are associated with increases in template bleeding time.

Recently, we and others have shown that inhibition of the serine protease, coagulation factor Xa, with potent and specific inhibitors can prevent platelet or fibrinogen-mediated vascular thrombosis (Neeper et al., 1990; Schaffer et al., 1991; Vlasuk et al., 1991; Sitko et al., 1992). Tick anticoagulant peptide (TAP), a 60-amino acid polypeptide originally isolated from the soft tick *Ornithodoros moubata* (Waxman et al., 1990), and antistasin, a 119 amino acid protein originally isolated from the leech *Haementeria officinalis* (Dunwiddie et al., 1989), effectively prevent platelet deposition in this baboon model of vascular graft thrombosis at doses which do not effect template bleeding time (Schaffer et al., 1991; Schaffer et al., 1992). The effect of TAP infusion is shown in Fig. 6. The antithrombotic effect of the factor Xa inhibitors probably results from inhibition of thrombin generation. Activated protein C, which inhibits thrombin formation by means of enzymatic cleavage and destruction of coagulation factors Va and VIIIa, also prevents occlusive thrombus formation at doses which do not elevate template bleeding time in this model (Gruber et al., 1990).

CONCLUSION

This baboon model of platelet-dependent vascular graft thrombosis represents a severe thrombotic challenge which is not responsive to standard heparin or aspirin. The positive responses in this model to RGD-based peptides and related compounds, direct inhibitors of thrombin, and inhibitors of thrombin generation such as factor Xa inhibitors and activated protein C, suggests that these novel, highly selective and potent agents may represent superior alternatives to clinically available therapies for the treatment and prevention of arterial thrombotic events in man. The methodology using radiolabeled platelets and fibrinogen provides a direct measure of the dynamics of platelet and fibrinogen deposition which is not available when using only a flow-probe.

REFERENCES

Atkins P and Hawkins LA (1965): Detection of venous thrombosis in the legs. *Lancet* 2: 1216-1219.

Bates ER, McGillem MJ, Mickelson JK, Pitt B and Mancini GJ (1991): A monoclonal antibody against the platelet glycoprotein IIb/IIIa receptor complex prevents platelet aggregation and thrombosis in a canine model of coronary angioplasty. *Circulation* 84: 2463-2469.

Bauman JM, and Landry AJ (1986): Practical platelet labeling with Indium^{111} oxine. *J Nucl Med Tech* 14: 15-17.

Beckerman C (1984): The fibrinogen uptake test and fibrinogen scintography for detection of deep vein thrombosis. In: *Imaging of the Peripeheral Vascular System*, Pinsky SM and Moss GS, eds. Orlando: Grune and Stratton, Inc.

Bousser MG, Eschwege E, Haguenau M, Lefaucconnier JM, Thibult N, Touboul D and Touboul PJ (1983): "AICLA" controlled trial of aspirin and dipyridamole in the secondary prevention of atherothrombotic cerebral ischemia. *Stroke* 14: 5-14.

Bush LR and Shebuski RJ (1990): *In vivo* models of arterial thrombosis and thrombolysis. *FASEB J* 4: 3087-3098.

Cadroy YC, Houston RA and Hanson SR (1989a): RGDV peptide selectively inhibits platelet-dependent thrombus formation *in vivo. J Clin Invest* 84: 939-944.

Cadroy Y, Harker LA and Hanson SR (1989b): Inhibition of platelet-dependent thrombosis by low molecular weight heparin (CY222): Comparison with standard heparin. *J Lab Clin Med* 114: 349-357.

Cadroy Y and Harker LA (1990): Platelets, thrombosis and antithrombotic therapies. In: *Cardiovascular Pharmacology*, Antonaccio M, ed. New York: Raven Press, Ltd.

Callow AD, Ledig CB, O'Donnell TF, Kelly JJ, Rosenthal D, Korwin S, Hatte C, Kahn PC, Vecchione JJ and Valeri CR (1980): A primate model for the study of the interaction of ^{111}In-labeled baboon platelets with Dacron arterial protheses. *Ann Surg* 191: 362-366.

Cerqueira MD, Stratton JR, Vracko R, Schaible TF and Ritchie JL (1992): Noninvasive arterial thrombus imaging with 99mTc monoclonal antifibrin antibody. *Circulation* 85: 298-304.

Collen D and Gold HK (1990): New developments in thrombolytic therapy. *Adv Exp Med Biol* 281: 333-354.

Coller BS, Folts JD, Smith SR, Scudder LE and Jordan R (1989): Abolition of *in vivo* platelet thrombus formation in primates with monoclonal antibodies to the platelet GPIIb/IIIa receptor. *Circulation* 80: 1766-1774.

DeNardo SJ, DeNardo GL, O'Brien T, Peek NF, Zielinski FW and Jungerman JA (1974): I-123 fibrinogen imaging of thrombi in dogs. *J Nucl Med* 15: 487.

Didisheim P (1972): Animal models useful in the study of thrombosis and antithrombotic agents. *Prog Hemostasis Thromb* 1: 165-197.

Dunwiddie C, Thornberry NA, Bull HG, Sardana M, Friedman PA,

Jacobs JW and Simpson E (1989): Antistasin, a leech-dervied inhibitor of factor Xa. *J Biol Chem* 264: 16694-16699.

Eidt JF, Allison P, Noble S, Ashton J, Golino P, McNatt J, Buja LM and Willerson JT (1989): Thrombin is an important mediator of platelet aggregation in stenosed canine coronary arteries with endothelial injury. *J Clin Invest* 84: 18-247.

Fressinaud E, Baruch D, Rothschild C, Baumgartner HR and Meyer D (1987): Platelet von Willebrand factor: Evidence for its involvement in platelet adhesion to collagen. *Blood* 70: 1214-1217.

Fuster V, Stein B, Ambrose JA, Badimon L, Badimon JJ and Chesebro JH (1990): Atherosclerotic plaque rupture and thrombosis. *Circulation* 82 (suppl II): II-47-II-59.

George, JN (1990): Hemostasis and fibrinolysis. In: *Internal Medicine*, Stein JH, ed. Little, Brown and Company.

Gold HK, Coller BS, Yasuda T, Saito T, Fallon JT, Guerrero JL, Leinbach RC, Ziskind AA and Collen D (1988): Rapid and sustained coronary artery recanalization with combined bolus injection of recombinant tissue-type plasminogen activator and monoclonal antiplatelet GPIIb/IIIa antibody in a canine preparation. *Circulation* 77: 670-677.

Gruber A, Hanson SR, Kelly AB, Yar BS, Bang N, Griffin JH and Harker LA (1990): Inhibition of thrombus formation by activated recombinant protein C in a primate model of arterial thrombosis. *Circulation* 82: 578-585.

Hanson SR and Harker LA (1985): Studies of suloctodil in experimental thrombosis in baboons. *Thromb Haemostasis* 53: 423-427.

Hanson SR, Kotze HR, Savage B and Harker LA (1985a): Platelet interactions with Dacron vascular grafts. A model of acute thrombosis in baboons. *Arteriosclerosis* 5: 595-603.

Hanson SR, Harker LA and Bjornsson TD (1985b): Effects of platelet-modifying drugs on arterial thromboembolism in baboons. *J Clin Invest* 75: 1591-1599.

Hanson SR and Harker LA (1988): Interruption of acute platelet-dependent thrombosis by the synthetic anti-thrombin D-phenylalanyl-L-prolyl-L-arginyl chloro-methyl ketone. *Proc Natl Acad Sci* 85: 3184-3188.

Hanson SR, Pareti FI, Ruggeri ZM, Marzec UM, Kunicki TJ, Montgomery RR, Zimmerman TS and Harker LA (1988): Effects of monoclonal antibodies against the platelet glycoprotein IIb/IIIa complex on thrombosis and hemostasis in baboons. *J Clin Invest* 81: 149-158.

Harker LA, Kelly AB and Hanson SR (1991): Experimental arterial thrombosis in nonhuman primates. *Circulation* 83 (suppl IV):

IV42-IV55.

Harker LA and Hanson SR (1979): Experimental arterial thromboembolism in baboons: Mechanism, quantitation and pharmacological prevention. *J Clin Invest* 64: 559-569.

Haskel EJ, Prager NA, Sobel BE and Abendschein DR (1991): Relative efficacy of antithrombin compared with antiplatelet agents in accelerating coronary thrombolysis and preventing early reocclusion. *Circulation* 83: 1048-1056.

Haskel EJ, Adams SP, Feigen LP, Saffitz JE, Gorczynski RJ, Sobel BE and Abendschein DR (1989): Prevention of reoccluding platelet-rich thrombi in canine femoral arteries with a novel peptide antagonist of platelet glycoprotein IIb/IIIa receptors. *Circulation* 80: 1775-1782.

Heras M, Chesebro JH, Penny WJ, Bailey KR, Badimon L and Fuster V (1989): Effects of thrombin inhibition on the development of acute platelet-thrombus deposition during angioplasty in pigs. Heparin versus recombinant hirudin, a specific thrombin inhibitor. *Circulation* 79: 657-665.

Heras HM, Chesebro JH, Penny WJ, Bailey KR, Lam JYT, Holmes DR, Reeder GS, Badimon L and Fuster V (1988): The importance of adequate heparin dosage in arterial angioplasty in a porcine model. *Circulation* 78: 654-660.

Heras M, Chesebro JH, Webster MW, Mruk JS, Grill DE, Penny WJ, Bowie EJ, Badimon L and Fuster V (1990): Hirudin, heparin, and placebo during deep arterial injury in the pig. *Circulation* 82: 1476-1484.

Hobb JT and Davies JWL (1960): Detection of venous thrombosis with ^{131}I-labeled fibrinogen in the rabbit. *Lancet* 2: 134.

Jakobsen E and Kierulf P (1973): A modified β-alanine precipitation procedure to prepare fibrinogen free of antithrombin III and plasminogen. *Thromb Res* 3: 145-149.

Jang IK, Gold HK, Ziskind AA, Leinbach RC, Fallon JT and Collen D (1990): Prevention of platelet-rich arterial thrombosis by selective thrombin inhibition. *Circulation* 81: 219-225.

Kelly AB, Marzec UM, Krupski W, Bass A, Cadroy Y, Hanson SR and Harker LA (1991): Hirudin interruption of heparin-resistant arterial thrombus formation in baboons. *Blood* 77: 1006-1017.

Knight LC, Budzynski AZ and Olexa SA (1981): Radiolabeling of fibrinogen using the iodogen technique. *Thromb Haemostasis* 46: 593-596.

Knight L (1990): Radiopharmaceuticals for thrombus detection. *Semin Nucl Med* 20: 52-67.

Koblik PD, DeNardo GL and Berger HJ (1989): Current status of immunoscintigraphy in the detection of thrombosis and thromboembolism. *Semin Nucl Med* 19: 221-237.

Lam JYT, Chesebro JH, Steele PM, Dewanjee MK, Badimon L and Fuster V (1986): Deep arterial injury during experimental angioplasty: Relationship to positive ^{111}Indium-labeled platelet scintigram, quantitative platelet deposition and mural thrombosis. *J Am Coll Cardiol* 8: 1380-1386.

Lotter MG (1985): Measurement of platelet survival and analysis of survival curves. In: *Platelet Kinetics and Imaging*, Vol 1, Heyns A, Badenhorst PN and Lotter MG, eds. Boca Raton, FA: CRC Press, Inc.

Mickelson JK, Simpson PJ and Lucchesi BR (1989): Antiplatelet monoclonal $F(ab')_2$ antibody directed against the platelet GPIIb/IIIa receptor complex prevents coronary artery thrombosis in the canine heart. *J Mol Cell Cardiol* 21: 393-405.

Muller DW, Ellis SG and Topol EG (1992): Experimental models of coronary artery restenosis. *J Am Coll Cardiol* 19: 418-432.

Neeper MP, Waxman L, Smith DE, Schulman CA, Sardina M, Ellis RW, Schaffer LW, Siegl PKS and Vlasuk GP (1990): Characterization of recombinant tick anticoagulant peptide. *J Biol Chem* 265: 17746-17752.

Owen J, Kvam D, Nossel HL, Kaplan KL and Kernoff PBA (1983): Thrombin and plasmin activity and platelet activation in the development of venous thrombosis. *Blood* 61: 476-482.

Plow EF, Marquerie G and Ginsberg M (1987): Fibrinogen, fibrinogen receptors, and the peptides that inhibit these interactions. *Biochem Pharmacol* 36: 4035-4040.

Schaffer LW, Davidson JT, Vlasuk GP, Dunwiddie CT and Siegl PKS (1992): Selective factor Xa inhibition by recombinant antistasin prevents vascular graft thrombosis in baboons. *Arteriosclerosis Thromb* 12: 879-885.

Schaffer LW, Davidson JT, Vlasuk GP and Siegl PKS (1991): Antithrombotic efficacy of recombinant tick anticoagulant peptide, a potent inhibitor of coagulation factor Xa in a primate model of arterial thrombosis. *Circulation* 84: 1741-1748.

Shebuski RJ, Ramjit DR, Sitko GR, Lumma PK and Garsky VM (1990): Prevention of canine coronary artery thrombosis with echistatin, a potent inhibitor of platelet aggregation from the venom of the viper, *Echis carinatus*. *Thromb Haemostasis* 64: 576-581.

Sitko GR, Ramjit DR, Stabilito II, Lehman DD, Lynch JJ and Vlasuk GP (1992): Conjunctive enhancement of enzymatic thrombolysis and prevention of thrombotic reocclusion with the selective factor Xa inhibitor, tick anticoagulant peptide. *Circulation* 85: 805-815.

Vlasuk GP, Ramjit D, Fujita T, Dunwiddie CT, Nutt EM, Smith DE and Shebuski JR (1991): Comparison of the *in vivo* anticoagulant properties of standard heparin and the highly selective FXa inhibitors antistasin and tick anticoagulant peptide (TAP) in a rabbit model of venous thrombosis. *Thromb Haemostasis* 65: 257-262.

Waxman L, Smith DE, Arcuri KE and Vlasuk GP (1990): Tick anticoagulant peptide (TAP) is a novel inhibitor of blood coagulation factor Xa. *Science* 248: 593-596.

Weitz JI, Hudoba M, Massel D, Maraganore J and Hirsh J (1990): Clot bound thrombin is protected from inhibition by heparin-antithrombin III but is susceptible to inactivation by antithrombin III-independent inhibitors. *J Clin Invest* 86: 385-391.

APPLICATION OF NUCLEAR IMAGING TO DRUG DELIVERY EVALUATION AND DEVELOPMENT: A REVIEW OF RADIOLABELED, INJECTABLE, COLLOIDAL SYSTEMS OF DELIVERY

Nancy J. Brenner, Christine Fioravanti and H. Donald Burns
Merck Research Laboratories

INTRODUCTION

The formulation or vehicle for a drug varies with the method of required administration (for example, oral, intravenous, intramuscular, or intraperitoneal) and, in some cases, according to which organs or tissues are targeted. Emulsions have been widely used as drug formulations for oral administration for many years, and recently, interest in emulsions as formulations for intravenous use has increased. This interest stems primarily from the fact that the distribution of a lipophilic drug can be altered if it is administered in an emulsion formulation. In addition, emulsions can be used to develop formulations of lipid soluble drugs which are not very soluble in aqueous solutions. Of particular interest are micro and nanocolloidal particles, liposomes, fat emulsions, and microfine emulsions.

The fate of these particulate emulsions after intravenous injection varies, depending on particle size, composition, surface characteristics (charge, hydrophobicity, etc.), and other factors. Large particles (larger than 5-10 μm) are trapped in the capillary bed of the lung while smaller

Nuclear Imaging in Drug Discovery, Development, and Approval
H.D. Burns, R. Gibson, R. Dannals, Peter Siegl, Editors

particles are rapidly cleared from the circulation by the macrophages of the reticuloendothelial system (RES) via phagocytosis, resulting in rapid accumulation in the liver and spleen. For detailed discussions of mechanisms of RES uptake and the effect on drug delivery and bioavailability, refer to Wilkins (1967), Napper and Netschey (1971), Van Oss et al. (1975), Douglas and Davis (1985; 1986), Davis et al. (1986) and Illum et al. (1986). This rapid uptake in the lungs and the RES has been used to target drugs to these sites. However, this process may result in an undesirable concentration of the drug in these tissues increasing the chances of toxicity to these organs and reducing the concentration of the drug at other targeted tissues to sub-therapeutic levels. Thus, in evaluating the efficacy of a new, particulate formulation of a drug, it is important to obtain information on the biodistribution of the drug and/or the colloidal or emulsion carrier. Since species differences can be important, it is useful to obtain information in an appropriate animal model or, when possible, in humans.

Plasma clearance rates for a drug in animals and humans can be obtained readily after intravenous administration. The tissue distribution of a drug and/or its colloidal or emulsion vehicle can be determined in small experimental animals by measuring tissue levels post-mortem. However, this is impractical in some animals (e.g., primates) and impossible in humans. Valuable information on the tissue distribution of drugs and/or drug vehicles can be obtained in animals and humans using non-invasive, nuclear imaging techniques and a suitable radiotracer. To accomplish this, either the drug or the drug vehicle can be labeled with an appropriate radionuclide and the distribution of the radiolabeled material can be followed using non-invasive planar, Single Photon Emission Computed Tomography, or Positron Emission Tomography (PET) imaging. Imaging of the distribution of the radiotracer after administration to the experimental subject (animal or human) provides dynamic, real-time information regarding the performance of the formulation when administered *in vivo*. This information can be used to evaluate the efficacy of the formulation with respect to drug delivery. Thus, suitably designed nuclear imaging studies can provide dynamic data on drug distribution in humans, revealing information which cannot be obtained by any other means.

Earlier reviews by Davis and others describe the general uses of nuclear imaging in drug delivery of injectable colloidal formulations (Wilson et al., 1982; Davis and Illum, 1983; Davis, 1985; Davis, 1986). Several articles have focused on Davis' work with particular drug targeting systems, such as emulsions, nanoparticles, and microspheres (Davis, 1976; Davis, 1982; Davis et al., 1985; Davis et al., 1986; Douglas and Davis, 1985; Douglas et al., 1987). These reviews provide excellent examples of the use of dynamic imaging in the design and improvement of injectable drug delivery systems.

This chapter will illustrate recent applications of nuclear imaging to the characterization of injectable dosage forms. Examples will be presented that have used imaging to evaluate the effects of surface characteristics and other modifications of drug vehicles on biodistribution. The selection of radionuclides and radiolabeling techniques (whether to label the drug, a drug analogue, or the drug vehicle), as well as validation techniques will be addressed. Methods for determination of the stability of various radiolabeled drug/vehicle complexes will also be highlighted. The injectable vehicles that are covered are micro and nanocolloidal particles, liposomes, fat emulsions, and microfine oil emulsions. In addition, recent work in our laboratories using imaging to evaluate an intravenous microfine emulsion formulation for lipophilic compounds will be presented.

RADIOTRACER DESIGN

For a drug delivery imaging study, either the drug or a suitable analogue of the drug can be radiolabeled. The radiolabel can also be incorporated as a vehicle component, for example, attached to the surface of a microsphere. In either case, the stability of the attachment of the radiotracer to the drug, drug analogue, or vehicle must be demonstrated in preliminary *in vitro* and *in vivo* experiments. *In vivo* stability can be predicted by challenging the attachment of the radionuclides to the drug, drug analogue, or vehicle in suitable *in vitro* models. Preliminary *in vivo* studies in appropriate animal models should be performed to provide direct analyses of blood level and tissue distribution concurrent with imaging data. Indirect evidence can also be obtained via controls in which biodistribution and clearance data for the radiolabeled drug/vehicle are compared with the distribution data for the free radiolabel in a solution (some examples involving labeled carriers are: Illum and Davis, 1982; Pimm et al., 1992; Gabizon et al., 1991; Rudolph et al., 1991; Davis et al., 1987).

Whether working with the radiolabel as a part of the drug, drug analogue, or vehicle, the investigator must demonstrate that the resulting radiolabeled drug/vehicle complex retains the desired physiochemical and pharmacokinetic properties, including *in vivo* stability, drug release rate, and biodistribution. Liposomes, for example, can be tagged either by entraping the radiotracer within the aqueous interior of the vesicles, or by addition of the radiolabel to an integral component during or after manufacture. Autoradiographic and imaging techniques have been very useful in the design and improvement of various liposomal drug delivery systems (Cohen et al., 1991; Liu et al., 1991; Yamazaki et al., 1992; Gabizon et al., 1991; Rudolph et al., 1991). Because of their generally rapid clearance from the blood by the RES and subsequent sequestration in the liver and spleen (with larger particles being trapped in the lungs), liposomes have been studied for targeting entrapped anti-tumor therapeutics to hepatic tumors.

Radionuclides

A variety of radionuclides are available for use in non-invasive, nuclear imaging studies (Table 1). The selection of a radionuclide for a particular study is determined by the length of the imaging study (which must cover the length of the biological event under consideration), the frequency of repetitions of dosing in the same subjects, the species serving as the *in vivo* model, the size and type of detector, the ease and

Major Emissions Radionuclide	Half-Life	Eenergy of photon, keV
Positron-emitters		
Carbon-11	20.5 min.	511
Fluorine-18	109.7 min.	511
Oxygen-15	2.07 min.	511
Nitrogen-13	9.96 min.	511
Gamma-emitters		
Selenium-75	118.5 days	136 + 265 + 280
Technetium-99m	6.02 hr.	141
Indium-111	2.81 days	171 + 245
Iodine-123	13.3 hr.	159
Iodine-131	8.05 days	365 + 637 + 284
Thalium-201	73 hr.	71 + 69 + 80 + 167

TABLE 1. Half-life of commonly used isotopes.

stability of radionuclide incorporation, and the radionuclide availability. The radiotracer should be labeled with an isotope of appropriate energy and half-life for the time required for synthesis, formulation, and imaging. Too long a half-life (for example, 120 days for Selenium-75), could result in undesirably high dosimetries for the subject, limiting administerable dosages, as well as lengthening the intervals between repetitive studies on the same individual. In the other extreme, a radiotracer incorporating the Carbon-11 isotope (with a half-life of 20 minutes) requires a short (optimally less than 30 minutes) synthesis and formulation and is suitable for PET imaging studies no longer than 90 minutes. The short half-life, however, allows more flexibility on dosing size and frequency.

Radionuclides that are useful for nuclear imaging include positron and gamma-emitting species. Some common positron-emitting species are Carbon-11, Fluorine-18, Oxygen-15, and Nitrogen-13, with half-lives of 20, 110, 2, and 10 min, respectively. An advantage of positron-emitting radionuclides (particularly Carbon-11) is that they can be used to isotopically label a drug. Disadvantages include the short half-lives (limiting synthesis, formulation, and imaging times) and the resulting requirement of an on-site cyclotron for these accelerator-produced radionuclides. For a detailed description of PET physics, see Links (Chapter 2), Wolf and Schlyer (Chapter 3) and Links (1990).

The gamma-emitting radionuclides most commonly used for imaging are Technetium-99m, Indium-111, Iodine-123, Iodine-131, and Thallium-201. Of these, Technetium-99m and Iodine-123 are the most versatile and suitable for *in vivo* imaging. Technetium-99m and Iodine-123 each exhibit simple, clean gamma spectra with ideal energies (141 and 159 keV, respectively) for good image resolution. Each provides suitable chemistries for radiotracer synthesis. Technetium-99m is generator-produced, with a half-life of 6 hrs. It is a transition metal which forms chelate complexes with coordinating ligands that can be incorporated into a drug delivery vehicle. Iodine-123 is accelerator-produced, with a 13.1 hr half-life, and is very useful for longer imaging sessions. This isotope exhibits typical halogen chemistry, forming covalent bonds with carbon, under appropriate conditions. Iodine-123 can, therefore, be covalently incorporated into many drugs or drug analogues, often resulting in relatively stable radiotracers with predictable physiochemical and pharmacological properties. Very few naturally occurring compounds contain iodine, however. The radioiodinated derivative is an analogue of the natural drug and must have appropriate properties for the particular study.

EXAMPLES OF NOVEL DRUG DELIVERY STUDIES

Radiolabeled Vehicles

There are many examples of drug vehicles that have been radiolabeled for a variety of clinical or research objectives. Depending on the goals of a particular study, certain considerations should be emphasized. One property, that of stability of the radiolabel-vehicle association, is so fundamental that it should be characterized and optimized in very early studies (preferably in *in vitro* studies, then in animal models).

As a part of the selection process for an appropriately radiolabeled drug vehicle, one must demonstrate suitable stability of the association of the radiotracer with the vehicle. Douglas and Davis (1986) performed an *in vitro* study with a radiolabeled poly(butyl 2-cyanoacrylate) nanoparticle system to determine whether the radiolabel-particle association could withstand exposure to serum protein for the entire length of an imaging study (at least 12 hrs). A Technetium-99m dextran complex was incorporated as a polymeric stabilizer during the synthesis of the nanoparticles, resulting in linkage of the radiolabel to the polymer matrix. The particles were studied *in vitro* in both phosphate-buffered saline (PBS) (pH 7.4) and in PBS with bovine serum albumin (BSA) to mimic plasma conditions. Although a slow, steady release of radioactivity occurred (probably because of degradation of the nanoparticle and release of Technetium-99m labeled dextran), the radioactivity release rate was relatively unaffected by the presence of the BSA and was not great enough to be of concern. This *in vitro* study is a good illustration of the importance of determining the stability of a radiolabel-vehicle association before performing *in vivo* experiments.

Another recent study compared two alternatives for labeling a biodegradable drug carrier, based on the *in vitro* stability of each in the presence of plasma (Willmott et al., 1991). Knowledge of the stability of the radiolabeled vehicle *in vivo* was critical for the selection of the radionuclide and for appropriate interpretation of imaging results. Biodegradable protein microspheres, potential delivery vehicles for cancer therapeutic agents, were designed to target the capillary beds of the liver, to increase exposure of tumor tissue to the drug, and to biodegrade, allowing for multiple dosing. These microparticles (10-50 μm in diameter) were radiolabeled with either Iodine-131 (covalently added to the tyrosine residues), or with Indium-111 (via diethylenetriaminepentacetic acid (DTPA) coupling to the protein and chelation of the radionuclide). *In vitro* stability studies in PBS and in human plasma showed good stability of the iodinated microspheres over several days, indicating the suitability of this radiolabeled system for *in vivo* studies of the distribution and biodegradation of this therapeutic carrier. The Indium-111 labeled microspheres, though stable in PBS, showed substantial loss of microsphere-associated radioactivity after several days in human plasma, rendering this radiolabeled system unsuitable for prolonged *in vivo* studies. This well-designed *in vitro* study led to the selection of an appropriately radiolabeled carrier, indicating good *in vivo* potential of that carrier.

Rudolph et al. (1991) used radiolabeled liposomes in both *in vitro* studies and *in vivo* animal experiments to characterize the distribution and stability of a potential erythrocyte substitute. Liposome-encapsulated hemoglobin (LEH) was radiolabelled with [^{99m}Tc]HMPAO and was then used as a probe for imaging the *in vivo* biodistribution of this oxygen delivery vehicle in the rabbit . *In vitro* extraction studies indicated that the [^{99m}Tc]HMPAO radiolabel of the LEH complex was internalized, possibly associated with the hemoglobin. Evidence that the radiotracer did not easily dissociate from the [^{99m}Tc]LEH complex was obtained by analysis of plasma fractions drawn throughout the experiment. No free hemoglobin or Technetium-99m was observed. In addition, no radioactivity was observed in the bile, kidney, or bladder, in contrast to the radioactivity observed in these organs of control animals given free [^{99m}Tc]HMPAO. This imaging data was validated by tissue sampling at the end of the study. Thus, these *in vitro* and *in vivo* studies revealed the kinetics of [^{99m}Tc]LEH distribution and indicated the potential for further development of this hemoglobin carrier in more advanced *in vivo* studies.

Gabizon et al. (1991) conducted pharmacokinetic and imaging studies in patients using adriamycin and liposome-entrapped radiolabeled deferoxamine to help elucidate the mechanism of antitumor drug delivery. Deferoxamine-containing liposomes were labeled using an Indium-111-oxine complex to label the deferoxamine. Demonstration of the *in vivo* stability of the radiolabel-liposome association was crucial for proper interpretation of imaging data. The location and physical association of the radiolabel with the liposome was suggested as existing either in association with the inner bilayers in the form of a lipophilic Indium-111-oxine complex, or bound to deferoxamine in the inner aqueous phase of the vesicles. When challenged by incubation with the Indium-111 chelator, DTPA, very little of the radiolabel was lost, indicating that it was incorporated into or internalized by the liposome.

The stability of the radiolabeled liposomes was tested *in vitro* by incubation with plasma. In comparison with the free radiotracer, [^{111}In]oxine, very little of the radiolabel became bound to plasma proteins. *In vivo* stability was checked by fractionization of patient plasma samples. The fraction of radiolabel bound to plasma proteins was very similar to that observed under the *in vitro* conditions. Preliminary *in vitro* results indicated that this radiolabeled liposome-entrapped drug complex was a suitable, stable probe for evaluating the drug delivery characteristics of this potential drug targeting system. The results of non-invasive imaging in patients indicated the importance of RES, liver function, and possible tumor cell interactions which, depending on the disease state and on the physical properties of the drug vehicle, could dominate or interfere with tumor-directed targeting (Gabizon et al., 1991; Perez-Soler, 1989; Richardson et al., 1979).

The previous examples illustrate *in vitro* and *in vivo* methods for characterization of the stability of radiolabeled drug carriers. These studies also provide some indication of the complexity of the design task for creating drug vehicles that target specific tissues and avoid often undesirable RES uptake. Radiotracers have been incorporated into a variety of injectable drug-carrier vehicles, including colloids such as biodegradable microspheres, synthetic polymeric particles, liposomes, and fat or oil emulsions, to evaluate the characteristics and mechanisms of distribution, drug release and potential interaction or interference with the RES. The *in vivo* distribution of these particulate drug vehicles is largely determined by their physical properties such as size, hydrophobicity, and surface charge. The macrophages of the RES remove most particles (greater than 150-300 nm in diameter and smaller than 5 μm,) via phagocytosis, resulting in rapid clearance from the blood circulation and uptake in the liver and spleen. The particles with the more hydrophobic surfaces are generally removed from circulation most rapidly. Several approaches have been taken to minimize undesirable RES uptake and improve drug targeting, including that of RES blockade.

APPLICATIONS FOR VEHICLE OPTIMIZATION

Effect of RES Blockade: Liposomes

An early imaging study of the effect of blockade of the RES in mice showed greater accumulation of radioactivity in tumors (and, therefore, successful targeting) with preblocked RES systems (Proffitt et al., 1983; Gregoriadis and Allison, 1980; Gregoriadis, 1976). Intravenous injections of unilamellar phospholipid vesicles (<0.1 μm in diameter), containing a 6-aminomannose derivative of cholesterol, were used to block the RES, reducing the clearance of subsequent injections of Indium-111-nitrilotriacetic acid loaded neutral liposomes. These blockade studies, though not clinically feasible, provided information on the performance of the RES in drug delivery studies in animal models and on the changes in liposomal distribution in the absence of RES interference.

Effects of Coatings on Distribution: Rigid Systems

Illum and Davis investigated the effects of colloidal particle coatings and varied surface hydrophobicities on their phagocytic uptake and entry into the RES (Illum and Davis, 1984; Illum et al., 1986). Coated and uncoated polystyrene microspheres (160 and 60 nm) were surface labeled with Iodine-131. Hydrophilic coatings (lecithin, SIgA, and non-ionic surfactants) showed reduced uptake of particles by the liver after intravenous administration. *In vivo* imaging results (and *in vitro* experiments using mouse peritoneal macrophages) confirmed that hydrophilic materials and non-ionic surfactants could be used to thwart phagocytosis and thereby modify the biodistribution of colloidal drug carriers.

Effects of Surfactants on Distribution: Emulsions

West used dual radiolabeling to study the *in vivo* behavior of a series of emulsions that were stabilized with surfactants of varying ionic properties (West, 1988). Each emulsion was labeled with Iodine-123 bound to the oil and [^{111}In]lasalocid sodium complex as an incorporated marker (for earlier examples of radiolabeled oils in emulsions, see Davis et al., 1987; Davis et al., 1990; and references therein.) A different surfactant was used to stabilize each emulsion. These were administered intravenously to rabbits for imaging, blood sampling, and eventual autopsy. Results showed similar *in vivo* behaviors for the Indium-111 and Iodine-123 radiotracers, indicating the stability of the emulsion droplets. The non-ionic emulsifiers remained in the circulation longer, reducing liver uptake. Dual labeling was used to make two determinations at once, comparing the *in vivo* stability of each emulsion and showing that the non-ionic emulsifiers were effective modifiers for avoidance or minimization of phagocytosis.

Radiolabeled Drug and Drug Analogues

As is evident from the range of studies reviewed, the vehicle is often the most convenient component of the drug/vehicle complex to radiolabel. The radiolabeling of the drug or a suitable analogue can involve complex chemistry, an often inconvenient challenge, particularly when dealing with large quantities of high energy, often short-lived and expensive radionuclides. In addition, the solubility characteristics of the radiolabeled drug or analogue must be appropriate for its effective incorporation into the delivery system.

Illum and colleagues used non-invasive nuclear imaging to compare a novel polymer-drug delivery system with ethanol/propylene glycol, and emulsion systems (Illum et al., 1985). The novel system, a globular, partially quaternized polymer, poly[thio-1-(N,N-diethylaminomethyl)-ethylene], had been used for intramuscular administration (to rabbits) of the radiolabeled steroid, [^{75}Se]norcholestenol. Imaging showed that the intravenously administered radiolabeled steroid polymer complex was taken up primarily in the lungs (blockage of vasculature occurred) whereas, in either the ethanol/propylene glycol solution or the emulsion, the greater activity was in the liver and spleen. These results quickly showed that this particular polymer system would be unsuitable as a carrier for intravenous administration of steroids.

MICROFINE EMULSION STUDY

Nuclear imaging was used in our laboratory to evaluate the *in vivo* distribution of a lipid soluble drug administered in a new emulsion formulation. MK-329 is a lipophilic (log P = 2.8), non-peptidal CCK-A antagonist that was under investigation as a potential treatment for a variety of gastrointestinal disorders. A microfine emulsion formulation of MK-329 was developed for use as a vehicle for intravenous administration. Since this formulation consists of small, MK-329 containing oil droplets, a key concern was would the drug be distributed throughout the body, as desired, or would the drug/vehicle behave as a colloid, becoming sequestered in the lungs or liver. If the formulation behaved as a colloid *in vivo*, it would be cleared from the circulation rapidly by the RES, resulting in approximately 80% of the radioactivity accumulating in the liver within 10 minutes post injection. For comparison, similar studies were conducted using a (polyethylene glycol)-based (PEG) formulation as a non-colloidal, control solvent. Since MK-329 is soluble in the PEG formulation, it would not be expected to behave as a colloid *in vivo*.

For this study, the drug, rather than the emulsion vehicle, was chosen for radiolabeling because distribution of the drug, not stability of the emulsion, was the critical issue. Two radiotracers were developed for use in these imaging studies, [^{11}C]MK-329 (**1**) and [^{123}I]L-366,012 (**2**). Initial work was conducted in our laboratories using [^{123}I]L-366,012, an iodinated analogue of MK-329, that has a high affinity for CCK-A receptors and is also lipophilic (log P = 3.9 compared to 2.8 for MK-329). These initial studies were designed to optimize formulation conditions and to obtain a preliminary indication of the ability of the

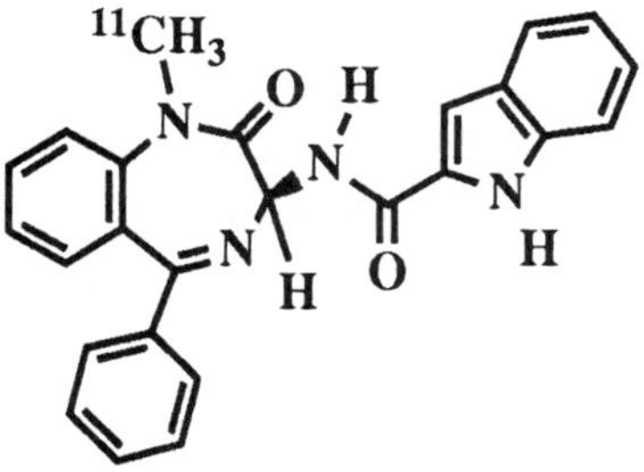

1 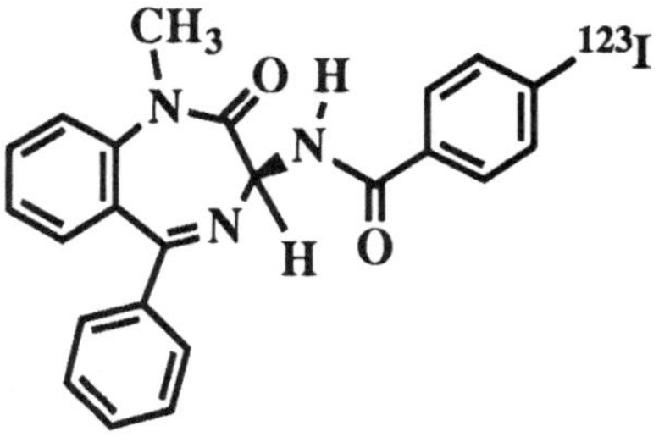 2

microfine emulsion to provide whole body distribution of a lipid soluble drug. The use of the iodinated analogue of MK-329 made it possible to obtain a preliminary evaluation of the new formulation in our own laboratories prior to conducting more costly and complex studies at a PET facility.

Since the results of these preliminary imaging studies (in rats and monkeys) with [^{123}I]L-366,012 indicated that the emulsion was giving the desired whole body distribution of the lipophilic radiotracer, similar studies using [^{11}C]MK-329 were conducted at Johns Hopkins Medical Institutions.

Radiotracer Synthesis and Formulation

The radioiodinated analogue of MK-329, [^{123}I]L-366,012, was synthesized via a melt exchange, replacing Iodine-127 with Iodine-123 as previously described (Burns et al., 1991). The tracer was extracted from the inorganic reaction mixture with ethanol and analyzed by High Performance Liquid Chromatography (HPLC). The typical product used in imaging studies had a radiochemical purity $\geq$ 99%. [^{11}C]MK-329 was synthesized by alkylation of desmethyl-MK-329 with [^{11}C]CH_3I (Burns et al., 1989). This tracer was purified by preparative HPLC and, after evaporation of the HPLC solvent, was dissolved in diethyl ether.

The microfine emulsion formulation of each radiotracer was prepared as follows. The solvent (ethanol for [^{123}I]L-366,012 and diethylether for [^{11}C]MK-329) was removed via rotary evaporation. Safflower and soybean oils (premixed with unlabeled MK-329) were added to the residue to dissolve the radiotracer. The aqueous phase, consisting of a mixture of water, egg lecithin, and glycerin, was added to the oils and the phases were mixed to form a coarse dispersion. The mixture was then processed into a microfine emulsion using a Microfluidizer® homogenizer (Microfluidics). The emulsion was immediately used in the imaging studies. A fraction of each emulsion was saved for subsequent measurement of particle size. Particle size determinations were performed with a Coulter N4MD laser-light scattering instrument using the photon correlation spectroscopy method of analysis. The average diameter of the particles used in these studies was less than 220 nm.

Formulation in the PEG-based solvent control was performed by evaporation of the ethanol or diethyl ether solvent from the radiotracer, followed by dissolving the radiotracer residue in the PEG-based solvent (45% polyethylene glycol, 21% glycerol, and 34% water).

Imaging Studies

Nuclear imaging was utilized to study the *in vivo* distribution of both formulations of [^{123}I]L-366,012 in African Green monkeys (*Cerupithecus Aethiups*). Prior to the imaging studies, the fasted monkeys were anesthetized with ketamine/xylazine via intramuscular injection. For each study, the monkey was placed in a supine position beneath a planar-gamma camera equipped with a low-energy, all-purpose collimator. Imaging began immediately after the intravenous injection of 0.5-1.0 mCi of the [^{123}I]L-366,012 solvent or microfine emulsion formulation. Pre- and post-injection activity of the [^{123}I]L-366,012 syringe was assayed to determine the net dose administered. Image analysis consisted of expressing the decay-corrected activity in the liver as a percentage of the injected dose over time.

A total of five experiments using the microfine emulsion formulation and two experiments using the PEG solvent formulation were conducted in three African Green monkeys. Two animals received both formulations and one received only the microfine emulsion. Images of the torso of a monkey injected with [^{123}I]L-366,012 in microfine emulsion are shown in Fig. 1. Distribution of the radioligand in the liver at peak activity (5 min) is visually greater than at the plateau activity (40 min). Gall bladder activity (included in the quantification of liver activity) becomes evident at approximately 30 min in both studies. Additionally, the diffuse activity throughout the field of view, other than the liver region, is evident with both the solvent and microfine emulsion formulations.

Uptake of the microfine emulsion formulation in the liver peaked at 18.1 ± 1.2% within 5 min. Similarly, the solvent form peaked at 15% within 5 min. The distribution curves for the two systems are quite similar (Fig. 2) and the differences between the groups were not significant.

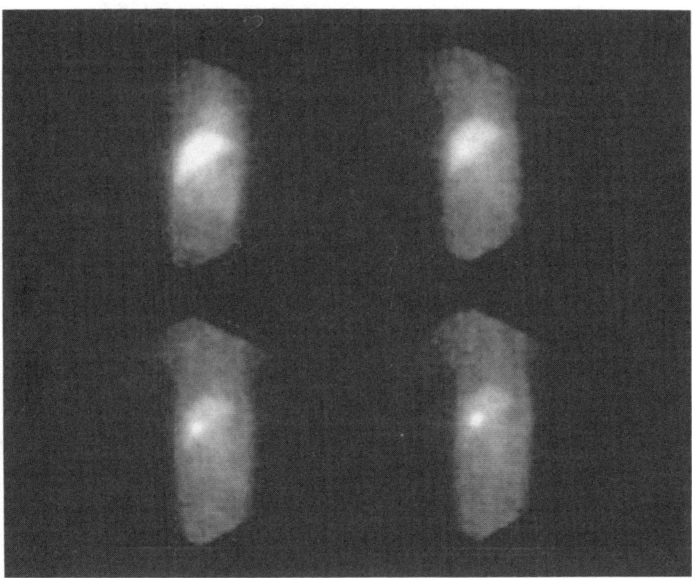

FIGURE 1. Anterior view of monkey torso 5 (upper left), 20 (upper right), 40 (lower left) and 60 min (lower right) post-injection of [^{123}I]L-366,012.

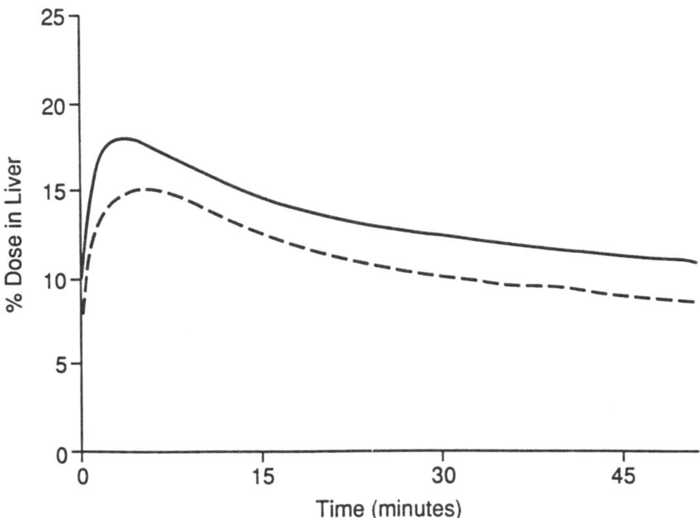

FIGURE 2. Distribution curves of [^{123}I]L-366,012 in the liver. Microfine emulsion (solid line) and PEG solvent (dashed line). The liver activity did not differ significantly. A radioalabeled colloid cleared by the RES would have given %-dose in liver 4- to 5-fold higher than observed in this study.

While imaging the distribution of the analogue, [^{123}I]L-366,012, provided a direct comparison between the two delivery systems, it did not provide information regarding the true compound of interest, MK-329. Thus, a second imaging study was conducted with the Carbon-11-labeled parent compound to confirm the results obtained with the iodinated analogue.

A planar gamma camera was utilized to examine the distribution of [^{11}C]MK-329 in two additional African Green monkeys. The [^{123}I]L-366,012 imaging procedure was modified to accommodate the higher energy and shorter half-life of Carbon-11. Imaging was performed with a high energy parallel hole collimator. Alternating anterior and posterior images of 5 min each were obtained for two hrs after the intravenous injection of 10-30 mCi of [^{11}C]MK-329 in solvent or microfine emulsion. Image analysis consisted of determining the activity in the liver as the geometric mean of the anterior and posterior counts expressed as a percentage of the injected dose [geometric mean = (anterior counts x posterior counts)$^{1/2}$]. The total injected dose was expressed as the geometric mean of the anterior and posterior whole body image.

The distribution of the [^{11}C]MK-329 closely follows the results obtained from the iodinated analogue (Figure 3). Noise in the graphic

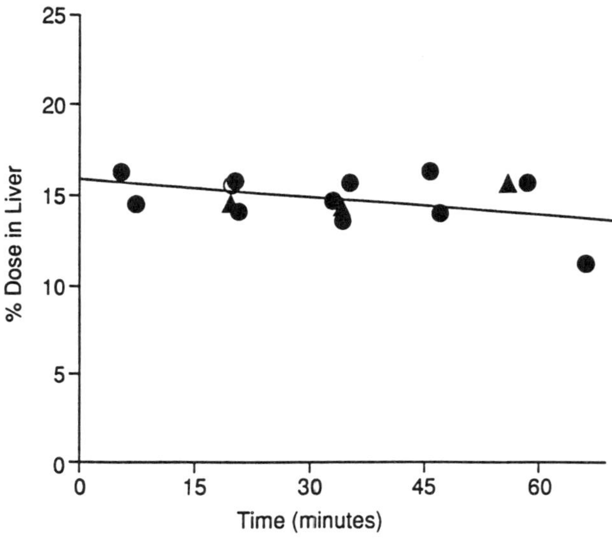

FIGURE 3. Representative distribution curves of [^{11}C]MK-329 in the liver. PEG solvent (●), microfine emulsion (▲).

analysis is due to difficulty in drawing the region of interest. The 12-16% uptake in the liver at early timepoints for both formulations of the parent compound closely parallels the results obtained with the iodinated analogue.

If the formulation behaved as a colloid *in vivo*, it would have been cleared from the circulation rapidly by the RES and approximately 80% of the radioactivity would have accumulated in the liver within 10 minutes of injection. The fact that less than 20% of the radioactivity was found in the liver for both tracers and the fact that both tracers behaved similarly in the microfine emulsion and PEG formulations indicates that, from a drug delivery perspective, the microfine emulsion is a suitable formulation for intravenous administration of MK-329 and is likely to be suitable for other lipid soluble drugs as well.

Thus, nuclear imaging was able to provide a direct, non-invasive *in vivo* evaluation of a new formulation as well as a comparison of two intravenous drug formulations. Results of these studies demonstrated that the microfine emulsion delivery system was as efficacious as the solvent system. Additionally, the second study with $[^{11}C]$MK-329 confirmed the information obtained with the iodinated analogue.

CONCLUSION

Nuclear imaging has been used in the design and *in vivo* characterization of various injectable drug delivery systems, including microparticles and microspheres, nanoparticles, liposomes, and emulsions. Radiolabeled drug/vehicle complexes must be challenged in *in vitro* and *in vivo* experiments to define the nature of association, stability, and localization of the radiotracer on and/or within the drug/vehicle complex. The stability of the radiolabel-drug or radiolabel-vehicle association should be demonstrated in animal models and, if possible, validated by plasma and/or tissue analysis or other biodistribution techniques. The technique of nuclear imaging can be extended to higher animal models and, eventually humans, to provide drug distribution information that cannot be obtained by other means. Although great care must be taken in the design, interpretation, and validation of nuclear imaging drug delivery studies, few alternative techniques provide dynamic, real-time information for the non-invasive, *in vivo* evaluation of novel drug delivery systems.

ACKNOWLEDGMENTS

Studies with $[^{11}C]$MK-329 and $[^{123}I]$L-366,012 were conducted in collaboration with J. Walkow, D. Karr, P. Fogel, H. Solomon, W. Eng,

Merck Research Labs and B. Dannals, J. Links, J. Frost, Johns Hopkins University and will be published in detail elsewhere. The authors are also grateful to R. Gibson for useful discussions and M. Bechtel for manuscript preparation.

REFERENCES

Burns HD, Brenner NJ, Gibson RE and Solomon HF (1991): Gamma Emitting, CCK-A antagonists for pancreatic imaging. *United States Patent No. 4,994,258.*

Burns HD, Eng W-s, Dannals RF, Frost JJ, Ravert HT, Wilson AA, Gibson RE, Wagner HN Jr and Solomon HF (1989): Synthesis of [C-11]MK-329: A potential tracer for imaging peripheral cholecystokinin receptors with PET. *J Nucl Med* 20: 931.

Cohen S, Bernstein H, Hewes C, Chow M and Langer R (1991): The pharmacokinetics of and humoral responses to antigen delivered by microencapsulated liposomes. *Proc Nat Acad Sci* 88: 10440-10444.

Davis SS (1976): The emulsion-obsolete dosage form or novel drug delivery system and therapeutic agent? *J Clin Pharm* 1: 11-27.

Davis SS (1982): Emulsion systems for the delivery of drugs by the parenteral route. In: *Optimization of drug delivery: Proceedings of the Alfred Benzon Symposium 17*, Bundgaard H, Bagger-Hansen A and Koford H, eds. Copenhagen: Munksgaard.

Davis SS (1986): Biopharmaceutical aspects of drug formulation. *ACTA PharmacSuecica* 23: 305-314.

Davis SS (1985): The use of scintigraphic methods for the evaluation of novel delivery systems. In: *Directed Drug Delivery*, Borchardt RT, Repta AJ and Stella VJ, eds. Clifton: Humana Press.

Davis SS and Illum L (1983): Drug delivery systems. *Practitioner* 227: 1537-1543.

Davis SS, Hunneyball IM, Illum L, Ratcliffe JH, Smith A and Wilson CG (1985). Recent advances in the use of microspheres for targeted therapy. *Drugs Under Exptl Clin Res* 11: 633-640.

Davis SS, Illum L, West P and Galloway M (1987): Studies on the fate of fat emulsions following intravenous administration to rabbits and the effect of added electrolyte. *Clin Nutr* 6: 13-19.

Davis SS, Illum L, Muller R, Landry F, Wright J and Harper G (1990): The effect of infused fat emulsions on reticuloendothelial function in the rabbit. *Clin Nutr* 9: 260-265.

Davis SS, Douglass SJ, Illum L, Jones PDE, Mak E and Muller RH (1986). Targeting of colloidal carriers and the role of surface properties. In: *Targeting of Drugs with Synthetic Systems, NATO Advanced Studies Institute Series A*, Gregoriadis G, Trouet A, Poste

G and Senior J, eds. New York: Plenum Press.

Douglas SJ and Davis SS (1985): The use of nanoparticles in drug targeting. *Chem Indust* 22: 748-751.

Douglas SJ and Davis SS (1986): Radiolabelling of poly(butyl 2-cyanoacrylate) nanoparticles with a TC-99m-dextran complex. *J Labelled Compd Radiopharm* 23: 495-504.

Douglas SJ, Davis SS and Illum L (1987): Nanoparticles in drug delivery. *Crit Rev Therap Drug Carrier Systems* 3: 233-261.

Gabizon A, Chisin R, Amselem S, Druckmann S, Cohen R, Goren D, Fromer I, Peretz T, Sulkes A and Barenholz Y (1991): Pharmacokinetic and imaging studies in patients receiving a formulation of liposome-associated adriamycin. *Br J Cancer* 64: 1125-1132.

Gregoriadis G (1976): The carrier potential of liposomes in biology and medicine. *N Engl J Med* 295: 704-710.

Gregoriadis G and Allison AC (1980): *Liposomes in Biological Systems.* New York: Wiley.

Illum L and Davis SS (1984): The organ uptake of intravenously administered colloidal particles can be altered using a non-ionic surfactant (Poloxamer 338). *FEBS Lett* 67: 79-82.

Illum L, Huguet J, Vert M and Davis SS (1985): Accumulation in the lung of Se-75 norcholestenol administered intravenously in a globular partially quaternized poly thio-1-(N, N-diethyl-aminomethyl)ethylene carrier system. *Int J Pharmac* 26: 113-121.

Illum L and Davis SS (1982): Cellulose microspheres as a sustained release system for parenteral administration. *Int J Pharm* 11: 323-327.

Illum L, Hunneyball IM and Davis SS (1986): The effect of hydrophilic coating on the uptake of colloidal particles by the liver and peritoneal macrophages. *Int J Pharm* 29: 53-65.

Links JM (1990): Physics and instrumentation of positron emission tomography. In: *Quantitative Imaging Neuroreceptors, Neurotransmitters, and Enzymes.* Frost JJ and Wagner HN Jr, eds. New York: Raven Press.

Liu D, Mori A and Huang L (1991): Large liposomes containing ganglioside GM-1 accumulate effectively in spleen. *Biochem Biophys ACTA* 1066: 159-165.

Napper DH and Netschey A (1971): Studies of the steric stabilization of colloidal particles. *J Colloid Interface Sci* 37: 528-535.

Perez-Soler R (1989): Liposomes as carriers of antitumor agents toward a clinical reality. *Cancer Treat* 16: 67-82.

Pimm MV, Clegg JA, Hudecz F and Baldwin RW (1992): Indium-111 labelling of a branched polypeptide drug carrier with a poly(L-lysine) backbone. *Int J Pharm* 79: 77-80.

Proffitt RT, Williams LA, Presant CA, Tin GW, Uliana JA andGamble RC (1983): Liposomal blockade of the reticuloendothelial system: Improved tumor imaging with small unilamellar vesicles. *Science* 220: 502-505.

Richardson VJ, Ryman BE and Jewkes RF (1979): Tissue distribution and tumor localization of 99m-Technetium-labelled liposomes in cancer patients. *Br J Cancer* 40: 35-43.

Rudolph AS, Klipper RW, Goins B and Phillips WT (1991): *In vivo* biodistribution of a radiolabeled blood substitute technetium-99m-labeled liposome-encapsulated hemoglobin in an anesthetized rabbit. *Proc Nat Acad Sci* 88: 10976-10980.

Van Oss CF, Gillman CF and Newman AW (1975): *Phagocytic Engulfment and Cell Adhesiveness.* New York: Dekker.

West PE (1988): Oil-in-water emulsions for intravenous drug delivery. Thesis, University of Nottingham, The British Library, Boston Spa, Wetherby, West Yorkshire, UK.

Wilkins DJ (1967): *The Reticuloendothelial System.* DiLuzio NR and Paoletti R, eds. New York: Plenum Press.

Willmott N, Murray T, Carlton R, Chen Y, Logan H, McCurrach G, Bessent G, Goldberg A, Anderson J, McKillop and McArdle CS (1991): Development of radiolabelled albumin microspheres: A comparison of gamma-emitting radioisotopes of iodine-131 and indium-111. *Nucl Med Biol* 18: 687-694.

Wilson CG, Hardy JG, Frier M and Davis SS (1982): *Radionuclide imaging in drug research,* London: Croom Helm.

Yamazaki N, Kojima S, Gabius S and Gabius H-J (1992): Studies on carbohydrate-binding proteins using liposome-based Systems I. Preparation of neoglycoprotein-conjugated liposomes and the feasibility of their use as drug-targeting devices. *Int J Biochem* 24: 99-104.

THE GASTROINTESTINAL TRANSIT AND SYSTEMIC ABSORPTION OF DILTIAZEM HCL FROM A MODIFIED RELEASE DOSAGE FORM

Donald L. Heald and John A. Ziemniak
Drug Disposition Department
Rhône-Poulenc Rorer

Ian R. Wilding
Pharmaceutical Profiles, Ltd

INTRODUCTION

The primary means of administration for the vast majority of therapeutic agents is via the oral route. Many of these oral dosage forms are designed to provide a rapid disintegration in the stomach, however, over the last 5-6 years there has been a growing trend within the pharmaceutical industry to develop complex pharmaceutical dosage forms designed to deliver drug in a controlled or timed released manner after oral administration. The design of these modified release (MR) drug products is intended to optimize the therapeutic regimen by providing slow continuous delivery of drug over the entire dosing interval while also providing greater patient compliance and convenience. The development of these MR dosage forms has also placed demands on the pharmaceutical industry to design and utilize new methodologies to assess the performance of these products *in vivo.* Likewise, regulatory agencies are faced with the arduous task of reviewing the performance of these new drug products to assure that the

Nuclear Imaging in Drug Discovery, Development, and Approval
H.D. Burns, R. Gibson, R. Dannals, Peter Siegl, Editors

dosage forms perform as expected. The task of demonstrating the performance of the dosage form is the responsibility of the sponsoring pharmaceutical company. This position is supported by the following recent quote presented at a workshop on MR dosage forms sponsored by the American Association of Pharmaceutical Scientists, the United States Pharmacopeia, and the United States Food and Drug Administration: "Sponsor should explain the rationale for the dose selected and the dosage form. It must be demonstrated that the dosage form performs *in vivo* according to its proposed rationale" (Wilding et al., 1991).

There are a limited number of methodologies currently available to provide this type of data. Wilding and co-workers have critically reviewed the relative merits of the different techniques available for assessing the *in vivo* performance of controlled release dosage forms (Wilding et al., 1991). The gastrointestinal (GI) transit of pharmaceutical dosage forms can be evaluated by a variety of techniques, but gamma scintigraphy is the most attractive technique since it offers a relatively simple method whereby a small amount of a gamma-emitting radionuclide is incorporated into a MR formulation and its passage through the different regions of the GI tract is monitored. The use of external gamma scintigraphy has become a popular method to investigate the GI performance of pharmaceutical dosage forms and the underlying principals have been reviewed in several review articles (Wilson and Hardy, 1982; Fell and Digenis, 1984; Jay et al., 1985; Wilson et al., 1988; Hardy, 1989; Digenis et al., 1989; Wilding et al., 1991). Typically, radiation photons (gamma particles) arising from the dosage form, or an anatomical site, pass through a collimator and strike the sodium iodide crystal of gamma camera detector head. The resultant flash of light is detected by a series of photomultiplier tubes. Most scintigraphic systems are able to separate emissions on the basis of different photopeak energies so that two isotopes can be monitored simultaneously and independently. This capability permits the simultaneous study, in a noninvasive manner, of two different dosage forms or two discrete components within the same delivery device (Davis et al., 1984; Digenis and Sandefer, 1991). A more in-depth review of the gamma scintigraphy procedure, choice of suitable radionuclide, and methods employed to remove inherent errors associated with the use of gamma scintigraphy is presented in Chapter 3 of this book and other literature (Wilding et al., 1991; Digenis and Sandefer, 1991).

There are two approaches often used for the conventional labeling of oral dosage forms: i) incorporation of a non-absorbable chelate of the radioactive isotope (Daly et al., 1982), and ii) incorporation of radiolabeled ion-exchange resin (Davis 1987). The use of these approaches, however, in the radiolabeling of novel drug formulations and in MR delivery systems may present several problems (Digenis and Sandefer, 1991; Wilding et al., 1991):

1) Synthesis time may be too long and preclude the use of short-lived radionuclides.
2) Equipment normally used to manufacture dosage form can become contaminated with radioactivity and thus prevent its use for other manufacturing purposes.
3) On-site small scale preparation of radiolabeled dosage forms may result in an altered dosage form relative to those manufactured under industrial scale conditions.
4) The labeling process itself may not be applicable to the formulation or may compromise the integrity of the dosage form.

These problems can be overcome by the use of stable isotopes and neutron activation methods (Parr et al., 1985; Parr et al., 1986; Parr et al., 1987; Hardy et al., 1991; Digenis et al., 1990; Wilding et al., 1992). In general, a stable isotope such as Erbium-170 or Samarium-152, as the non-absorbable isotopically enriched oxide, can be incorporated into the dosage form in very low amounts. Since the amounts used are minimal, performance of the dosage form is usually not compromised; however, rigorous *in vitro* assessment is still required to ensure the manufacturing integrity of the formulation. The dosage form is then irradiated in a neutron source for a period of time in order to convert the isotope into a gamma emitting material (Erbium-171 or Samarium-153). The radiation exposure to a subject following dosing with these radionuclides is comparable to that received using conventional radionuclides such as Technetium-99m and Indium-111. The radioactive decay half-lives for Erbium-171 and Samarium-153 are 7.5 hrs and 47 hrs, respectively, eliminating the time restraints of other markers. Potential benefits of this activation method would include being able to use the same production equipment at the same site of manufacture thus improving quality assurance while guaranteeing that the formulation to be tested is representative of the commercial product (Wilding et al., 1991; Digenis and Sandefer, 1991).

In order to maintain prolonged delivery, an oral MR formulation must release drug throughout the different regions of the GI tract, including the small and large intestine. Knowledge of the relevant characteristics of the differing environments and GI transit in various regions of the GI tract are important factors when addressing the *in vivo* performance of the formulation. The relevant physiologic features of the

GI tract and the process of gastric emptying in man has been reviewed in the literature (Wilding et al., 1991; Digenis and Sandefer, 1991). In general, the stomach behaves in two very discrete ways, depending on whether it contains food or is empty (Richter, 1988). Generally under fasting conditions, the stomach is void of any material aside from swallowed saliva, mucus and cellular debris, however, there may be large particles of indigestible food left from the previous meal. Expulsion of these materials and the routine cleansing of the stomach is the purpose of the migrating myoelectric complex (MMC). The MMC is divided into four consecutive phases of activity (Phillips, 1988):

Phase 1 - a period of quiescence lasting about sixty minutes with no contractions.
Phase 2 - about 40 minutes of intermittent action potentials and contractions that gradually increase in intensity as the phase progresses.
Phase 3 - a short phase of intense distal and proximal gastric contractions lasting between five and fifteen minutes. It is the ability of Phase 3 to so completely empty the stomach that has led to the name "housekeeper wave".
Phase 4 - a short transitory period between the intense activity of Phase 3 and quiescence of Phase 1.

The different phases of the cycle migrate distally from the stomach to the terminal ileum over a two hr period. Therefore, when one Phase 3 reaches the terminal ileum, another one is beginning in the stomach. On feeding, the cycle is immediately broken with a normal meal disrupting the activity for about 3-4 hours (Vantrappen et al., 1977).

METHODS

Once-a-Day Formulation of Diltiazem HCl

Rhône-Poulenc Rorer (RPR) has developed a new once-a-day formulation of diltiazem HCl (DTZ) for the treatment of hypertension utilizing the Geomatrix™ technology of a tri-layered tablet. DTZ is classified as a calcium channel blocker or calcium antagonist. The mechanism of action of this group of drugs is through the inhibition of transmembrane influx of calcium ions into cardiac muscle and vascular smooth muscle. This inhibition causes a negative inotropic effect on the heart, diminished myocardial oxygen consumption and dilated coronary and peripheral arteries allowing for the use of DTZ in angina and hypertension (Buckley, 1990; Weiner, 1988).

Figure 1 presents a diagram of the novel Geomatrix system. The hydrophilic core of the tablet contains the active drug and is designed to swell over time as the tablet hydrates. *In vitro* data determined that after twenty-four hrs the tablet increased from approximately 7 mm to a diameter of 14-15 mm in size. The rate and duration of the release of the active drug from the Geomatrix system is regulated by the exposed surface area and the volume dimension of the center core. Further adjustments in the release rate is accomplished by 2 external layers which serve to control the rate of hydration of the core and restrict the surface area available for drug diffusion. By controlling matrix porosity and swelling, the formulation maintains a constant and sustained release (SR) profile.

Gamma-Scintigraphy

RPR utilized gamma scintigraphy to monitor the GI transit and release of a SR form of DTZ, under normal physiological conditions in healthy male volunteers while also examining the influence of a high fat breakfast on the transit and release characteristics of the formulation. The commercial formulation consists of either a 180 or 240 mg capsule of diltiazem which is composed of either three or four, 60 mg Geomatrix tablets, respectively. Tablets for this study were either labeled in the core only or both the core and the layers were labeled with different non-absorbable gamma-emitting radionuclides prior to encapsulation. Frequent monitoring with a gamma camera as well as simultaneous blood sampling were used to provide information on the location, transit rate, extent of dispersion of the individual tablets, obstruction of the GI tract, and the *in vivo* absorption of DTZ. The pharmacokinetic profile of

FIGURE 1. Geomatrix System.

the drug in the systemic circulation was determined from measurements of blood samples obtained at frequent intervals over the 48 hrs post-dosing period. Relationships between GI transit and *in vivo* release behavior, with resultant pharmacokinetic data were explored.

DTZ MR tablets (Dilacor XR™) were manufactured by Rhône-Poulenc Rorer Central Research, Collegeville, Pa, on a small laboratory scale of 350 tablets using the nonabsorbing stable isotopes Samarium-152 and Erbium-170. Each 60 mg tablet contained a small amount of [^{152}Sm]samarium oxide blended into the formulation making up the central core of the tablet and accounted for less that 0.5% of the overall tablet weight of the active layer. Likewise, a second series of tablets were also prepared where in addition to the Samarium-152 in the core a very small amount of [^{170}Er]erbium oxide was blended into the coating layers or caps accounting for approximately 5% of the total weight of the external layer formulation. Tablets were shipped to Pharmaceutical Profiles Limited, Nottingham, UK, who conducted the clinical portion of the study.

In vitro dissolution testing of the commercial Dilacor XR tablets and those tablets containing the nonabsorbable tracers confirmed that the addition of the samarium oxide did not alter the *in vitro* dissolution profiles of the tablets. Similar studies on irradiated tablets showed that the release properties of the tablets was not affected by the neutron activation process. On the day prior to dosing (24 hrs prior to dosing), the single and dual labeled tablets were irradiated in a neutron flux of 10^{12}n/cm^2/s^1 for five and fifteen min, respectively. The tablets were assayed for radioactive content and encapsulated so that each capsule contained one dual labeled and three single labeled Geomatrix 60 mg tablets for a total dose of 240 mg DTZ.

According to the open label, randomized, two-period crossover design, eight healthy male volunteers received one capsule of the test formulation of DTZ 240 mg administered with 150 ml of water at 8 am following either an overnight fast starting at 10:00 pm on Day -1 or a high fat breakfast in an effort to determine what effect food has on the transit and integrity of the Geomatrix tablets. The two study periods were separated by a total of 7 days (washout). During each period, subjects were required to stay at the study center for 3 nights in a dormitory room.

Each volunteer was provided with a glass of apple juice or a cup of caffeine free coffee at both 2 hours and 7 hours post-dose. A standard lunch with a cup of caffeine free coffee was provided 4 hours post-dose, and dinner at 10 hours post-dose. Water was allowed *ad libitum*. The dietary regimen was repeated over the second study period except all volunteers received a standard breakfast and lunch. The standard high fat breakfast consisted of 2 eggs (scrambled), 2 strips of bacon, 2 slices of toast, 2 oz. of butter, 1 slice of fried bread, and 250 ml of milk.

Subjects were imaged using a gamma camera having a 40 cm diameter field of view, fitted with an appropriate parallel hole collimator. Anterior scintigraphic images of the abdomen were taken at regular intervals throughout the study period and stored on magnetic tape for analysis at a later stage. Images were taken every 20 min for the first 10 hrs and every hour after dinner for the remainder of the day (up to 16 hrs). Subjects were then permitted to sleep and imaging resumed at hourly intervals beginning at 24 hrs post-dose and ending at 36 hrs, and a final image taken at 48 hrs post-dose. Feces was collected throughout the study and examined for emitter prior to release at 48 hrs. Blood samples (8 mls) were obtained either via an indwelling cannula, irrigated with heparin, or by direct venipuncture, at the following time intervals: 0 (pre-dose), 0.5, 1, 1.5, 2, 3, 4, 6, 8, 10, 12, 14, 16, 24, 28, 32, 36, and 48 hrs. Samples were collected into heparinized glass tubes and gently mixed several times, plasma harvested and then transferred to two labeled polypropylene tubes prior to freezing at -80° C prior to HPLC analysis.

Analysis

Pharmacokinetic (PK) analyses for DTZ was performed using noncompartmental techniques in order to determine if any relationship exists between the PK performance of the capsule and its behavior in the GI tract. The maximum plasma drug concentrations (Cmax) and the times at which they occurred relative to the dose (Tmax) were determined from the plasma drug concentration-time profile. Terminal elimination rate constants (Ke) were determined by linear regression of the terminal linear phase of the log plasma DTZ concentration versus time profile. Areas under the plasma DTZ concentration-time curve through the last detectable concentration C_{LDC} [$AUC_{(LDC)}$] were calculated by numerical integration using the trapezoidal rule. Areas under the plasma DTZ concentration versus time curve from zero extrapolated to infinity, AUC_{inf} were calculated by the trapezoidal rule and extrapolated to infinity using the following equation where C_{LDC} represents the last detectable concentration:

$$AUC_{inf}=AUC_{(LDC)} + C_{LDC}/Ke$$

The ANOVA for a 2-way crossover design was used to test for differences in the means for each parameter among all treatments at an alpha level of 0.05.

RESULTS

The gamma scintigraphic data from this study were analyzed to determine the following parameters:

1) GI transit of each of the four individual once-a-day tablets.
2) Anatomical position of loss of tablet integrity.
3) Times for loss of tablet integrity.
4) The integrity of the dual labeled tablet with respect to the interaction of the swelling core and the impermeable layers.

GI transit was defined as the time for 50% of the tablets to leave the stomach, 50% to arrive at the colon, and by the difference, the 50% value which describes small intestinal transit time. Loss of tablet integrity or disintegration time is defined as the time at which the impermeable caps separate from the central core of the tablets. Figure 2 presents a diagram of the GI tract which provides a reference to the various segments of the tract as they are discussed elsewhere in the chapter.

Biopharmaceutics

Semilogarithmic plots of the mean plasma DTZ concentration versus time are presented in Figure 3. In general, the mean plots indicate that coadministration of a high fat breakfast does not result in "dose dumping" of DTZ into the general circulation. The mean plots also suggest that a high fat breakfast may slightly delay the overall absorption of diltiazem from 0 to 2 hrs and then enhance absorption of the drug between 6 and 12 hrs. Table 1 presents a summary of the mean biopharmaceutic and statistical analyses. A detailed analysis of the biopharmaceutic/pharmacokinetic analysis is beyond the focus of this chapter and will be presented elsewhere (Heald et al., 1992).

Gamma Scintigraphy

In general, an individual's scintigraphic results coincided with that individual's pharmacokinetic data and provided an opportunity to understand the functionality of the new SR dosage formulation. More specific trends were masked by the large degree of inherent variability and first pass biotransformation associated with this drug. The results from the gamma scintigraphy determined no apparent evidence of inter-tablet adhesion, therefore, it would appear that *in vivo*, each of the four

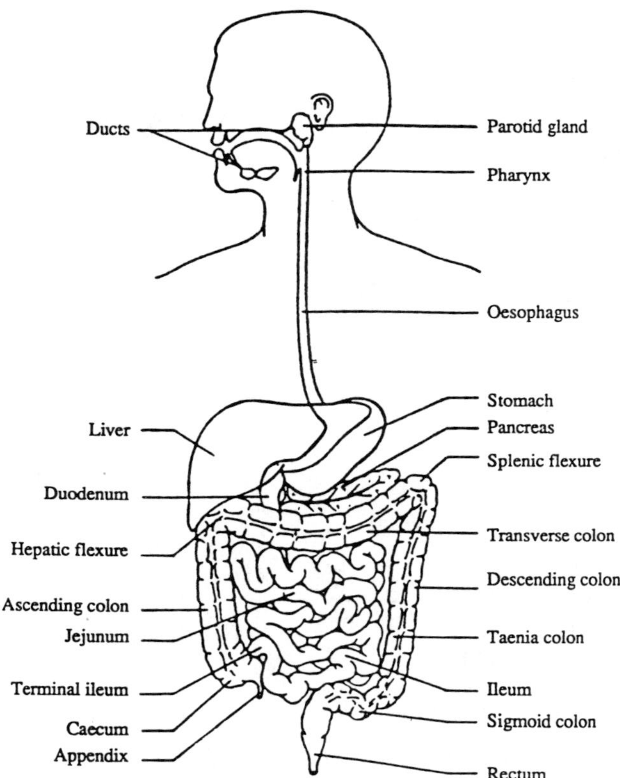

FIGURE 2. Diagram of the gastrointestinal tract.

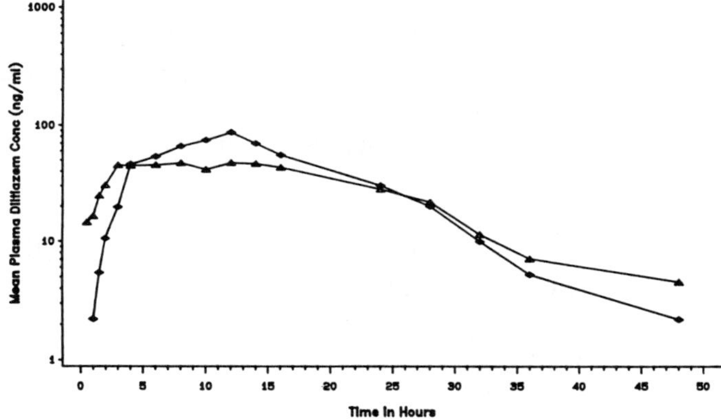

FIGURE 3. Semilogarithmic plots of the mean plasma diltiazem concentration versus time in subjects given 240 mg diltiazem HCL SR in the fasted (▲) and fed (◆) states.

Parameter	Fasted Condition	Fed Condition	ANOVA[1] p > F
AUC_{inf} (ng * hr/ml)	1201.8 (0.35)[2]	1422.0 (0.47)	0.168
Cmax (ng.ml)	59.8 (0.37)	92.9 (0.45)	0.006
Tmax (hr)	8.5 (0.53)	10.3 (0.26)	0.497
Ke (hr^{-1})	0.16 (0.30)	0.16 (0.15)	0.959
$T_{1/2}$ (hr)	4.33	4.33	

1 Two factor analysis of variance.

2 coefficient of variation.

TABLE 1. Mean bioavailability parameters following oral administration of 240 mg Dilacor XR to 8 healthy male volunteers under fasted and fed conditions.

tablets behaved as a discrete SR dosage form. Fig. 4 presents a typical gamma scintigraphy image, which was obtained from Subject 6 at nine hrs post-dose. The figure illustrates the position of the reference marker on the subject's upper right abdomen and the position of four individual tablets determined to be in the transverse colon. The outline of the gut wall was traced on either side of the radioactive trail to help visualize the transverse colon.

The labeling of the impermeable caps for one of the four tablets provided information on the mechanism of tablet disintegration. It had been postulated that separation of the inactive layers from the core might precede tablet disintegration. This was only observed in one volunteer under fasted conditions in one of the study periods (Subject #7) and did not alter the plasma DTZ concentration versus time profile when compared to the profiles from the other subjects. In the remaining cases, it would appear that tablet integrity was lost as a consequence of nonspecific structural failure of the unit as a whole and not due to any specific mechanism.

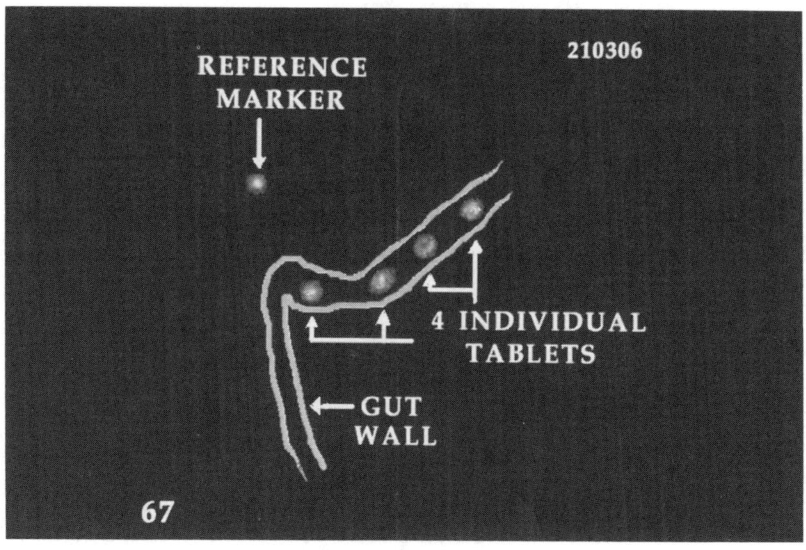

FIGURE 4. Typical gamma scintigraphic image.

The results from the gamma scintigraphy study also determined that the SR formulation did not obstruct nor bind to the intestinal tract. No tablets were defecated intact during the 72 hr study period. Individual plots of the plasma drug concentration versus time coupled with the scintigraphic data determined that DTZ is absorbed throughout the GI.

Fasted Conditions

Individual and mean results from the gamma scintigraphy study describing GI transit times and loss of tablet integrity profiles under fasted conditions are presented in Tables 2 and 3. In general, the average gastric emptying time was 52 min and varied greatly, range 7 min to 250 min. The erratic gastric emptying times may be due to the time of arrival of the dosage form in the stomach in relation to contractual activity of the MMC. For example, if the dose was administered close to a housekeeper wave, the formulation would most likely be propelled out of the stomach into the small intestine. The data indicate that the SR formulation rapidly leaves the stomach with all the tablets arriving in the small intestine by the time lunch is administered 4 hrs post dose. The erratic transit of the tablets through the GI tract does not appear to affect the SR properties of the dosage form following oral administration under fasted conditions suggesting that DTZ is uniformly absorbed throughout the GI tract.

Subject No.	Fasted Condition			Fed Conditions		
	Gastric Emptying	Small Intestinal Transit	Colon Arrival	Gastric Emptying	Small Intestinal Transit	Colon Arrival
1	31	409	441	*	-	-
2	73	145	218	582	-	-
3	250	304	554	*	-	-
4	19	227	246	*	-	-
5	7	174	181	232	141	373
6	19	184	203	191	183	374
7	8	197	205	*	-	-
8	7	244	251	578	-	-
Mean (min)	51.75	235.5	287.4	395.8	162.0	373.5
Mean (hrs)	0.86	3.9	4.8	6.6	2.7	6.2
Range "	0.12-4.2	2.4-6.8	3.0-9.2	3.2-9.7	2.3-3.1	6.2-6.2

* Tablets disintigrated in the stomach, therefore no gastric emptying times are reported.

TABLE 2. Gastrointestinal transit times (in minutes) for Dilacor XR.

In the case of one volunteer, Subject 3, the capsule was retained in the oesophagus for nearly two hours post-dose. An attempt was made to dislodge the capsule by having the volunteer drink some water, however, this attempt failed to promote transit. Plasma DTZ concentrations were very low (< 10 ng/ml) for this subject over the first two hours. The low plasma DTZ concentrations may be due to the inability to either dissolve the capsule covering or hydrate the individual tablets. It is likely that the onset of the MMC activity coincided with the prolonged oesophageal transit of the capsule and that the tablets were finally emptied from the stomach when the next housekeeper wave occurred two hours later. The onset of absorption coincided with transit of the capsule from the oesophagus into the GI tract. The use of gamma scintigraphy provided the company with the opportunity to determine that the "outlier" in this case was not due to a problem with the geomatrix formation, but rather due to an apparent adhesion to oesophageal tissue. In a recent study carried out by the Nottingham group, capsules and tablets were observed

Subject Number	Fasted Condition Tablet 1	Tablet 2	Tablet 3	Tablet 4	Fed Condition Tablet 1	Tablet 2	Tablet 3	Tablet 4
1	923 DC	923 SC	923 SC	1195* HF	472 ST	552 ST	592 ST	632 SI
2	633 HF	633 TC	671 TC	671 SF	625 SI	625 SI	625 SI	924 AC
3	1829 AC	1829 AC	1829 AC	1829 AC	722 ST	722 ST	722 ST	722 ST
4	1204* TC	1204* SF	1204* SF	1776 AC	513 ST	513 ST	513 ST	513 ST
5	927 SF	927 SF	1201* SF	1201* DC	1205* TC	1205* SC	1205* SC	1476 SC
6	626 TC	722 SF	722 SF	722 SF	934 SF	934 SF	934 SF	1207* AC
7	473 TC	807 SF	869 SF	869 SF	582 ST	582 ST	582 ST	582 ST
8	716 SC	716 SC	1210* DC	1210*DC	627 SI	628 SI	628 SI	628 SI
N	32				32			
Mean (min)	1037.3				747.7			
Mean (hrs)	17.3				12.5			
Range (hrs)	7.9 - 30.5				7.9 - 24.6			

* Tablet disintegration occurred between 16 and 24 hours post-dose. Recorded value equals the mid-point of the range.

ST = **ST**omach, **SI** = **S**mall **I**ntestine, **AC** = **A**scending **C**olon, **HF** = **H**epatic **F**lexure, **TC** = **T**ransverse **C**olon, **SF** = **S**plenic **F**lexure, **DC** = **D**escending **C**olon, **SC** = **S**igmoid **C**olon

TABLE 3. Loss of tablet integrity for Dilacor X. Results expressed in minutes and location. In some cases, time and position of tablet disintegration was unclear and recorded values can only provide an estimate of tablet integrity.

to lodge in the oesophagus in two out of ten (20%) of the subjects (Davis et al., 1986). Channer and Virjee in 1986 reported similar results reporting 37 capsules lodged in 175 subjects (21%). Hey et al. (1982) reported that 160 out of 726 (22%) subjects had either tablets or capsules lodge in their oesophagus.

Under fasted conditions, the average small intestinal transit time was approximately 3.9 hrs (range 2.4-6.8 hrs). The Nottingham group have shown previously that the transit of different pharmaceutical dosage forms through the small intestine is independent of feeding conditions or physical properties of the system and that the average time is 3 hrs with a standard deviation of 1 hr (Davis et al., 1986). The small intestinal transit times obtained in the current study are very similar to the results reported earlier.

The tablets that spread in the small intestine were observed to regroup at the ileo-caecal junction, before entering the caecum and ascending colon. This (stagnation) effect is probably related to the suggested reservoir function of the terminal ileum and has been observed previously for mini-tablets (Khosla, 1989).

In general, the average loss of tablet integrity, defined as the time in which the impermeable caps separated from the core tablets, in the fasted condition occurred in the large bowel at 17.3 hrs (range 7.8 hrs to 30.4 hrs). The anatomical site of loss of tablet integrity ranged from the ascending colon to the sigmoid colon. Although tablet integrity was lost, the core of each tablet continued to release drug in a controlled manner. Continual absorption, as well as the presence of diffuse regions of high counts, continued for 20 to 24 hrs post-dose. This diffuse area was observed to disperse over relatively large areas of the intestine, but distinct regions of high versus no activity were observed for up to 20 hrs post-dose. Individual plasma DTZ versus time profiles indicate that the SR formulation continued to release the drug in a controlled manner and that DTZ is readily absorbed from the colon. Feces were examined for regions of high activity and analysis determined that no tablets were defecated intact during the 72 hr study period.

In the case of Subject 3, the tablets remained intact for up to 30 hrs post-dose with little dispersion of the individual units in the ascending colon. Subject 3 had a very flat plasma profile with net change in plasma diltiazem concentrations of only 5.4 ng/ml (range of 17.2 ng/ml to 22.6 ng/ml) from 16 hrs to 36 hrs (5 sampling times). Interestingly, this volunteer has a bowel movement frequency of once every three days and the slow GI movement for this subject is supported by the scintigraphic results. The imaging of the labeled tablets determined that the gastric emptying time was approximately 4.2 hrs compared to the average (n=7) of 24 min (range 7 min to 73 min). The longer duration of the tablets in the stomach may be responsible for the largest Cmax determined from this subject relative to the other subjects, since mechanical grinding may facilitate an erosion of the tablet or may affect the swelling of the hydrophilic polymer. Interestingly, the four hr residence time of the tablets in the stomach is also the period of fasting between administration and lunch. The thought of consuming lunch may be responsible for the actual purging of the stomach contents, including the four tablets.

Fed Conditions

Individual and mean results from the gamma scintigraphy study showing GI transit and loss of tablet integrity profiles of the tablets under fed conditions are presented in Tables 2 and 3. Khosla and coworkers have reported that indigestible solids up to 7 mm in diameter can empty from the fed stomach of man in a similar manner to that observed for multiple unit pellet systems (Khosla et al., 1989). However, in this study, the individual tablets emptied as a bolus from the fed stomach at times roughly corresponding to the onset of interdigestive activity which is consistent with data for undigestible tablets of greater than 12 mm diameter. As a consequence, since retention in the stomach was observed for up to 400 min it is reasonable to assume that the Geomatrix tablets must swell *in vivo* to at least 12 mm in diameter.

Administration of the tablets after a high fat breakfast led to prolonged periods of gastric residence. This observation is characteristic of the continuous feeding regimens as utilized in this study in which the stomach always remains in the fed state, with the onset of interdigestive motility patterns being arrested by lunch and in some individuals, by dinner. In all cases of prolonged gastric residence (Subjects 1, 3, 4, and 7), loss of tablet integrity occurred in the gastric contents and the remnants were emptied from the stomach along with the solid phase of the food. No gastric emptying times could therefore be calculated.

Based on this pilot study in eight individuals, both the overall extent of drug absorption (mean AUC_{inf} increased 18%) and maximum peak drug concentrations (mean Cmax increased 55%) of DTZ increased under fed conditions when compared to fasted administration. A possible explanation for the noted increases may be explained by the grinding of solid food into fine particulate suspension by the stomach. This mechanical grinding may also be responsible for facilitating a physical erosion of the microenvironment of the SR formulation or simply could affect the rate of hydration of the hydrophilic polymer. The gamma scintigraphy and pharmacokinetic data indicates that food may erode some of the core into small particles, increasing the surface area and thus promoting an increased dissolution of DTZ. The drug, dissolved in stomach secretions, passes through the pyloric sphincter into the small intestines and is readily absorbed. Fig. 5 presents a linear regression plot of individual Cmax concentrations versus gastric emptying times from all 8 subjects in both fed and fasted conditions (n=16). Under fed conditions, gastric emptying times could not be determined for Volunteers 1, 3, 4, and 7 because tablet integrity was lost while residing in the stomach. Gastric emptying times for each of these four volunteers was set equal to the mean time of loss of tablet integrity for the four MR tablets. The plot suggests that the longer the tablets reside in the stomach (delayed gastric emptying time) the larger the plasma diltiazem Cmax values (R^2=0.40, p=0.009).

The mean time for loss of tablet integrity occurred at 12.5 hrs, approximately 5 hrs earlier relative to fasted administration. Interestingly, in subjects having a relatively rapid gastric emptying of the tablets from the fed stomach (Subjects 5 and 6), tablet integrity and disintegration times were similar to those observed in the fasted regime. It appears that in these individuals, probably due to chance, the tablets emptied into the duodenum four hrs after the administration of the dose which coincides closely to the lunch time meal served four hours post-dose. Likewise, a similar occurrance may have been responsible for the emptying of tablets from the stomach for Subjects 2 and 8, immediately prior to the supper time meal, 8 hrs post-dose. Emptying of the formulation from the stomach immediately prior to a meal may be responsible for the generally lower peak plasma concentration noted with these four subjects when compared to the individual Cmax concentrations of DTZ determined from the four subjects who retained the tablets in their stomachs for longer periods of time. No tablets were

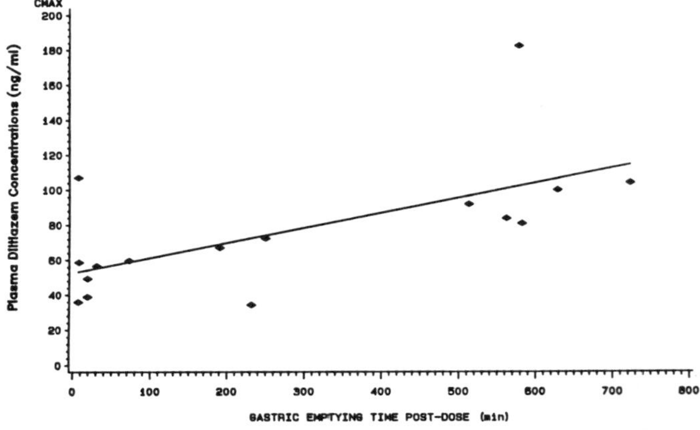

FIGURE 5. Linear regression plot of individual Cmax versus gastric emptying times.

defecated intact during the 72 hr study period.

Finally, an effort was made to integrate the plasma drug concentration versus time data with the corresponding gamma scintigraphic image. An image was matched with each corresponding plasma drug concentration verses time data and the progression over 48 hrs was illustrated in a concentration versus time profile as presented in Figure 6. The left panel of Fig. 6 incorporates the same scintigraphic image as presented in Fig. 4 and the upper right panel includes a description of the elapsed time (9.1 hrs), position of the four tablets (4 in the Transverse Colon), and the plasma DTZ concentration of 32.8 ng/ml. The bottom right panel presents the entire plasma drug concentration versus time profile with a solid line representing the profile through 9.1 hrs. Images were sequenced from time zero through the 48 hr study period for each subject under fasted and fed conditions and then output to video tape every five seconds to create a motion picture of the events through the GI tract. The video was well received by various regulatory agencies and helped the company demonstrate that this unique MR dosage form of DTZ performed *in vivo* according to the proposed rationale.

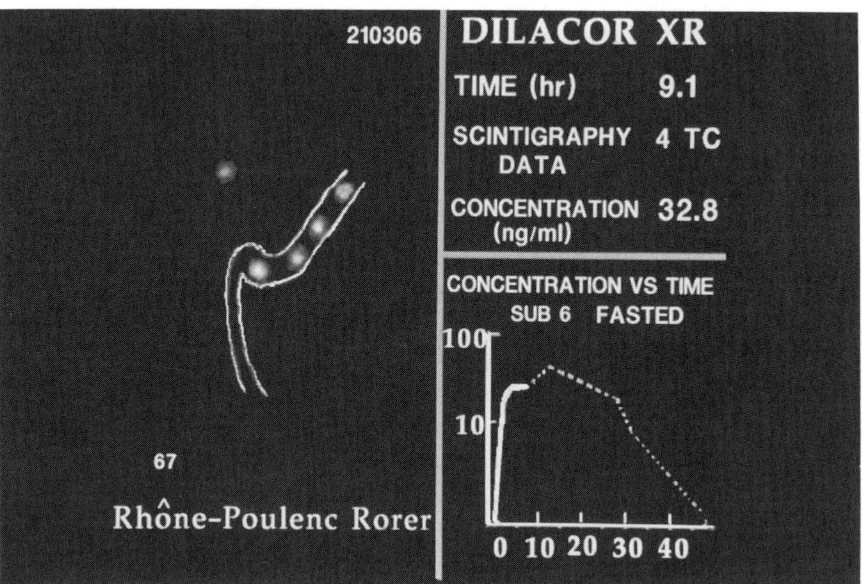

FIGURE 6. Combined scintigraphic and pharmacokinetic data for the Food and Drug Administration.

CONCLUSION

In conclusion, the use of gamma scintigraphy was an invaluable tool in the development of the final dosage form. Outliers determined in the pharmacokinetic analysis were explained utilizing the scintigraphic data and the data supported the rationale of the Geomatrix technology. The GI transit of the dosage form was consistent with the proposed once-a-day use. Individual tablets apparently swelled *in vivo* and traversed the intestine as discrete and separate entities. No aggregation of the tablets was noted, nor was evidence seen for systemic blockage or binding to the GI tract. Consistent with the biodegradable nature of the tablets, fecal recovery studies failed to detect any evidence of excretion of intact or fragmented tablets. The use of this tool can aid the pharmaceutical industry in the development of various dosage forms via the oral, intranasal, inhalation, suppository routes of administration and support the regulatory filing of the NDA.

REFERENCES

Buckley M, Grant S, Goa K, McTavish D and Sorkin E (1990): Diltiazem: A reappraisal of its pharmacological properties and therapeutic use. *Drugs* 39: 757-806.

Channer K and Virjee J (1986): The effect of size and shape of the tablets on their oesophageal transit. *J Clin Pharmacol* 26: 141-146.

Daly PB, Davis SS, Frier M, Hardy, JG, Kennerley JW and Wilson CG (1982): Scintigraphic assessment of the *in vivo* dissolution rat of a sustained release tablet. *Int J Pharmacol* 10: 17-24.

Davis S, Hardy J, Taylor M, Whalley D and Wilson C (1984): A comparative study of the gastrointestinal transit of a pellet and a tablet formulation. *Int J Pharm* 21: 167-177.

Davis S, Hardy J and Fara J (1986): Transit of pharmaceutical dosage forms through the small intestine. *Gut* 27: 886-892.

Davis SS (1987): The design and evaluation of controlled release dosage forms for oral drug delivery. *STP Pharmac* 3: 412-417.

Digenis G, Parr A and Jay M (1989): Neutron activation methods for evaluation of pharmaceutical dosage forms. In: *Drug Delivery to the Gastrointestinal Tract,* Hardy J, Davis S and Wilson C, eds. Chichester: Ellis Horwood.

Digenis G and Sandefer E (1991): Gamma scintigraphy and neutron activation techniques in the *in vivo* assessment of orally administered dosage forms. *Crit Rev Therap Drug Carrier Systems* 7: 309-345.

Digenis G, Sandefer E, Parr A, Beihn R, McClain C, Scheinthal B, Ghebre-Sellassie I, Iyer U, Nesbitt R and Randinitis E (1990):

Gastrointestinal behavior of orally administered radiolabeled erythromycin pellets in man as determined by gamma scintigraphy. *J Clin Pharmacol* 30: 621-631.

Fell J and Digenis G (1984): The *in vivo* behavior of solid oral dosage forms. *Int J Pharm* 22: 1.

Hardy J (1989): Radionuclide imaging in drug formulation, In: *Radiopharmaceuticals Using Radioactive Compounds in Pharmaceutics and Medicine*, Theobald A, ed. Chichester: Ellis Horwood.

Hardy J, Lamont G, Evans D, Haga A and Gamst O (1991): Evaluation of an enteric-coated naproxen pellet formulation. *Aliment Pharmacol Therap* 5: 69-75.

Heald D, Ziemniak J and Wilding I. (1992): manuscript in preparation.

Hey H, Jorgensen F, Sorensen K, Hasslebach H and Wamberg T (1982): Oesophageal transit of six commonly used tablets and capsules. *Br Med J* 285: 1717-1719.

Khosla R, Feely LC and Davis S (1989): Gastro-intestinal transit of non-disintegrating tablets in the fed state. *Int J Pharm* 53: 107-117.

Jay M, Beihn R, Digenis G, DeLand F Caldwell L and Mlodozeniec A (1985): Disposition of radiolabeled suppositories in humans. *J Pharm Pharmacol* 37: 266.

Parr A, Jay M, Digenis G and Beihn R (1985): Radiolabelling of intact tablets by neutron activation for *in vivo* scintigraphic studies. *J Pharm Sci* 74: 590-591.

Parr A, Beihn R and Jay M (1986): *In vivo* scintigraphy evaluation of enteric coated tablets using neutron activation of erbium-170. *Int J Pharm* 32: 251-256.

Parr A, Beihn R, Szpunar G and Jay M (1987): Correlation of ibuprofen bioavailability with gastrointestinal transit by scintigraphic monitoring. *Pharm Res* 4: 486-489.

Phillips S (1988): Small bowel. In: *An Illustrated Guide to Gastrointestinal Motility*, Kumar D and Gustavsson S, eds. Chichester: John Wiley and Sons.

Richter H (1988): Stomach and duodenum. In: *An illustrated Guide to Gastrointestinal Motility*, Kumar D and Gustavssoer S, eds. Chichester: John Wiley and Sons.

Vantrappe G, Janssens J, Hellemans J and Ghoos Y (1977): The interdigestive motor complex of normal subjects and patients with bacterial overgrowth of the small intestine. *Pharm Res* 4: 486-489.

Weiner D (1988): Calcium Channel Blockers. *Med Clin North Am* 72: 83-113.

Wilding I, Coupe A and Davis S (1991): The role of γ-scintigraphy in oral drug delivery, *Adv Drug Delivery Rev* 7: 87-117.

Wilding I, Davis S, Sparrow R, Bloor J, Hayes G and Ward G (1992): The effect of food on the *in vivo* behavior of a novel sustained release formulation of tiaprofenic acid. *Int J Pharm*: in press.

Wilson C, Washington N, Norman S, Greaves J, Peach J and Pugh K (1988): A gamma scintigraphic study to compare oesophageal clearance of 'Expidet' formulations, tablets and capsules in supine volunteers. *Int J Pharm* 46: 241-246.

Wilson C and Hardy J (1982): *Radionuclide Imaging in Drug Research*, London: Croom Helm.

THE POTENTIAL USES OF RADIOPHARMACEUTICALS IN THE PHARMACEUTICAL INDUSTRY

Raymond E. Gibson and H. Donald Burns
Department of Radiopharmacology
Merck Research Laboratories.

William C. Eckelman
PET Department, Clinical Center
National Institutes of Health

INTRODUCTION

The earliest pharmaceuticals were discovered based on observations of the favorable effects of natural products. Some of the better known examples are digitalis-containing foxglove, quinine-containing cinchona bark, salicylate-containing willow bark, and morphine-containing opium. Many of today's new drugs are also the result of natural products screening, e.g., avermectins and mevacor. Merck Research Laboratories has recently purchased the rights to certain South American rain forests to continue the search for natural products.

The second phase of pharmaceutical development also depends on screening for physiologic effect. This has been deemed the era of rational design (Waldrop, 1990). Medicinal chemists synthesize a large number of synthetic derivatives based on natural products or endogenous hormones and neurotransmitters which are screened for effect using isolated enzymes or receptor preparations *in vitro*. Tritium labeled ligands have played an important part in the identification and use of

Nuclear Imaging in Drug Discovery, Development, and Approval
H.D. Burns, R. Gibson, R. Dannals, Peter Siegl, Editors

these isolated binding sites. Characterization of these binding sites provides the basis for the use of radioligands *in vivo*.

The third phase was initiated by the introduction of the powerful techniques of molecular biology. Now, receptors and enzymes from human tissues are cloned, expressed in measurable quantities, and their drug interactions characterized. Although most three dimensional structures of receptor-ligand complexes are still conjecture, crystal structures of ligand-receptor interaction such as the growth hormone-receptor complex are now being obtained. The hope is that smaller ligands that have optimal pharmaceutical properties can be designed based on these structures.

THE MAJOR USES OF SPECT AND PET

Today, most pharmaceuticals are designed using second and third phase techniques with validation in animals prior to initiation of human studies. Physiologic effect is still the major end-point, but autoradiographic studies and pharmacokinetic modeling are carried out using Carbon-14 and Tritium labeled compounds. However, given the long history of species differences, this approach is certainly not fool proof. Methodologies which permit the determination of drug interactions *in vivo* in man should circumvent many of the difficulties which arise in applying results obtained from animal studies to the clinic. The methodologies described in these chapters, common to the field of Nuclear Medicine, provide techniques which can be applied not only to the study of drug interactions in man, but can be used in the earlier stages of drug development using animal models. This is particularly valuable since novel uses of imaging technologies may require testing in animals prior to tests in man. The authors have provided an overview of the techniques and methods needed to use imaging technologies in drug development, though a volume of this size cannot be inclusive.

The major uses of gamma scintigraphy in Nuclear Medicine are the measure of perfusion and biochemical processes by external imaging. Anatomic imaging is more effectively accomplished with x ray computed tomography or Magnetic Resonance Imaging (MRI), the latter providing no apparent radiation dose to the subject. The examples provided in the previous chapters of applications of planar, Single Photon Emission Computed Tompgrahy (SPECT) and Positron Emission Tomography (PET) techniques are a small number of the potential studies available using Nuclear Medicine methodologies. The radiopharmacopea used for the diagnosis of different diseases or pathologies is quite large (Table 1). Most of the agents in Table 1 are taken from a text edited by Bernier, et al. (1989) or are presented in this text; but, Table 1 is not inclusive of the field. Among those areas not

Targets	Radiopharmaceutical	Physiologic or Pharmacologic Parameter Measured
Heart	[^{201}Tl]Thallium chloride	Perfusion/ischemia
	[^{99m}Tc]Red Blood Cells	Cardiac function
	[^{99m}Tc]Sestamibi	"
	[^{99m}Tc]Teboroxime	"
	[^{99m}Tc]pyrophosphate	Acute infract
	[^{99m}Tc]anti-fibrin Ab	"
Brain	[^{99m}Tc]HMPAO	"
	[^{99m}Tc]ECD	"
	[^{123}I]iodoamphetamine	"
	[^{99m}Tc]DTPA	Lesions of the blood-brain-barrier (tumors, stroke, etc.)
Liver	[^{99m}Tc]sulfur colloid	Blood flow/RES fn
	[^{99m}Tc]albumin colloid	"
	[^{99m}Tc]NGA	Hepatocyte fn /Receptor Conc.
Gallbladder	[^{99m}Tc]HIDA	Hepatobiliary fn
Lungs	[^{99m}Tc]MAA	Perfusion
	[^{133}Xe]Xe	Ventilation perfusion /pulmonary emboli
	[^{99m}Tc]aerosols	"
Kidney	[^{123}I]Iodohippuran	Renal plasma flow
	[^{99m}Tc]MAG$_3$	"
	[^{99m}Tc]DTPA	GFR
Bone	[^{99m}Tc]MDP	fractures/tumors
	[^{68}Ga]Gallium citrate	inflammation
Thyroid	[^{123}I]Iodide	nodules/tumors
	[^{99m}Tc]TcO$_4^-$	tumors
Adrenals	[^{131}I]MIBG	tumors

TABLE 1. Commonly used Radiopharmaceuticals in Nuclear Medicine and their uses.

covered in this text are radiolabeled antibodies used to image or treat tumors, to image infarcts, thrombi and plaque, Also, many experimental agents such as [^{123}I]4IQNB, a muscarinic receptor antagonist which may find use in the differential diagnosis of dementias (Weinberger et al., 1991), are not included in Table 1. PET radiotracers are available which can be used for some of the functions presented in Table 1. Cerebral blood flow can be determined using [^{15}O]H_2O (Herscovitch et al., 1983; Raichle et al., 1983) or [^{15}O]butanol (Berridge et al., 1991). The very short half-life of Rubidium-82 (75 sec) makes it very useful for blood flow studies in CNS and heart (Mullani et al., 1983). The tracer for glucose metabolism, [^{18}F]2-fluoro-2-deoxy-D-glucose ([^{18}F]FDG), has been used in the CNS to evaluate tumors, dementias, and epileptic foci. In the heart, [^{18}F]FDG provides a measure of ischemic metabolism, which, coupled with a perfusion tracer such as [^{13}N]NH_3, provides an excellent evaluation of the damage to ventricular muscle from acute infarcts (Krivokapich et al, 1990; Krivokapich et al., 1991). However, many of the targets listed in Table 1 do not have PET radiotracers since the single-photon agents have proved so useful. The most exciting uses of PET remain in the areas of defining receptor and enzyme function or interactions *in vivo*.

Any of the above radiopharmaceuticals (including those described in previous chapters), either single-photon or positron emitting, can be applied to preclinical studies using animals models. However, preliminary studies to establish appropriate protocols may be necessary. For example, we were interested in determining the effect of a drug on gallbladder function using a non-human primate. The extensive literature on the use of radiopharmaceuticals for gallbladder imaging in humans did not provide us with the knowledge that isoflurane anesthesia inhibits the gallbladder contraction one normally expects when challenged with cholecystokinin. A study which was considered to be quite easy in concept required several months of effort to adequately characterize in the non-human primate. Thus, previous extensive clinical experience with a radiopharmaceutical does not guarantee ease of application to pre-clinical models. Nonetheless, the data obtained from the gallbladder study was non-invasively obtained with no harm or stress to the experimental animals involved in the study.

PET Versus SPECT

Scientist who use Nuclear Medicine technology to investigate drug interactions and/or mechanisms, or basic physiology and pathology, generally express a preference for either PET or SPECT, with each camp providing sound reasons for their preferences. Realistically, both technologies have advantages and disadvantages and should be considered complementary areas, rather than competitors. The selection of either PET or SPECT technologies for a study is determined by numerous factors. PET appears to have more advantages than SPECT since i) drugs can be isotopically labeled; ii) the half-lives of PET isotopes are shorter which reduces dose delivered per mCi injected and permits repeat studies on the same day; and, iii) resolution and quantitation with currently available instrumentation is better than with SPECT. For most biochemical probes, PET has the advantage of the positron-emitting radionuclides of carbon, nitrogen, and oxygen. The pharmaceutical scientist can therefore determine the behavior of the drug of interest, rather than an analogue. However, as mentioned in previous chapters, methods of incorporating positron-emitting nuclides into drugs are more difficult because of the short synthetic times required for an isotope such as Carbon-11. Lastly, PET exhibits greater sensitivity than SPECT since it does not require collimation of radiation to provide resolution.

Perhaps the greatest disadvantage of PET is that most studies require a cyclotron on the same site as the PET scanner. This is particularly true for studies with Oxygen-15, and to a lesser extent, Carbon-11. Several PET isotopes are, however, available through generator systems, specifically, Gallium-68 (Loc'h et al., 1980) and Rubidium-82 (Bonow et al., 1991). Establishing a cyclotron-PET facility within a pharmaceutical company (purchase of the cyclotron and PET instrument, construction of the cyclotron vault with power supply, radiochemistry laboratory with hot-cells, etc.) is quite expensive. An attractive alternative is to pursue PET studies in collaboration with existing facilities at medical institutions and national laboratories, though the pace of research in an academic environment is frequently somewhat slower than the needs of pharmaceutical research.

The advantages of SPECT are those that counter the difficulties encountered in PET. The radioisotopes have sufficiently long half-lives that they can be purchased from a variety of commercial firms. The most commonly used isotope in Nuclear Medicine, Technetium-99m, is available from a generator which can be replaced weekly. Many studies which do not involve tracing the drug of interest can be pursued using existing radiopharmaceuticals (e.g., Table 1). Instrumentation for obtaining and processing planar images, or a rotating gamma-camera which can be used for either planar or SPECT, are relatively inexpensive compared to PET instrumentation. Recently, multi-slice, multi-head or ring machines for single photon emitting radiotracers have become available (e.g., the TRIAD, the PRISM, the Strichman SPET machine, the MUMPI, the ASPECT, the CERTO96, and the TOMOMATIC). Although the sensitivity of SPECT is less than that of PET, there is no inherent limit to the resolution of the SPECT methodology. Unlike PET in which the inherent resolution is determined by the distance traveled by a positron before annihilation, the limits of SPECT are determined by the number of detectors and the collimator. It is therefore theoretically possible to construct an instrument with resolution less than 1 mm, but, this resolution would be at the expense of sensitivity and require construction of a collimator with matching resolution. In studies in humans, the limits imposed by dosimetry and time required to acquire tomographic data result in a compromise with resolutions of 7-8 mm for the best available instruments. However, in animal studies higher doses of radiopharmaceuticals can be administered, e.g., 5 mCi of a Technetium-99m radiopharmaceutical to a 250 g rat, and a properly anesthetized animal can be imaged for many hours. High resolution SPECT instrumentation for animal studies are under development and may become commercially available.

The selection of methodology, PET or SPECT, is frequently determined by the type of study needed which may specify the sensitivity and therefore the speed of the detection devices. Cerebral blood flow (CBF) can be rapidly and repeatedly determined using $[^{15}O]H_2O$. The influence of a drug candidate on blood flow can be easily determined by a $[^{15}O]H_2O$ CBF study before, during, and after drug intervention. Not only does Oxygen-15 have a short physical half life, but water has an even shorter biological half-life in tissue. Although a radiopharmaceutical with the characteristics of water cannot be designed for use with SPECT, CBF can be determined using one of the new Technetium-99m labeled agents which behave like molecular microspheres; but, repeat studies even in animals are limited by the longer half-life (6 hrs). A protocol has been reported in which a control CBF is determined using a low dose of the single-photon radiopharmaceutical followed by drug intervention and a subsequent high-dose determination of CBF (Fioravanti et al., 1992). By contrast, cerebral glucose metabolism (and potential drug effects) can only be determined using PET since there is currently no SPECT equivalent of $[^{18}F]$2-fluoro-2-deoxy-D-glucose.

Drug disposition in man may be demonstrated using the isotopically labeled drug (usually with Carbon-11). Since metabolism of the radioligand is frequently very rapid, results on either the gross distribution or paths of excretion may be quite misleading. The specific localization of a drug with a target organ, through binding to an enzyme or receptor, may be the only method of demonstrating that the drug interacts at the appropriate site. These studies must be, however, validated with preliminary studies in animals to insure that the detected signal is the radiolabeled drug, not a metabolite. By contrast, SPECT studies of drug interactions are usually confined to analogues since few pharmaceuticals naturally contain even an iodine. Thus drug interactions can only be inferred via *in vivo* competition studies with the radiolabeled analogue.

MAGNETIC RESONANCE IMAGING

We also did not include a chapter on the uses of magnetic resonance imaging (MRI), a rapidly growing area which also deserves an entire volume. Although total clinical Nuclear Medicine procedures have stayed flat over the years, this is the result of decreases in anatomical imaging procedures, such as liver imaging and brain imaging for blood-brain-barrier abnormalities. Computed tomography (CT) and MRI, also known as nuclear magnetic resonance imaging, prompted this shift because of their higher resolution anatomical images (approx. 2 mm). Since the information is obtained from the proton signals in tissue, no functional information is provided in a typical MRI scan, i.e., the brain image from MRI (or CT) may be the same whether or not the subject is alive. In order to provide information on function, particularly blood flow, new agents (MRI contrast media) have been developed using techniques developed for Technetium-99m.

MRI contrast media change the relaxation times of the protons primarily in water in the tissue. This feature allows the new MRI contrast media to better define anatomical changes and functional changes in major organ systems such as the kidney and liver. For example, the iron chelate, Fe-EHDG (Lauffer and Brady, 1989), and the gadolinium chelate, Gd-BOPTA (Vittadini et al., 1988), are being studied as potential hepatobiliary agents. Although MRI offers increased resolution, this is counterbalanced by the need for 1 to 100 µM concentrations of the MRI contrast agent (Hattner and White,1990). The high concentrations needed to measure capillary perfusion and biochemical changes is an important concern for toxicity. However, diffusible contrast media such as gadolinium-DTPA have shown different pharmacokinetics in normal and infarcted myocardial and cerebral tissue (Rosen et al.,1989, Bradley et al., 1989); but, biochemical probes are still in the early stages of development (Josephson et al., 1990). MRI contrast agents covalently linked to antibodies may be useful for tumor imaging. Calculations of the concentration of gadolinium bound to antibodies necessary to enhance the signal at tumors indicate that either increases in relaxivity or in the labeling approach will be necessary (Eckelman et al., 1988). In addition to the metal chelates, other nuclei for MRI are being suggested, e.g., Fluorine-19 for 2-fluoro-2-deoxyglucose which has the potential for use similar to that of the [^{18}F]FDG studies now carried out using PET. However, the concentration of [^{19}F]FDG required with present-day machines saturates the hexokinase system (Bolo et al., 1987). MRI spectroscopy also has the potential to measure biochemical changes using either endogenous signals from nuclei such as phosphorus or from exogenous "contrast" agents. Although at an earlier stage of development than other nuclear medicine methodologies, applications of MRI to drug development will be an area to track for the future.

CONCLUSION

The non-invasive imaging techniques of nuclear medicine provide a powerful research tool that can be used in a variety of ways in the discovery and development of new drugs. Basic studies with properly designed and characterized, high specific activity, receptor specific radioligands can be used to evaluate the interaction of new drug candidates with their target receptors *in vivo*, providing information which cannot be obtained by any other means. Information on the relationship of dose to receptor occupancy, the kinetics of interaction of drugs with receptors and the relationship between receptor occupancy and pharmacological response can be obtained non-invasively in experimental animals and humans. In these experiments, the receptor specific radiotracer functions as a probe that allows one to follow the interaction of an unlabeled drug with its intended receptor. This is especially important in diseases for which there are no suitable animal models, particularly CNS disorders such as schizophrenia. Similar studies with these receptor specific radioligands can reveal changes in receptor distribution or concentration and allow researchers to gain a better insight into the role that specific receptors play in various diseases or disorders. Studies of this type can be conducted using either PET or SPECT imaging with appropriate, receptor specific radiotracers, e.g. [^{11}C]NMSP and PET imaging or [^{123}I]IBZM and SPECT imaging for drugs involving the dopamine D_2 receptor.

Many drug candidates can also be studied using isotopically labeled compounds (primarily Carbon-11 and, in some cases, Fluorine-18) and PET. Studies with these labeled drugs can reveal information on the kinetics of the drug in the target tissue or accumulation of the drug in other tissues (suggesting potential toxic side effects in that tissue). Similar studies can be conducted to evaluate the effect that various factors such as formulation or route of administration have on the biodistribution of a drug. Information obtained by imaging with compounds isotopically labeled with positron emitting radionuclides can be used to complement information from more traditional studies conducted with Carbon-14 and Tritium labeled tracers. These imaging studies offer the advantage that they can be conducted in humans, thus eliminating any concerns regarding potential species differences.

Other nuclear medicine techniques can be used to assess the

physiological and/or toxicological effects induced by drugs or drug candidates. Routine clinical nuclear medicine procedures are available that can be used for evaluating the effect that a drug has on a wide variety of physiological parameters (e.g., on blood flow, cardiac ejection fraction, hepatobiliary function, etc.). These procedures can be conducted either in animal models early in the drug development process or subsequently in humans associated with phase II and III IND studies.

We have presented examples of some of these applications of imaging technology in various chapters of this text. The first two chapters describe the instrumentation requirements for planar, PET and SPECT imaging, the special requirements for preparation of PET radiotracers and availability of radioisotopes for planar and SPECT imaging. The next three chapters provide a description of the chemistry of positron and single-photon emitting radiotracers and an outline describing the application of these chemistries to the development of new radiotracers. We also provide two chapters on applications of autoradiography, a field ancillary to *in vivo* imaging, and techniques frequently necessary in the characterization of new radiotracers prior to *in vivo* imaging studies. This text then provides a number of varied examples of the application of radiotracers to *in vivo* imaging. The application of nuclear imaging does require specific skills which are not necessarily available within a pharmaceutical company and may require the development of such a department. Alternatively, academic centers and national laboratories are available with quality investigators who can help the pharmaceutical researchers in all aspects of these studies. Many of these individuals have contributed to this text. The efficient utilization of nuclear imaging will, in many cases, accelerate drug development. Imaging technologies can also provide data which cannot be obtained by any other means. This utility is singularly obvious in the application of imaging to non-invasive studies in humans.

REFERENCES

Bernier DR, Christian PE, Langen JK and Wells LD (1989): *Nuclear Medicine Technology and Techniques*, St. Louis: C.V. Mosby Co.

Berridge MS, Adler LP, Nelson AD, Cassidy EH, Muzic RF, Bednarczyk EM and Miraldi F (1991): Measurement of human cerebral blood flow with [^{15}O]butanol and positron emission tomography. *J Cereb Blood Flow Metab* 11: 707-715.

Bolo NR, Brennan KM, Jones RM and Budinger TF (1987): Fluorodeoxyglucose brain metabolism studies by NMR and PET. *Ann N Y Acad Sci* 508: 451-459.

Bonow RO, Berman DS, Gibbons RJ, Johnson LL, Rumberger JA, Schwaiger M and Wackers FJ (1991): Cardiac positron emission tomography. A report for health professionals from the committee on advanced cardiac imaging and technology of the council on clinical cardiology, American Heart Association, *Circulation*, 84:447-454.

Bradley RH, Kent TA, Eisenberg HM, Quast MJ, Ward GA, Campbell GA and Hillman G (1989): Middle cerebral artery occlusion in rat studied by magnetic resonance imaging. *Stoke* 20: 1032-1036.

Eckelman WC, Tweedle M and Welch MJ (1988): NMR enhancement with Gd labeled antibodies. In: *Radiolabeled Monoclonal Antibodies for Imaging and Therapy*, Srivastat SC, ed. New York: Plenum Press.

Fioravanti C, Burkholder D and Gibson RE (1992): Global cerebral blood flow in rhesus monkeys determined using Tc-99m ECD and planar imaging. J Nucl Med 33 (suppl): 979.

Hattner RS and White DL (1990): Gallium-67/stable gadolinium antagonism: MRI contrast agent markedly alters the normal distribution of gallium-67. *J Nucl Med* 31: 1844-1846.

Herscovitch P, Markham J and Raichle ME (1983): Brain blood flow measured with intravenous $H_2{}^{15}O$: I. Theory and error analysis. *J Nucl Med* 24: 782-789.

Josephson L, Groman EV, Menz E, Lewis JM and Bengele H (1990): A functionalized supermagnetic iron oxide colloid as a receptor directed MR contrast agent. *Magnet Resonance Imgng* 8: 637-646.

Krivokapich J, Barrio JR, Huang SC and Schelbert HR (1990): Dynamic positron tomographic imaging with nitrogen-13 glutamate in patients with coronoary artery disease: comparison with nitrogen-13 ammonia and fluorine-18 fluorodeoxyglucose imaging. *J Am Coll Cardiol* 16: 1158-1167.

Krivokapich J, Huang SC, Ratib O and Schelbert HR (1991): Noninvasive detection of functionally significant coronary artery stenoisis with exercise and positron emission tomography. *Am Heart J* 122: 202-211.

Lauffer RB and Brady TJ (1989): Iron ethylene bis(2-hydroxyphenylglycine) as a hepatobiliary MRI contrast agent. In: *Magnetic Resonance Imaging*, Partain CL, Price RR, Patton JA, Kulkarni MV and Jasmes AE Jr, eds. Phiadelphia: Saunders.

Loc'h C, Mazèré B and Comar D (1980): A new generator for ionic gallium-68. *J Nucl Med* 21: 171-173.

Mullani NA, Goldstein RA, Gould KL, Marani SK, Fisher DJ, O'Brien HA Jr and Loberg MD (1983): Myocardial perfusion with Rubidium-82. I. Measurement of extraction fraction and flow with external detectors. *J Nucl Med* 24: 898-906.

Raichle ME Marten WRW, Herscovitch P, Mintunu MA and Markham J (1983): Brain blood flow measured with intravenous $H_2{}^{15}O$: II. Implementation and validation. *J Nucl Med* 24: 790-798.

Rosen BR, Belliveau JW and Chien D (1989): Perfusion imaging by nuclear magentic resonance. *Magnet Resonance Q* 5: 263-281.

Vittadini G, Felder E, Tirone P and Lorusso V (1988): B-19036, a potential new hepatobiliary contrast agent for MR proton imaging. *Invest Radiol* 23 (suppl 1): S246-S248.

Waldrop MM (1990): The reign of trial and error draws to a close. *Science* 247: 28-29.

Weinberg DR, Gibson R, Coppola R, Jones DW, Molchan S, Sunderland T, Berman KF and Reba RC (1991): The distribution of cerebral muscarinic acetylcholine receptors in vivo in patients with dementia. *Arch Neurol* 48: 169-176.

INDEX

Related Birkhäuser Titles

Emerging Strategies in Neuroprotection
Edited by Paul J. Marangos and Harbans Lal
1992, 320 Pages
ISBN 0-8176-3544-0

Molecular Biology of G-Protein-Coupled Receptors
Mark R. Brann, Editor
ISBN 0-8176-3465-7, 1992, 256 Pages

Cholinergic Basis for Alzheimer Therapy
Edited by Robert Becker and Ezio Giacobini
1991, 494 Pages
ISBN 0-8176-3566-1

Genetically Defined Animal Models of Neurobehavioral Dysfunctions
Edited by Peter Driscoll
1992, 328 Pages
ISBN 0-8176-3460-6

The Encyclopedia of Neuroscience
Edited by George Adelman
1987, 1,565 Pages, 2 Volumes
ISBN 0-8176-3335-9

Vincent van Gogh: Chemicals, Crises, and Creativity
By Wilfred N. Arnold
1992, 330 Pages
ISBN 0-8176-3616-1

Cortical Memory Functions
By C. M. Fair
1992, 210 Pages
ISBN 0-8176-3615-3

New Approaches for Antifungal Drugs
Edited by Prabhavathi B. Fernandes
1992, 201 Pages
ISBN 0-8176-3602-1

Slow Potential Changes in the Brain
Edited by Wolfgang Haschke, E.-J. Speckmann, and Alexander I. Roitbak
1992, 334 Pages
ISBN 0-8176-3583-1

Machinery of the Mind
Edited by E. Roy John
1990, 566 Pages
ISBN 0-8176-3461-4

Neurodevelopment, Aging, and Cognition
Edited by Ivica Kostović, Stevo Knezević, Henryk M. Wisniewski, and George J. Spilich
1992, 396 Pages
ISBN 0-8176-3599-8

Central Neural Mechanisms of Cardiovascular Regulation
Edited by George Kunos and John Ciriello
1991, 282 Pages
ISBN 0-8176-3545-9

Central Neural Mechanisms of Cardiovascular Regulation, Vol. 2
Edited by George Kunos and John Ciriello
1992, 400 Pages
ISBN 0-8176-3569-6

Neurotransmitter Interactions and Cognitive Function
Edited by Edward D. Levin, Michael Decker, and Larry L. Butcher
1992, 376 Pages
ISBN 0-8176-3617-X

Neuropharmacology of Ethanol: New Approaches
Edited by Roger E. Meyer, George F. Koob, Michael J. Lewis, and Steven M. Paul
1991, 271 Pages
ISBN 0-8176-3463-0

Electromagnetic Fields and Circadian Rhythmicity
Edited by Martin C. Moore-Ede, Scott S. Campbell, and Russel J. Reiter
1992, 210 Pages
ISBN 0-8176-3552-1

Kindling and Synaptic Plasticity: The Legacy of Graham Goddard
Edited by Frank Morrell
1991, 304 Pages
ISBN 0-8176-3466-5

Activation to Acquisition:
Functional Aspects of the Basal Forebrain Cholinergic System
Edited by Russell T. Richardson
1991, 392 Pages
ISBN 0-8176-3467-3

The Neurosciences: Paths of Discovery, II
Edited by Fred Samson and George Adelman
1992, 352 Pages
ISBN 0-8176-3503-3 (Softcover: 0-8176-3629-3)

Why We Nap: Evolution, Chronobiology,
and Functions of Polyphasic and Ultrashort Sleep
Edited by Claudio Stampi
1992, 350 Pages
ISBN 0-8176-3462-2

Oxygen Free Radicals in Tissue Damage
Edited by Merrill Tarr and Fred Samson
1992, 256 Pages
ISBN 0-8176-3609-9

Cortical Circuits: Synaptic Organization
of the Cerebral Cortex Structure, Function, and Theory
By Edward L. White
1989, 250 Pages
ISBN 0-8176-3402-9

The Cholinergic Neuron and its Target
By V. P. Whittaker
1992, 574 Pages
ISBN 0-8176-3553-X

The Neurosciences: Paths of Discovery, I
Edited by Frederic G. Worden, Judith P. Swazey,
and George Adelman
1992, 622 Pages
ISBN 0-8176-3506-8 (Softcover: 0-8176-3621-8)

Basic Mechanisms of the EEG
Edited by Stephan Zschocke and E.-J. Speckmann
1993, 300 Pages (Tent.)
ISBN 0-8176-3596-3